A.S. Marfunin

Spectroscopy, Luminescence and Radiation Centers in Minerals

Translated by V. V. Schiffer

With 170 Figures

Springer-Verlag
Berlin Heidelberg New York 1979

Dr. Arnold S. Marfunin
IGEM Academy of Sciences
Staromonetniy 35
Moscow 109017/USSR

The original Russian edition was published by Nedra, Moscow, in 1975.

ISBN 3-540-09070-3 Springer-Verlag Berlin Heidelberg New York
ISBN 0-387-09070-3 Springer-Verlag New York Heidelberg Berlin

Library of Congress Cataloging in Publication Data. Marfunin, Arnold Sergeevich. Spectroscopy, luminescence and radiation centers in minerals. Translation of Spektroskopiia, liuminesfsenfsiiâ i radiafsionnye fsentry v mineralakh. Bibliography: p. Includes index. 1. Mineralogy, Determinative. 2. Spectrum analysis. 3. Luminescence. 4. Holes (Electron deficiencies). 5. Color centers. I. Title. QE369.S65M3713   549'.125   78-26367.

Offsetprinting and bookbinding by Brühlsche Universitätsdruckerei, Lahn-Gießen

2132/3130-543210

Preface

The development of mineralogy, the evolutionary changes in comprehending the mineral substance of the earth are closely associated with the progress of research methods.

Over a space of more than two and half centuries, from the goniometry of the mineral crystals to microscopic petrography and optical mineralogy, to crystal structure determinations, electron microscopy and electron diffraction and finally investigations into their electrical, magnetic and mechanical properties, all this has led to the formation of the existing system of mineralogy, its notions, theories and to a proper description of minerals.

However, no matter how great the variety of methods employed in mineralogy, they all come to a few aspects of substance characteristics. These are methods of determining the composition, structure and properties of the minerals. Thus the X-ray microanalyzer, the atom-absorption, neutron-activation, chromatographic and other analyses open up new opportunities for determining nothing else but the elementary composition of minerals.

In the last 10 to 15 years the scope of these analytical possibilities has been considerably enlarged by an addition of the spectroscopy of solids methods which help expose a new facet in the cognition of the substance in general and of the mineral one, in particular. The spectroscopy of solids includes electron paramagnetic resonance, nuclear magnetic and nuclear quadrupole resonance, Mössbauer (γ-resonance) spectroscopy (all of them owing their origin to the discovery of new physical phenomena), X-ray and electron spectroscopy (that began developing after construction and improvement of new spectrometer types), optical spectroscopy and luminescence (which have assumed quite a new significance due to use of crystal field theory).

Embracing all regions of the spectrum, from the gamma and X-ray and throughout the ultraviolet, visible, infrared, to the superhigh-frequency and radio-frequency, being based on the nuclear, electron, vibrational and spin transitions and making use of radically different types of spectrometers, these methods have one major common feature that distinguishes them from all other research procedures.

The salient feature common to all branches of the spectroscopy of solids is the observation of atoms in a state of chemical bonding within concrete crystals and compounds. This gives a complete characterization of crystals and compounds, viz. their composition, the arrangement of atoms in the structure, properties of the atoms in a given crystal structure (electronic structure), and properties of the crystal as a whole, determined by the composition, atomic and electronic structure.

Spectroscopic parameters present the only direct possibility to measure and define the actual states of chemical bonding, a natural experimental groundwork forming a basis for the current theories of chemical bonding, such as the crystal field, molecular orbitals, and energy band theories which, in their turn, are at the same time the spectroscopic theories.

The spectroscopy of solids not only opens up a new chapter in the understanding of the substance, but within the framework of the visible "physical oecumen", if one may put it this way, it appears as conclusive, synthetizing, and unique in describing causal relationships among the composition, structure, and properties, and enables it to comprehend and assess these properties. The crystals themselves and compounds are a product subsequent to self-consistency of the electronic properties of the atoms, the one that has led to a given arrangement of the atoms in the crystal structure and has imparted specific features to them in the crystal, as well as shaped the properties of the crystals themselves.

In all directions of these spectroscopic methods, the detailed and thorough investigations have been effected in mineralogy at the most advanced technical and theoretical level. Both fundamental data and those relevant to individual rock-forming and ore minerals have been already obtained.

Thus, in addition to a change in the general approach to the chemical binding in minerals, X-ray and electron spectroscopy helped trace molecular orbital diagrams for rock-forming silicates.

The nuclear-magnetic resonance spectra furnished a basis for measuring intracrystalline fields, while from the electron paramagnetic resonance spectra the degree of ionicity – covalency – has been determined. In the case of sulfides, the bonding states have been estimated according to the Mössbauer and optical reflectance spectra and for the arsenic, antimony, bismuth sulfosalts – on the ground of the nuclear quadrupole resonance data.

Spectroscopic methods are of aid in establishing the distribution of cations in nonequivalent positions within the structures of olivines, amphiboles, mica, and other minerals, with some of these serving as a means for setting apart geothermometers, determining types and localization of water in minerals, obtaining all the information about the impurity elements, interpreting causes accounting for colors and luminescence of minerals and helping describe in great detail the state of iron (its valent states, magnetic properties, supermagnetism of ultra-fine particles, properties of diamond and apatite with implanted iron, etc.).

The results thus made available are used in subdividing intrusive massifs, sedimentary and metamorphic complexes, separating facial rock varieties, determining relative ages, temperature conditions attending their formation and educing prospective indications for exploration.

It was the electron paramagnetic resonance alone that helped discover in minerals an abundance and a great variety of radiation electron-hole centers and to interpret their models. By this time such centers have been shown to be of geological and prospective value, to influence flotation of minerals, determine the photochromatism of some crystals (of sodalites in particular), utilized in T.V. screens, to define the mechanism of thermoluminescent dosimetry, specify the choice of radiation-stable lasers, piezoelectrics, the service-life of reactor materials, to intensify and change the coloration of gems. The entire complex of spectroscopic methods occupies a place of prime importance in continuing investigations of lunar rocks and minerals.

In many branches of the spectroscopy of solids, natural minerals and their synthetic analogs represent the best-studied systems, such, for instance, as those of fluorite, scheelite, apatite, ruby in the electron paramagnetic resonance and luminescence, spinels-chromites-ferrites in the Mössbauer spectroscopy.

The introduction of new methods into mineralogy pre-supposes each time their adaptation conformably to features specific for mineral matter and to problems of mineralogy, petrology, lithology, ore deposits geology, prospecting and enrichment of raw mineral materials. In the past decade the potentialities and trends of each one of the spectroscopic methods in this cycle of research have taken their main shape; the notions, methods of measurement and interpretation have been selected and data covering major mineral groups accumulated. An attempts at systematic presentation of all this material has been made in this book.

A general theoretical introduction to all sections of the book is the crystal field theory and molecular orbital theory, discussed in the preceding book of the author[1], where subject to consideration was also the optical absorption spectroscopy of minerals.

The bibliography includes selected works which, however, reflect all the principal trends of research in the domain of spectroscopy of the minerals. References to works cited in *Physics of Minerals and Inorganic Materials* are almost omitted here and, therefore, one should look there for literature sources touching upon related problems (spectroscopy and chemical bonding, spectroscopy and crystal field, molecular orbital theories, etc.).

The author wishes to express his indebtness and gratitude to L.V. Bershov, V.M. Vinokurov, I.N. Penkov, A.N. Platonov, A.N. Taraschan, S.A. Altshuler, M.M. Zaripov, M.I. Samoilovich, V.I. Nefedov, A.M. Bondar, V.O. Martirosyan, M.L. Meilman, R.M. Mineeva, A.R. Mkrtchyan, M.Ya. Scherbakova, A.V. Speransky for fruitful discussions of many issues concerned with the spectroscopy of minerals which proved of great help to the author in writing this book.

Moscow, April 1979   A.S. Marfunin

1 A.S. Marfunin: Physics of minerals and inorganic materials. An introduction. Berlin, Heidelberg, New York: Springer 1979.

Contents

1   Mössbauer (Nuclear Gamma-Resonance) Spectroscopy

1.1   Basic Principles and Experimental Arrangement   1
1.1.1   Isomer Nuclear Transitions and Gamma-Ray Emission .   1
1.1.2   Resonance Fluorescence   1
1.1.3   Mössbauer Effect: a Recoilless Gamma-Fluorescence . . .   3
1.1.4   Experimental Arrangement for Observing the Nuclear
        Gamma-Resonance (Mössbauer Spectrometer)   4
1.1.5   Development of the Method   5
1.2   Mössbauers Nuclei   6
1.3   The Mössbauer Spectra Parameters   7
1.3.1   Isomer (Chemical) Shift   7
1.3.2   Quadrupole Splitting   12
1.3.3   Magnetic Hyperfine Structure   16
1.4   The Mössbauer Spectra of Minerals   19
1.4.1   Distribution of Fe^{2+} and Fe^{3+} in Rock-Forming Silicates .   20
1.4.2   The Spectra of Sulfide Minerals   28
1.4.3   Spectra of Ferric Oxides and Hydroxides   32
1.4.4   Spectra of Iron in Different Classes of Minerals, in Meteorites,
        Tektites, and Lunar Samples   35

2   X-Ray and X-Ray Electron Spectroscopy

2.1   Basic Concepts   38
2.1.1   Development Stages   38
2.1.2   Systematics of X-Ray and X-Ray Electron Spectroscopy
        Types   39
2.1.3   X-Ray Emission Spectra   44
2.1.4   X-Ray Absorption Spectra. The Quantum Yield Spectra.
        Reflection Spectra. Isochromatic Spectra   50
2.1.5   X-Ray Electron Spectroscopy (Electron Spectroscopy for
        Chemical Analysis). Auger-Spectroscopy   56
2.2   Application of X-Ray and X-Ray Electron Spectroscopy to
        Study of Chemical Bonding in Minerals   59
2.2.1   Determination of Molecular Orbital and Energy Band
        Schemes from X-Ray and X-Ray Electron Spectra   60
2.2.2   Chemical Shifts in the X-Ray and X-Ray Electron Spectra
        and Determination of Effective Charges   72

3   Electron Paramagnetic Resonance

3.1   The Substance of the Electron Paramagnetic Resonance
       (EPR) Phenomenon   76
3.1.1 The Scheme of Obtaining EPR Spectra   76
3.1.2 The Energy Levels Scheme   82
3.1.3 Substances Which Can Be Investigated by Using EPR . . .   85
3.1.4 EPR, Lasers, and Masers   87
3.2   Physical Meaning of the EPR Spectra Parameters   88
3.2.1 The Order of Measurements; Experimental and Calculated
       Parameters   88
3.2.2 g-Factor and Splitting of the Spin Levels in a Magnetic Field   91
3.2.3 Fine Structure of the EPR Spectra; Fine Structure B_n^m
       (or D, E) Parameters; Initial Splitting   94
3.2.4 Parameters of Hyperfine Structure. Interaction with Magnetic
       Nuclei of Paramagnetic Ions   104
3.2.5 Spin-Hamiltonian; Order of Calculation the Paramagnetic
       Ions Spectra   111
3.3   Investigation of Minerals by EPR Spectra   114
3.3.1 Principal Applications of EPR Spectroscopy in Mineralogy
       and Geochemistry   114
3.3.2 Survey of EPR Data for Paramagnetic Impurity Ions in
       Minerals   115

**4   Nuclear Magnetic Resonance (NMR) and Nuclear
       Quadrupole Resonance (NQR)**

4.1   The Principle of the Phenomenon and Types of Interaction
       in NMR   119
4.1.1 Types of Nuclei Viewed from the Standpoint of NMR . .   122
4.1.2 Two Types of NMR-Investigations   123
4.1.3 Principal Mechanisms of Interactions in NMR   124
4.1.4 Spectra of Nuclei with $I = 1/2(H^1, F^{19})$ in Solids   126
4.1.5 Spectra of Nuclei with $I \geqslant 1$ in Solids   128
4.1.6 High-Resolution NMR-Spectroscopy in Solids   130
4.2   Nuclear Magnetic Resonance in Minerals   130
4.2.1 Types and Behavior of Water in Minerals; Structural
       Position of Protons   130
4.2.2 Structural Applications of NMR   134
4.2.3 Experimental Estimations of the Crystal Field Gradient .   135
4.3   Nuclear Quadrupole Resonance   136
4.3.1 The Energy Levels Diagram and the Resonance Condition
       in NQR   136
4.3.2 Quadrupole Nuclei and Requirements on the Study Sub-
       stance   138
4.3.3 NQR Spectra Parameters   139
4.3.4 Minerals Investigated and Data Obtained   140

5 Luminescence

5.1   Major Steps in the Development and the Present-Day State 141
5.1.1   Applications of Luminescence in Mineralogy 141
5.1.2   Major Steps in the History of Luminescence 143
5.2   General Concepts, Elementary Processes, Parameters . . . 146
5.2.1   Theoretical Bases Necessary for Understanding the Processes
        of Luminescence   147
5.2.2   Absorption, Luminescence, Excitation Spectra: the Scheme
        of the Experiment and Energy Levels 148
5.2.3   Energy Level Patterns and Configuration Curves Diagrams 151
5.2.4   Kinetics of Ion Luminescence in a Crystal; Fluorescence
        and Phosphorescence. 153
5.2.5   Transfer of Energy in Luminescence: Sensitization and
        Quenching 160
5.2.6   Representation of Luminescence in the Band Scheme and the
        Luminescence of Crystallophosphors 176
5.2.7   Methods of Luminescence Excitation 188
5.3   Types of Luminescent Systems in Minerals 194
5.3.1   Transition Metal Ions; the Crystal Field Theory and
        Luminescence Spectra 194
5.3.2   Rare Earths; Absorption and Luminescence Spectra . . . 196
5.3.3   Actinides; Absorption and Luminescence Spectra 211
5.3.4   Mercury-Like Ions; Pb^{2+} in Feldspars and Calcites . . . 215
5.3.5   Molecular Ions S_2^-, O_2^-, and F-Centers 216
5.3.6   Crystallophosphors of the ZnS Type; Natural Sphalerites
        and Other Sulphides 217
5.3.7   Luminescence of Diamond   220

6 Thermoluminescence

6.1   Mechanism and Parameters of Thermoluminescence . . . 224
6.1.1   The Nature of Emission Centers 224
6.1.2   The Nature of Trapping Centers 225
6.1.3   Determination of the Thermoluminescence Parameters . . 228
6.2   Experimental Data, Their Interpretation and Application
        in Geology 230
6.2.1   Alkali Halide Crystals 231
6.2.2   Fluorite   232
6.2.3   Anhydrite. 235
6.2.4   Quartz 236
6.2.5   Feldspars 237
6.2.6   Calcite and Dolomite. 238
6.2.7   Zircon 238
6.2.8   Geological Applications. 238
6.2.9   Crystallochemical Factors. 239
6.2.10 Physicochemical Factors 239
6.2.11 Geological and Geochemical Factors 239
6.2.12 Geological Age Dependences 240

7   Radiation Electron-Hole Centers (Free Radicals) in Minerals

7.1   Basic Principles and Methods 242
7.1.1   Discovery of Free Radicals in Minerals and Their Wide
         Distribution 242
7.1.2   Defects and Centers 244
7.1.3   Free Radicals in Crystals 246
7.1.4   Molecular Orbital Schemes and EPR Parameters 247
7.1.5   The Way to Identify the Electron-Hole Centers from the
         EPR Spectra 251
7.1.6   Systematics of the Electron-Hole Centers in Minerals and
         Inorganic Compounds 254
7.2   Description of the Centers 256
7.2.1   Oxygen Centers: O^-, O_2^-, O_2^{3-}, O_3^- 257
7.2.2   Carbonate Centers: CO_3^{3-}, CO_3^-, CO_2^- 262
7.2.3   Sulfate and Sulfide Centers: SO_4^-, SO_3^-, SO_2^-, S_2^-, S_3^- . . . 263
7.2.4   Silicate Centers: SiO_4^{5-}, SiO_4^{3-}, SiO_3^-, SiO_2^- 264
7.2.5   Phosphate Centers: PO_4^{4-}, PO_4^{2-}, PO_3^{2-}, PO_2^{2-}, PO_2^0 . . 265
7.2.6   Impurity Cation Centers 265
7.2.7   Hole Center S^- 266
7.2.8   Atomic Hydrogen in Crystals 267
7.3   Models of Centers in Minerals 268
7.3.1   Prevalence and Significance of Centers in Minerals . . . 268
7.3.2   Features Specific to the Structural Type and Models of the
         Centers in Minerals 269
7.3.3   Quartz 270
7.3.4   Feldspars 278
7.3.5   Framework Aluminosilicates with Additional Anions:
         Scapolite, Cancrinite, Sodalite, Ussingite Groups 279
7.3.6   Zeolites 283
7.3.7   Zircon 283
7.3.8   Beryl, Topaz, Phenakite, Euclase, Kyanite, Danburite,
         Datolite 283
7.3.9   Calcite 284
7.3.10   Anhydrite 284
7.3.11   Barite and Celestine 284
7.3.12   Apatite 286
7.3.13   Sheelite 286
7.3.14   Fluorite 287
7.4   Electron-Hole Centers in Alkali Halide Crystals 289
7.4.1   F Center 289
7.4.2   F Center in Compounds of Other Types 294
7.4.3   F Aggregate Centers 298
7.4.4   V Centers and Molecule Ions Hal_2^-, Hal_2^{3-} 301

References 304
Subject Index 347

1. Mössbauer (Nuclear Gamma-Resonance) Spectroscopy

1.1 Basic Principles and Experimental Arrangement

1.1.1 Isomer Nuclear Transitions and Gamma-Ray Emission

A nucleus, like an atom, has discreet (quantized) ground and excited levels, the transition from the upper to the lower level being accompanied by gamma-ray emission. Nuclear gamma emission line spectrum is similar in this respect to atomic optical emission spectrum occurring as a result of the transition from the upper to the lower electronic level.

The period during which the nucleus continues to stay in any excited state determines its mean lifetime. This is the time necessary for the transition from a given to the ground, or some other lower-lying excited level.

Two nuclei with equal charge and mass number, but in different excitation states with easily measurable lifetimes, are called isomer nuclei. By isomer states are understood excited states of a nucleus with a fairly long lifetime (typical values of the mean lifetimes of the isomer states of the nuclei used in the Mössbauer spectroscopy are 10^{-6}–10^{-10} s).

The isomer transitions that are used in Mössbauer spectroscopy are shown in Figure 1 for nuclei of iron, iridium and tin. Thus, gamma-emission of a Fe^{57} nucleus with energy of 14,400 eV occurs as a result of an isomer transition from the excited state of the Fe^{57} nucleus with a nuclear spin $I = 3/2$ (for this excited state), mean lifetime $\tau = 1.4 \cdot 10^{-7}$, and energy $E = 14,400$ eV to the ground state of a Fe^{57} nucleus with a nuclear spin $I = 1/2$. The nuclear spins of the ground and excited states determine the splitting of nuclear levels in crystalline fields, as well as in external electric and magnetic fields; the lifetimes determine the natural width of the lines.

Since in the case of Fe^{57} an isomer with energy of 14,400 eV has a lifetime of only about 10^{-7} s, in practice a radioactive cobalt isotope Co^{57} with a half-life of 270 days is taken as a source of gamma radiation; then through electron capture, Co^{57} transforms into an excited isomer Fe^{57}. Thus, following the cobalt decay, an iron isomer emitting gamma-rays emerges directly in the spectrometer.

Similarly, upon decay of radioactive osmium-191, iridium isomers emerge. In the case of tin, its isomer with a long lifetime (250 days) is used.

1.1.2 Resonance Fluorescence

1. Atomic (optical) resonance fluorescence. Should a vessel containing Na vapors under low pressure be irradiated by the light of the yellow line of the spectrum of

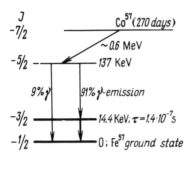

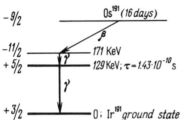

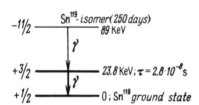

Fig. 1. Isomer transitions of the Fe^{57}, Ir^{191}, Sn^{119} nuclei. Shown are transitions whose gamma-emission is used in the Mössbauer spectroscopy: for $Fe^{57} = 14.4$ keV, for $Ir^{191} = 129.0$ keV and for $Sn^{119} = 23.8$ keV; τ is transition time determining the natural line width; I is nuclear spins in ground and excited states

the same Na (589 nm), absorption of this radiation by the Na vapors will then take place, owing to transition from the lower level $^2S_{1/2}$ to the upper one $^2P_{1/2,3/2}$, followed by emission of yellow light as a result of the reverse transition. This lies at the base of the phenomenon of resonance fluorescence (equal frequencies of primary and secondary emission).

2. Nuclear gamma-resonance fluorescence in the case of free atom nuclei. If Fe^{57} gamma emission, occurring as a result of an isomer transition from the level 3/2 with energy of 14,400 eV to the level 1/2 of the ground state (see Fig. 1), is directed against the free atom nuclei of the same Fe^{57}, there should then be observed the resonance fluorescence, due to transition from the ground state 1/2 to the same level of 3/2. This, however, does not happen.

3. The reason for the difference in conditions under which atomic (optical) and nuclear (gamma) resonance fluorescence are observed is as follows. The emission of a gamma quantum by the nucleus is followed by a nuclear recoil (as a barrel recoil after a shot), in which a part of the energy of the transition of the nucleus from the excitation to the ground level is spent. An equal amount of energy is spent in accelerating the absorption nucleus (absorber). Hence, the gamma-quantum energy fails to equal the difference between the absorption nucleus levels, but is lesser by

the amount of double energy of recoil. The latter is much greater than the width of the line, which is determined by the diffuseness of the excited level and, for this reason, conditions for resonance (equal energy difference between levels of the emitting and absorbing nuclei) are not complied with, and in this case no resonance gamma fluorescence is observed.

Thus for a Fe^{57} nucleus with emission energy of $E = 14,400$ eV, the recoil energy R is $0.19 \cdot 10^{-2}$ eV, and the line width[1] $\Gamma = 4.6 \cdot 10^{-9}$ eV.

In the case of the Na atom for the D line with $\lambda = 589$ nm (16.978 cm^{-1}) $E = 2.1$ eV, $R = 10^{-10}$ eV, $\Gamma = 4.4 \cdot 10^{-8}$. Accordingly, here the line width is much smaller than the recoil energy ($2R \ll \Gamma$), and recoil does not stand in the way of resonance fluorescence. For the Fe^{57} nucleus that yields emission of a far greater energy, the recoil energy $2R$ is accordingly much higher than for Na atom (though it constitutes only a negligent part of the total emission energy), being by far in excess of the line width, i.e., $2R \gg \Gamma$. For this reason no resonance fluorescence occurs here.

1.1.3 Mössbauer Effect: a Recoilless Gamma-Fluorescence

It has been possible to eliminate nuclear recoil during emission and absorption of gamma quanta. The essence of the Mössbauer effect consists in placing a nucleus in a crystal, when the recoil impulse is borne not by a single nucleus but by the whole of the crystal lattice, the emission and absorption of gamma-quanta by nuclei thus fixed taking place with practically no energy losses in recoil. This permits it to obtain resonance gamma fluorescence as mentioned above and as exemplified in the case of Fe^{57}. To this end it suffices for the emitting nucleus (Fe^{57}, Os^{191}, Sn^{119} and others) and the absorber of the same composition to form only part of a solids.

Here, a major property of recoilless gamma emission becomes manifest, i.e., an extremely small width of the line by comparison with emission energy. Thus for Fe^{57} $\Gamma = 4.6 \cdot 10^{-9}$eV with gamma-emission energy $E = 14,400$ eV. The Γ/E ratio equals $3.2 \cdot 10^{-13}$. This means that the energy of the emission can be determined to an accuracy of the order of 10^{-13}. Because of this, recoilless gamma emission can be measured most exactly of all regions of the electromagnetic spectrum.

Doppler Velocity. Owing to the apparent difference in the energy of emitting and absorbing nuclei and disturbed resonance conditions, it is possible to record the slightest deviations in the chemical state of absorbers. The conditions of resonance have to be re-established in order to assess and measure all these deviations. This is accomplished by using the Doppler effect, whose essence consists in the emission source moving toward the observer at a velocity of v, and the emission energy increasing, while with the source moving away from the observer it diminishes by a value of $\Delta E = \pm v/c \cdot E$.

---

[1] Conformable to Heisenberg's uncertainty relation $\tau\Gamma = h$, where τ is mean lifetime of the state; Γ is uncertainty of the excitation state energy that determines the width of the excitation state energy that determines the width of the excitation level, and consequently that of the line; $\hbar = h/2\pi$, where h is Planck's constant. For Fe^{57} $\tau = 1.4 \cdot 10^{-7}$ s (Fig. 1); $\hbar = 1.05 \cdot 10^{-27}$ erg. Hence $\Gamma = \hbar/\tau = 0.75 \cdot 10^{-20}$ erg $= 4.6 \cdot 10^{-9}$eV.

The difference in the emission and absorption nuclear energy can be offset by displacing the specimen (absorber) with varying speed relative to the source.

For the Fe^{57} nucleus, the speed of the relative motion of the absorber and the source $v = 1$ mm s^{-1} corresponds to an energy difference:

$$\Delta E = \pm \frac{1 \text{ mm s}^{-1}}{3 \cdot 10^{11} \text{ mm s}^{-1}} \cdot 14{,}400 \text{ eV} = 4.8 \cdot 10^{-8} \text{ eV}.$$

For Sn^{119} ($E = 23{,}800$ eV) 1 mm $s^{-1} = 7.9 \cdot 10^{-8}$ eV; for I^{191} ($E = 129{,}000$ eV) 1 mm $s^{-1} = 4.3 \cdot 10^{-7}$ eV.

1.1.4 Experimental Arrangement for Observing the Nuclear Gamma-Resonance (Mössbauer Spectrometer)

As with any other spectra, Mössbauer spectra are measured by using spectrometers consisting of an emission source, a specimen (absorber), and a detector of the resonant gamma-rays transmitted through (or emitted, or scattered by) the specimen (absorber). A necessary component for the Mössbauer effect is a system capable of moving the specimen (absorber) relative to the source (Fig. 2).

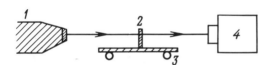

Fig. 2. Experimental arrangement for measuring the Mössbauer absorption spectrum. *1*, Source of gamma-emission; *2*, specimen containing the same nucleus as the source; *3*, the motlon system; *4*, gammaquanta counter

The discovery of the Mössbauer effect has provided researchers with a source of monochromatic recoilless gamma emission. Radioactive isotopes, Co^{57}, for example, are embedded into metallic iron, stainless steel, platinum and other host matrices suitable for preparation of a given source of emission. According to patterns shown in Figure 1, they produce recoilless gamma emission.

A distinctive feature of Mössbauer spectra is the fact that a test specimen (absorber; in the form of powder or a monocrystal) reveals resonance fluorescence of the same nucleus (the same isotope) which is incorporated in the source and emits gamma-rays. Thus, by means of the Fe^{57} emission, it is possible to investigate only the ironmaking part of the specimen-absorber (Fe^{57} isotopic abundance is 2.17%), whereas to observe the spectrum of tin, a Sn^{119} source is needed, and so on.

The spectra of absorption, emission or scattering of gamma-rays by the specimen, can be measured but it is the absorption spectrum that is commonly measured.

Detecting is done by means of a gamma counter, whose basic component is usually NaI (Tl) scintillation crystal mounted on a photomultiplier tube, which converts weak light flashes into sufficiently strong electric pulses.

The specimen-absorber motion relative to the source (to equalize, by means of the Doppler effect, the energy of gamma quanta emitted by the source's nuclei and absorbed by the specimen), and the measurement of Doppler velocity of this motion are performed by using a carriage traveling at controlled speed. Depending on the latter, two types of spectrometer are distinguished: with constant velocity, and with constant acceleration. In the first the absorber moves with constant speed, and during displacement of the absorber, gamma quanta that enter the detector on traversing the absorber are counted. Thereupon, a new value for the velocity is recorded, the counting is repeated, and so on.

In the other type of spectrometer, the motion proceeds at a variable velocity, gradually and smoothly, increasing and decreasing in a set regularity. Special radio-technical devices—multichannel analyzers—accept the pulses and record the number of gamma quanta transmitted by the absorber during successive velocity intervals. Mechanical and electromechanical spectrometers are also distinguished, depending upon the mode of the motion.

Thus, the *Mössbauer spectrum is the velocity spectrum* and represents the number of gamma quanta recorded by the counter at different Doppler velocities of the absorber relative to the source.

The minimum of the gamma-quanta transmission and, accordingly, their maximum absorption by the specimen are in line with and correspond to the Doppler velocity at which the resonance absorption occurs. The velocity matching the maximum of absorption characterizes the degree of difference in the environment of the specimen's nucleus, and that of the source (when absorber and source nuclei are identical, the velocity at which maximum absorption is observed is nil). The values obtained for the maximum absorption shifts are relative and in this connection the source against which measurements are effected should be mentioned (for Fe^{57} in relation to a Fe^{57} source in stainless steel, for instance). The position of the absorption lines is usually indicated in terms of velocity in mm s^{-1}, or cm s^{-1}, but this can be easily rescaled in eV.

1.1.5 Development of the Method

The Mössbauer discovery was preceded by research into the possibilities offered by gamma-fluorescence of the free atom nuclei. First, gamma fluorescence could be obtained by means of Doppler acceleration of the emission source relative to the absorber, by centrifugation of the latter at a velocity of hundreds of m/s. Broadening of lines due to thermal vibrations (leading to Doppler shifting of the emission frequency and, consequently, to broadening of the lines) were also studied. Thus, the magnitude of nuclear recoil R with emission of a gamma quantum of Ir^{191} amounts to 0.05 eV, whereas the Doppler width of the line (due to thermal vibrations) is as great as 0.1 eV. Here, resonance fluorescence occurs in the case of free iridium atom nuclei.

While investigating resonant scattering of this isotope gamma emission, Mössbauer discovered in 1958 the phenomenon of nuclear recoilless gamma-resonance fluorescence, achieved by fixing the source and absorber nuclei in solids. Lamb's theory dealing with the interaction of slow neutrons constrained in a crystal lattice was used in interpreting these observations.

The low intensity of the effect in the case of Ir^{191} and the need for using low temperatures complicate experimental work with it. The Mössbauer effect found genuine recognition and wide application in 1959, on measurement in the Fe^{57} nuclei, where its observation was easy because of its extremely high intensity, small natural width of the lines, an extraordinarily low magnitude of Γ/E, and also due to the possibility of making investigations at room and even elevated temperatures. A number of physical experiments (gravitational displacement of gamma-quanta frequency, thermal shifts, measurements of the nuclear excited states lifetime, etc.) were carried out by using recoilless Fe^{57} gamma radiation. After observing the effect in the Fe^{57}, Sn^{119} and other nuclei, this technique found application in solid-state physics and later also in the chemistry of solids, and mineralogy.

Fundamental points of, and the theory underlying the phenomena discussed in this section, and also the application of the method in nuclear physics, in the theory of relativity and solid-state physics are dealt with in many published works [6, 8, 23, 42, 45, 46, 71, 133–135, 156, 171, 172].

1.2 Mössbauers Nuclei

The possibilities which Mössbauer spectroscopy offers in extensive studies of minerals and inorganic compounds are, naturally, determined by the number of elements that can be analyzed by this technique.

Limitations imposed upon the method are due to the following factors:

1. For a given element a source of gamma emission must exist—a parent nucleus, whose decay (see Fig. 1) gives rise to the appearance of isomer nuclei.

2. The Mössbauer emission energy lies within the range of from a few keV up to lower hundreds of keV. With rising emission energy, the recoil energy increases, and the probability of observing the effect exponentially diminishes. For this reason, the nuclei of light elements with gamma transitions of the order of meV have to be excluded from the number of even potentially possible Mössbauer nuclei.

3. The utilized radioactive isotope (parent nucleus) must have a fairly long half-life.

4. The lifetime of the isomer transition excited level should be within the range of $\tau = 10^{-6} - 10^{-3}$ s. The lifetime is a factor determining the width of the lines ($\Gamma = \hbar/\tau$) and, accordingly, the ratio of the line width to transition energy Γ/E (sometimes it is the E/Γ ratios that are compared). Most suitable is the ratio of $\Gamma/E = 10^{-10} - 10^{-14}$ (for Fe^{57} Γ/E is $3.2 \cdot 10^{-13}$). With a longer lifetime, reduced Γ and lower Γ/E (for Zn^{67} $\Gamma/E = 5 \cdot 10^{-16}$) Doppler velocity becomes too low and measurable only with difficulty (for Zn^{67} 0.15 mm s^{-1} and the spectrometer vibration with an amplitude of the order of 1 Å may upset the resonance conditions (however, for Zn^{67} the Mössbauer effect could still be recorded). Conversely, with rising Γ/E, the resonance selectivity diminishes.

Hence, suitable energy values and the lifetime of the first excited state are among major requirements which the Mössbauer nucleus must meet.

Among the most typical Mössbauer nuclei are those of Fe^{57}, Sn^{119}, Te^{125}, I^{127}, $Au,^{197}$ $Ir^{191,193}$, many rare earths, and others.

The properties of all the Mössbauer nuclei are cited in a number of survey works and monographs [45, 71].

Here we shall list the properties of the Fe^{57} nucleus, the most important in Mössbauer spectroscopy.

Isotopic abundance	2.17%
Energy of the first excited level E	14.30 keV
Level lifetime	$1.4 \cdot 10^{-7}$ s
Line width	$4.6 \cdot 10^{-12}$ keV
Ratio of line width to transition energy Γ/E	$3 \cdot 10^{-13}$
Ground state spin	1/2
First excited state spin	3/2
Ground state magnetic moment	+0.09 nm
First excited state magnetic moment	−0.15 nm
Quadrupole moment of ground state	0
Quadrupole moment of the first excited state	0.29 barn
Recoil energy R	$1.9 \cdot 10^{-3}$ keV
Conversion coefficient	9.00
Mössbauer absorption cross-section	$2.4 \cdot 10^{-18}$ cm²
Parent nucleus	Co^{57}

1.3 The Mössbauer Spectra Parameters

1.3.1 Isomer (Chemical) Shift

Displacement of Nuclear Levels and Isomer Shift of the Mössbauer Spectra Lines. The difference in the source and sample (absorber) nucleus environment results in changes in the distance between the excited and ground nuclear levels in the sample as compared to the distance of the same nucleus of the source.

To compensate for this change, Doppler velocity is imparted to the absorber (or to the source); the magnitude of the velocity determines the sample absorption line position relative to the source in the Mössbauer velocities spectrum. The displacement of the absorption line in the velocities spectrum in relation to the position which this line would have occupied if the "chemical" environment of the sample's nucleus had been the same as that of the source nucleus is called the isomer or chemical shift.

Graphically this is shown in Figure 3a (in Fig. 3b and c the isomer shift is seen, together with the splitting of the levels and corresponding changes in the spectrum).

Isomer Shift Values Relative to Different Sources. The source of Fe^{57} recoilless gamma emission is usually prepared by incorporating radioactive Co^{57} in the stainless steel, but other sources, such as Fe^{57} in chromium, platinum, iron, copper, sodium nitroprusside, and others are also used. In all these sources, the Fe^{57} nucleus

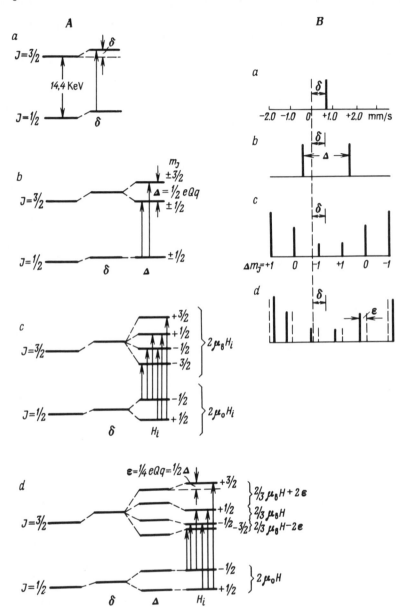

Fig. 3 A and B. Fe^{57} energy levels splitting **A**, and schemes of the corresponding spectra **B**. *a*, Isomer shift δ; *b*, quadrupole splitting–Δ; *c*, magnetic hyperfine structure H_i; *d*, quadrupole splitting and magnetic hyperfine structure

environment is dissimilar. To compare the isomer shifts, cited in various works, the positions of the Fe[57] line in diverse sources relative to Fe[57] in sodium nitro-prusside: Na_2 Fe(CN)$_5$NO 2H$_2$O, taken as a reference [36] are shown in Figure 4.

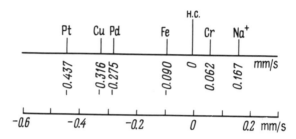

Fig. 4. Fe[57] isomer shift in different sources relative to Fe[57] in stainless steel (s.s.)

Chemical Nature of the Isomer Shift. In order to form an idea as to how the changes in the electron structure of atoms (when these enter different compounds) can affect the displacement of nuclear levels, one has to bear in mind that the nucleus is not only surrounded by electron clouds, but the electron charge passes through (permeates) the nucleus itself[2].

Thus an electrostatic interaction occurs between the nucleus potential and the electron charge (that yielding the well-known Hamiltonian term Ze^2/r). However, it is only s-electron density at the nucleus that differs from zero which causes the interaction of the nucleus with the electrons to take place. The intensity of such an interaction is determined by the s-electron density on the nucleus (at zero point), and is denoted by $\psi_s^2(0)$.

The isomer shift is thus a measure of the s-electron density at the nucleus, as compared with their density at the source nucleus. For the Fe[57] nucleus, the positive isomer shift corresponds to a decreased electron density on the nucleus. The s-electron density, however, becomes subject to the effect produced by the 3d-electrons of iron. This effect manifests itself by shielding the s-electrons on the nucleus. There-fore, an increase in the number of the d-electrons leads to a reduction of the s-electron density at the nucleus and, consequently, to a greater isomeric shift.

Calculation shows that differences in numbers and density of the 3d-electrons more strongly affect the density of the 3s-electrons than that of the 2s- and 1s-elec-trons. The isomer shift is more sensitive to a change in the s-electrons' density than to that of the d-electrons, the s- and d-electrons influencing the isomeric shift in an opposite direction. Hence, the diminution in density of the s-electrons and a mount-ing number of 3d-electrons tend to increase the isomeric shift.

The following observations become therefore quite understandable: (1) by com-parison with metallic iron (electron configuration $3d^64s^2$ or $3d^74s^1$) in the Fe^{2+}($3d^6$)

---

[2] In this way, as a result of the interaction of electrons with their nucleus (EPR spectra hy-perfine structure), and with that of the adjoining atom (superhyperfine structure), splitting of the electronic spin levels takes place (in electronic paramagnetic resonance). In the Mössbauer spectra a reverse effect is in evidence, that of the electron density on the nuclear levels.

and Fe^{3+} ($3d^5$) compounds, the density of the s-electrons diminishes, and a positive isomeric shift is always in evidence; (2) in the Fe^{2+} ($3d^6$) compounds, the number of the $3d$-electrons is greater than in the Fe^{3+}($3d^5$) compounds and, therefore, the s-electron densities at the nucleus for Fe^{2+} is smaller, and the isomer shifts are generally larger; (3) both in the Fe^{2+} and Fe^{3+} compounds, an increasing of ionicity of the bond entails lowering of density of the s-electrons and, accordingly, a larger isomer shift. The characteristic isomer shift values for Fe^{3+} are 0.4–0.5 mm s^{-1} for for Fe^{2+} 1.3–1.4 mm s^{-1} (relative to stainless steel).

Physical Interpretation and Estimation of the Isomer Shift. In considering the Mössbauer spectra of a number of compounds (74) the expression for the isomer shift is deduced as follows:

$$\delta = k \frac{\Delta R}{R} \Delta \psi_s^2(0),$$

where $k = 4\pi(5)Ze^2R^2$; R is mean radius of the nuclear charge distribution; ΔR is difference between the nuclear charges radii in excited (isomeric) and ground states; $\Delta \psi_s^2(0)$ is difference between the electron densities at the specimen (absorber) and source nuclei.

This expression for the isomeric shift emerges from the following model concepts.

The difference in energies of the ground and excited states of the nucleus is due to the difference of the nucleus radii in these states, determined relative to the radius of a hypothetic point nucleus. The electrostatic potential of the point nucleus is Ze/R, and of the actual nucleus is $(Ze/R)(3/2-r^2/2R^2)$. The displacement of any level energy (of the excited or ground states, of the source or of the specimen) relative to the point nucleus may be presented as follows:

$$\delta E = \frac{2\pi}{5} Ze^2 R^2 \psi_s^2(0) = kR^2\psi_s^2(0).$$

The difference in such displacements for the excited level δE_{ex}, and for the ground level δE_{rg} is

$$\delta E_{ex} - \delta E_{gr} = k(R_{ex}^2 - R_{gr}^2) \, \psi_s^2(0).$$

The isomer shift is determined by the difference of this disparity between the absorber and the source,

$$(\delta E_{ex} - \delta E_{gr}) \text{ abs} - (\delta E_{ex} - \delta E_{gr}) \text{ src} = k(R_{ex}^2 - R_{gr}^2) \times [\psi_s^2(0) \text{ abs} - \psi_s^2(0)\text{src}]$$

or $\quad \delta = k \dfrac{\Delta R}{R} [\psi_s^2(0)\text{abs} - \psi_s^2(0)\text{src}].$

This expression for the isomer shift has two factors: one of these, $\Delta R/R$, includes only nuclear parameters, and is assumed to be constant for the Fe^{57} nucleus in different sources and samples, whereas the second factor is an atomic parameter linked with the difference in the distribution of the electron density in concrete samples relative to the source. The values for $\psi^2(0)$ are calculated for different $3d^n$ configurations. By assuming typical δ values for the Fe^{3+} and Fe^{2+} compounds (0.55 and 1.40 mm s^{-1}), it is possible to find the magnitude of $\Delta R/R$. Considering the screening effect of the 4p-electrons with respect to 4s-electrons, this magnitude will be

$$\frac{\Delta R}{R} = -0.9 \cdot 10^{-3}.$$

The negative sign of $\Delta R/R$ for Fe is due to a positive isomer shift, since reduced density of the s-electrons in the Fe^{2+} and Fe^{3+} compounds with removed outer s-electrons by comparison with Fe corresponds to negative values of $\psi^2(0)$ abs and $\psi^2(0)$ src. Moreover, in contrast, the negative shifts in the Sn^{4+} compounds relative to Sn accord with the positive value for $\Delta R/R$.

Diagram of Relation Between the Fe^{57} Isomer Shift and Electron Density. The atom of iron has the following electron configuration:

$$1s^2 \quad 2s^2 \quad 2p^6 \quad 3s^2 \quad 3p^6 3d^6 \quad 4s^2.$$

When analyzing the effect of a change in the density of the s-electrons on the isomer shift, e.g., the influence of the $\psi_s^2(0)$ term on the δ in the above expression for the isomer shift, it is opportune to consider separately the contributions from: (1) 1s , 2s and 3s-electron density on the nucleus; (2) the number of 3d-electrons ($3d^6$ in Fe^{2+} and $3d^5$ in Fe^{3+}) which influence the isomer shift through the medium of the s-electrons, by shielding them against the nucleus; (3) the density of the valence 4s-electrons destined to accomplish chemical binding in the compounds, their fraction of influence on the isomer shift depending on the degree of ionicity-covalency, and denoted in the case of a partially covalent bond by $4s^x$ and in the pure ionic bonding to $4s^0$.

These contributions are shown on the graph in Figure 5 [36]. The summary densities of the 1s , 2s- and 3s-electrons at the nucleus for different numbers of the 3d-electrons ($3d^6$ for Fe^{2+} and $3d^5$ for Fe^{3+}, in particular) and without participation of the 4s-electrons, are indicated on the ordinate axis to the left, viz. $2\Sigma\psi_{ns}^2(0)$ for pure $3d^5$ and $3d^6$ and other $3d^n$ configurations (without 4s-electrons). The greater the number of the d-electrons, the lower the density of the s-electrons on the nucleus, "pushed up" from inside the atom because of the presence there of an additional electron density produced by continuing penetration of the d-electron wave functions into it.

Values adopted for the isomer shift for pure $3d^n$ configurations are the maximum ones observed, i.e., those corresponding to the uppermost ionic state. They are (in mm s^{-1}): for Fe^{3+}: in $Fe_2(SO_4)$ $6H_2O = 0.52$, in $Y_3Fe_3^{3+}O_{12} = 0.57$ (in the octahedral position); for Fe^{2+}: in $FeSO_4$ $7H_2O = 1.40$, in $FeFe_2 = 1.40$ and in $KFeF_3 = 1.39$.

Since the isomer shift–electron density dependence is linear, two δ values (one for $3d^5$ and one for $3d^6$) suffice to plot a scale of isomer shifts (see Fig. 5, ordinate axis to the right).

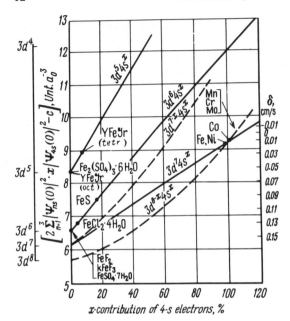

Fig. 5. Diagram showing relationship between electron density on the nucleus and isomer shift

The straight lines originating from the points $3d^5$, $3d^6$, $3d^7$ on the ordinate axis fit configurations $3d^5 4s^x$, $3d^6 4s^x$, $3d^7 4s^x$, where the 4s-electrons' contribution is indicated on the abscissa axis. The scale of the summary electron density of all the s-electrons is shown on the ordinate axis to the left (for the points on the axis itself $x = 0$, i.e., $3d^n 4s^0$).

Hence, the isomer shift enables it to assess both the total density of the s-electrons on the nucleus and the contribution to it of the valence 4s-electrons.

The chart in Figure 5 shows the electron configuration of metallic iron to be close to $3d^7 4s^1$.

This diagram gives only an approximate appraisal of the isomer shift dependences. An analogous diagram has been plotted by Goldansky and co-workers [74] by taking into consideration the effective value of the $3d^n$-electrons. A diagram of similar type was used by Nefedov [290] in interpreting data on isomer shifts and the X-ray line displacements obtained for the same compounds. The approximation here is the purely ionic nature of the compounds for which the isomeric shift values were taken in plotting the scale in Figure 5. Moreover, the δ values exceeding the used values were measured later (see Tables 1–4).

1.3.2 Quadrupole Splitting

Nuclear level scheme and quadrupole splitting in the Mössbauer spectra. The notions of the nuclear quadrupole moment and of the electric field gradient are discussed in Chapter 4. In the Mössbauer spectra quadrupole splitting becomes manifest in the presence of two conditions: (1) when the Mössbauer nucleus displays the quadrupole moment in the ground or excited (isomeric) states; (2) when the symmetry of the intracrystalline field at the site of the nucleus position is lower than the cubic one, e.g., when there is electric field gradient.

Figure 3b illustrates diagramatically the splitting of energy levels for a Fe^{57} nucleus. Figure 6 presents the spectrum of berthierite, $FeSb_2S_4$, with a distinctly marked quadrupole splitting. In its ground state, the Fe^{57} nucleus has a spin $I = 1/2$ and is therefore devoid of the quadrupole moment and its ground level does not split. In the excited state with the Fe^{57} spin $I = 3/2$, however, a quadrupole moment emerges and the isomer level 3/2 (14,400 eV) splits into two sublevels with magnetic quantum numbers $m = 3/2$ and $m = 1/2$. The energy of the sublevels is:

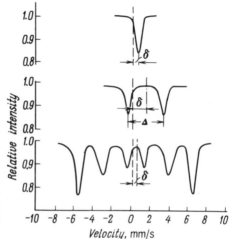

Fig. 6. Mössbauer spectra parameters. δ, Isomer shift (bornite spectrum); Δ, quadrupole splitting (berthierite spectrum); magnetic hyperfine structure (chalcopyrite spectrum)

$$E_Q = e^2qQ \frac{3m^2 - I(I + 1)}{4I(2I - 1)} \text{ eV.}$$

The energy difference between transition from the ground state ($m = 1/2$) to the two excited state sublevels ($m = 1/2$ and 3/2) makes up a quadrupole splitting

$$\Delta = E_Q(\pm 3/2) - E_Q(\pm 1/2) = \frac{e^2qQ}{2},$$

e.g., the difference of transitions $\pm 1/2 \rightleftarrows \pm 1/2$ (σ transition) and $\pm 1/2 \rightleftarrows \pm 3/2$ (π transition) equals $\Delta = 1/2\ e^2qQ$. When the energy of the $\pi =$ transition is greater than the energy of the σ transitions, Δ is then positive, while with a reverse relation, it is negative.[3]

---

[3] Different values characterizing the quadrupole splitting are cited: $\Delta = 1/2e^2qQ$ is distance between quadrupole doublet lines; ε is $1/2\ \Delta = 1/4e^2\ qQ$; $2\Delta = e^2\ qQ$.

The Sn^{119} nucleus, like Fe^{57}, has a ground state spin 1/2 and excited state 3/2 and, therefore, its pattern of quadrupole splitting is analogous to that of Fe^{57}. The energy levels of other Mössbauer nuclei are characterized by different spin values, and the picture of the quadrupole splitting in them varies (with a spin $I = 5/2$, for example, three sublevels emerge: $\pm 5/2$, $\pm 3/2$, and $\pm 1/2$).

The relative probabilities of the transitions $\pm 3/2 \rightarrow \pm 1/2$ and $\pm 1/2 \rightarrow \pm 1/2$ are identical, but display an angular dependence. For this reason, the relative intensities of the quadrupole doublet lines in a monocrystal depend upon the orientation, whereas in a polycrystalline specimen they are identical. For transitions of $\pm 1/2 \pm 1/2$, the angular dependence is in the form of $3/2 (1 + \cos^2\theta)$, for transition $\pm 1/2 \rightarrow \pm 1/2$, it is written in the form of $(1 + 3/2 \sin^2\theta)$, where θ is the angle between the electric field symmetry axis and the direction of the gamma radiation.

Comparison with Nuclear Quadrupole Resonance. One and the same theory and an identical splitting of the levels lie at the root and explain the spectra of the nuclear quadrupole resonance (NQR) and the quadrupole splitting in the Mössbauer spectra. The difference between them can be gleaned from Figure 7. In the

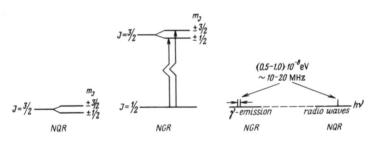

Fig. 7. Comparison of the quadrupole splitting in the Mössbauer spectra (NGR) and in the quadrupole resonance spectra (NQR)

NQR the ground level alone is considered, while in the Mössbauer spectra it is both the ground and excited levels (in the case of Fe^{57}, Sn^{119} only the excited level is subject to splitting). In the NQR, the transition takes place directly between two sublevels, whereas in the Mössbauer spectra, the transition proceeds from the ground level to the quadrupole sublevels of the excited level. The NQR is possible with the absorption of a radio-frequency quantum of the order of tens or hundreds MHz, which equals energy of the order of 10^{-7}–10^{-8}eV. In the Mössbauer spectra the quadrupole splitting accounts for only an insignificant part (1 mm s^{-1} = 4.8 · 10^{-8} eV = 11.61 MHz for Fe^{57}) of the gamma-emission energy (14,400 eV for Fe^{57}) · measurable only because of an extremely small width of the Mössbauer spectrum lines. The displacement of the levels by 2–3 mm s^{-1}, i.e., by 1–1.5 · 10^{-7} eV, or by 10–30 MHz, is of the same order as the distance between the NQR levels.

The nuclei studied by the NQR technique (As, Sb, Bi and others) are either not Mössbauer nuclei, or those on which no effect has been observed so far. Conversely, the most important Mössbauer nuclei: Fe^{57}, Sn^{119} have no quadrupole moment in their ground state.

Mechanism of the Quadrupole Splitting. The asymmetry of the electric charge about the nucleus, which gives rise to the electric field gradient q, is composed of two constituents:

$$q = (1 - R)q_{val} + (1 - \gamma_\infty)q_{lat},$$

where q_{val} is the contribution to the gradient of the field produced by the electron shell proper of the ion; q_{lat} is the contribution coming from the "lattice", i.e., due to the distribution of the charge produced by the ligand ions (or by all the ions of the "lattice"); R and γ_∞ are the Sternheimer antishielding and shielding factors, respectively, which characterize the polarization of the electron shell by the gradient of the field brought forth through contribution of the valence electrons and the lattice.

The contribution of the surrounding atoms q_{lat} is the same as that from the crystalline field, which gives rise to the splitting of orbital electron levels in the optical absorption spectroscopy and to the splitting of the spin electron sublevels in EPR (see Chap. 7).

However, in the case of Mössbauer spectra electrical field gradient is of importance at the site of the nucleus, differing from the gradient produced by the ligand electrons because of the interaction of these ligand electrons with intrinsic electrons of the atom. This type of interaction, when the intrinsic electrons of the atom create an electric field gradient additional to the outer one, is called antishielding.

Since precisely the electrical field gradient (difference of values) at the nucleus site is needed for the two contributions, both from the outer and intrinsic electrons, a deviation of the symmetry from cubic at the site of the atom is necessary. A lowering of symmetry leads to the appearance of the ligand field gradient, and lifts the degeneracy of the intrinsic electron levels.

The value of q_{lat}, that provides for the contribution of the structure distortion to the quadrupole splitting, is proportional to coefficient A_2^0 in the lattice sums. Among the intrinsic atom electrons, the s-electrons display a spherical symmetry, and make no contribution to the field gradient, the latter being provided by the p- and d-electrons alone.

In the case of Fe^{57}, various electron configurations of Fe^{2+} and Fe^{3+} are factors causing differences in these ions relative to the quadrupole splitting.

In Fe^{3+} five $3d$-electrons (term $^6S_{1/2}$) form a half-occupied shell (in a high-spin state) with a spheric symmetry and, for this reason, the intrinsic Fe^{3+} electrons fail to contribute to the quadrupole splitting, which here is determined by the ligand field (but in the case of the low-spin state of Fe^{3+} the spherical symmetry is not retained).

In Fe^{2+} the field gradient is produced largely by the intrinsic electrons, but on condition that the axial or a lower ligand field removes degeneration of the cubic symmetry levels. Therefore, the quadrupole splitting in the Fe^{3+} compounds is commonly considerably less than in the case of the Fe^{2+} compounds. The value of the quadrupole splitting in Fe^{2+} depends on the temperature, for it is associated with changes in the occupancy of the intrinsic electron levels at different temperatures. In contrast, the quadrupole splitting in Fe^{3+} is only slightly temperature-dependent.

A comparison of the value of the quadrupole splitting with the degree of a coordination polyhedron distortion [119, 130] generally shows no unequivocal interdependence. The contribution not only of the first coordination sphere, but also of remote ions, and the overlapping of wave functions of outer valence electrons and occupied shells should also be taken into account. Due consideration of all these interactions helped in calculating the field gradients (in Al_2O_3 and Fe_2O_3) which are in good agreement with the experimental ones [195].

1.3.3 Magnetic Hyperfine Structure

The Scheme of Magnetic Splitting of Nuclear Levels and Hyperfine Structure in Mössbauer Spectra. In magnetically ordered compounds (ferro- and antiferromagnetic) because of an interaction of the magnetic dipole moment of the nucleus μ with the local magnetic field H_i created by the atom's own electrons, Zeeman splitting of the nuclear levels leading to the appearance of a magnetic hyperfine structure in the Mössbauer spectra occurs.

The energy arising subsequent to this splitting of the levels is

$$E_m = - \frac{\mu}{I} H_i \cdot m_I = -g_N \mu_N H_i \, m_I,$$

where the first term describes the properties of the nucleus, i.e., the magnetic moment μ and spin I, or the nuclear g-factor (g_N) multiplied by the nuclear magneton μ_N; the second is the magnetic field of the atom; the third determines the magnetic level $m_I = I, I - 1 \ldots - I$, i.e., a total of $(2I + 1)$ levels arising out of the nuclear level with spin I.

The energy difference ΔE_m of transitions to magnetic sublevels is $\frac{\mu}{I} H_i$.

The level with spin $I = 1/2$ splits in the inner magnetic field into $(2I + 1) = 2$ magnetic sublevels with $m_I = +1/2$ and $m_I = -1/2$. The level with spin $I = 3/2$ splits into $(2I + 1) = 4$ sublevels with $m_I + 3/2, +1/2, -1/2, -3/2$. The selection rule for this magnetic dipole transition is $\Delta m_I = 0, \pm 1$.

In Figure 3c, the splitting of the levels for Fe^{57} and the pattern of a characteristic spectrum of six hyperfine structure lines are shown. The relative probability of these transitions is 3:2:1:1:2:3, this relation of intensities, however, being observed in polycrystalline samples, whereas in monocrystals it is determined by the angular dependence: for the transition $\pm 3/2 \rightarrow \pm 1/2$ in the form $9/4 \, (1 + \cos^2\theta)$; for the transition $\pm 1/2 \rightarrow 1/2$ in the form $3 \sin^2\theta$; for $\pm 1/2 \rightarrow \pm 1/2$ in the form $3/4 \, (1 + \cos^2\theta)$ where θ is the angle between the magnetic field direction and that of the gamma radiation.

The spectrum of a polycrystalline cubic chalcopyrite, $CuFeS_2$, may serve as an example of a typical hyperfine structure (see Fig. 6). For monocrystals, the intensity of transition $\pm 1/2 \rightarrow \pm 1/2$, with $\theta = 0$, i.e., with the gamma-ray passing along the magnetic field axis, is zero.

Further complications in Mössbauer spectra occur with a simultaneous magnetic dipole and electric quadrupole interaction (see Fig. 3d). As compared to the magnetic interaction, the quadrupole splitting is rather small, and one may assume that it displaces magnetic sublevels to one and the same value, but the energy of the $+1/2$ levels declines, and that of the $3/2$ increases.

The magnetic hyperfine structure has been found to exist in many Fe^{57} compounds, whereas in other Mössbauer nuclei (Sn^{119}, rare earths, etc.) it could be observed only with these elements incorporated in ferromagnetic alloys.

The splitting of the nuclear levels in the local magnetic field described above is quite similar to the splitting of the levels in nuclear magnetic resonance (NMR). However, as distinct from the Mössbauer spectra, the magnetic field in NMR is a superimposed outer field generated by the magnet. In NMR the magnitude of the outer magnetic field usually amounts to a few kilooersteds (often 10 Oe) whereas the intensity of the atom magnetic field is commonly of the order of 200–600 kOe. In NMR it is the ground state alone that comes under consideration, while in Mössbauer spectroscopy, one considers both the ground and excited states. NMR results from transition between neighboring magnetic sublevels ($\Delta m_I = \pm 1$) during absorption of a radio-frequency quantum of the order of tens of MHz ($10^{-7} - 10^{-8}$ eV). In the Mössbauer spectrum, on the other hand, a gamma-transition from magnetic sublevels of the ground level to magnetic sublevels of the excited level ($\Delta m_I = 0$, ± 1) occurs.

Local Magnetic Field at the Nucleus. An effective magnetic field at the nucleus H_i is a local one, i.e., it does not extend over the whole of the crystal, but arises due to interaction of the nucleus itself with its own electrons. This interaction includes three components: the interaction of the nucleus with the s-electrons, with the orbital magnetic moment of the atom and with the spin of the atom. If the orbital moment is zero (in Fe^{3+} with $3d^5$ and ^6S) then the interaction with it obviously fails to make any contribution to the local magnetic field.

In actual fact this field is almost completely linked with the interaction of the nucleus with the s-electrons, this being known as the contact Fermi interaction. However, as in the case of the isomer shift, the s-slectrons are subject to the action of unpaired d-electrons, which polarize the s-electrons even of the occupied shells. The greatest number of unpaired d-electrons ($3d^5$ in Fe^{3+}) causes a most strong polarization of the s-shells and largest magnetic fields (of the order of 400–600 kilooersted in Fe^{3+}). Decrease in the number of unpaired d-electrons (four unpaired electrons of the six $3d^6$ in Fe^{2+}) leads to decrease of the magnetic fields (of the order of 200–400 kilooersted in Fe^{2+}).

Let us compare three expressions that describe the basic parameters of the Mössbauer spectra:

Isomer shift: $$\delta = k \frac{\Delta R}{R} \cdot \Delta \psi_s^2(0)$$

Quadrupole splitting: $$\Delta = e^2 Q q$$

Energy difference of transitions to $\Delta E_m = \frac{\mu}{I} H_i.$
magnetic sublevels:

In all cases the first term includes the nuclei parameters, the second atomic (electronic) ones, in conformity with the nature of Mössbauer spectroscopy relating nuclear transitions to electronic structure of the compounds.

The magnetic moments of the nuclei in the ground states are known from experiments with NMR and, therefore, by measuring the magnetic hyperfine structure, it becomes possible to determine the magnetic moment of the excited state of the nucleus and the local magnetic field H_i.

Hyperfine Structure in Paramagnetic Compounds. Local magnetic fields at the nuclei exist also in paramagnetic compounds, but as distinct from ferro- and antiferromagnetics, the hyperfine structure can be observed only in some paramagnetics, and this in very small concentrations and at low temperatures. These differences can be traced by comparing the hyperfine structure in the antiferromagnetic hematite, Fe_2O_3, and in the paramagnetic corundum with an admixture of iron, Al_2O_3: Fe^{3+} [229]. In both cases Fe^{3+} has a spin of the atom $S = 5/2$. The ion levels are, however, split in different ways by the crystalline fields of Fe_2O_3 and Al_2O_3: Fe^{3+}. In the first of these, the exchange interaction of the iron ions results in the state with the spin number of the atom $S = 5/2$ being much lower than the other states, and the only occupied one, from which state alone the hyperfine structure arises.

Quite the reverse, with low Fe^{3+} concentrations in Al_2O_3:Fe^{3+} no exchange interaction occurs; the ion levels are split by the crystalline field to small distances only and all three electronic states, $m_s = 1/2, 3/2, 5/2$, become occupied. Each of these states supplies a hyperfine structure of its own in the Mössbauer spectrum which, therefore, appears as a superposition of three systems of lines.

Why then is the magnetic hyperfine structure not observed in all iron compounds, para-, ferro- and antiferromagnetic?

The hyperfine structure in the Mössbauer spectrum of paramagnetics can be observed under the condition of a large spin-lattice relaxation time (see Chap. 3) of the atomic spin. It is dissimilar for different ions (for Fe^{3+} the spin-lattice relaxation time is long, for Fe^{2+} it is short) and depends upon the concentration and temperature (small concentrations down to $0.n\%$ and low temperatures increase the relaxation time). Thus a distinct hyperfine structure, more complex than in Fe_2O_3, is observed in the Mössbauer spectra of Al_2O_3:Fe^{3+} with the iron content of 0.04% at 78 K, indistinct with 0.4%, and only a single central line being apparent with the iron content of 0.9%.

Probability of the Gamma-Quanta Recoilless Absorption f (or emission) describes the efficiency of absorption (or emission) of the Mössbauer nuclei bounded differently in the lattice (4, 36, 88, 159):

$$f = \exp\left[\frac{-6E_R}{k} \cdot \frac{T}{\theta_D^2}\right] \text{ for } T > 1/2\theta_D,$$

where E_R is the recoil energy of the free atom ($1.9 \cdot 10^{-2}$eV for Fe^{57}, see Sect. 1.2), k is the Boltzmann constant (0.862 eV/K), θ_D is the Debye temperature (depending on the vibrational spectrum of the solid, in particular that related to a given cation site). T is the temperature K of the sample. The f values must be taken into account in quantitative determination of the Mössbauer nucleus content: tin contents in ores by Sn^{119} gamma resonance, Fe^{3+}/Fe^{2+} ratios, quantitative estimations of iron ions distribution between structural positions (see Sect. 1.4).

1.4 The Mössbauer Spectra of Minerals

In mineralogy, the Mössbauer technique has been used mainly since the second half of the sixties. By now the characteristics for major classes and groups of minerals and basic trends have been obtained, and potentialities of the Mössbauer investigations of rocks, ores, meteorites, and lunar regolith have taken a definite shape.

Until very recently Mössbauer spectroscopy in mineralogy could be designated as a method of a single element, iron, with some few isolated works published on tin. This is due not so much to experimental advantages inherent in the gamma resonance of the Fe^{57} nucleus, but rather to the diversity of manifestations of this element in the earth's crust, that find a clearcut reflection in the Mössbauer spectra. It must be expected, however, that within the next years some mineralogical and analytical work will be also done on the nuclei of Au, Pt, Te, Sb, and other elements.

In applying the Mössbauer technique to geology, it is very important that the samples should be investigated in powdered form, and without separation of the monomineral fraction, i.e., both as nondestroyed monocrystals and as grounded samples of rocks and ores, in amounts of up to 200 mg, in the form of intimate to electron-microscopic intergrowths, in fine-dispersed and roentgen-amorphous phases, vitreous formations, adsorbed films, at any concentration, starting from the major component, and ending with impurities (hundredth part of a percent). The information obtained from these spectra also does not diminish in its scope when passing from minerals to monocrystalline rocks, to ochers, clays, fine-grained ores, and earthy formations. In case of the latter this technique in fact represents the most complete characterization of their phase composition.

The following trends in research work are distinguished:

1. Complete characterization of the Fe state in the mineral or in minerals of rock and ores; its valency, coordination, high- or low-spin state, ionic or covalent type of the chemical bonding. In this way one also obtains the characteristic of the Fe-bearing mineral itself and, apart from the features mentioned above, it becomes possible to determine its actual magnetic state, i.e., para-, ferro-, antiferro- or supermagnetic state.

2. Determining the form in which iron is found in dispersed phases and in the state of impurity; each mineral has its own typical spectrum, and this determines the diagnostic importance of the method, especially in fine-grained mixtures.

3. Determining the distribution of Fe^{2+} and Fe^{3+} in nonequivalent structural positions and ascertaining the disordered or ordered state.

4. Detailization of chemical bonding in Fe-containing minerals (effective charges, significance of radical groups formation in sulfides, correlation with in-

teratomic distances, etc.); the Mössbauer nuclei may be employed as a probe introduced into the mineral structure by way of diffusion (into a diamond, or apatite) so as to obtain information on distinctive features of the crystalline fields.

5. Investigations by means of the Mössbauer technique of phase transformations, particularly under high pressures, magnetic transformations in cooling and heating, during processes of oxidation and decomposition.

6. Express analyses and control of the content (for Fe, Sn, Au, and others).

These trends are indicated for Fe, but are valid also in the case of other nuclei.

Let us now consider the application of Mössbauer spectroscopy to different classes and groups of minerals.

1.4.1 Distribution of Fe^{2+} and Fe^{3+} in Rock-Forming Silicates

In the structures of major groups of Fe-Mg minerals, such as olivines, pyroxenes, amphiboles, mica, and also other silicates, there are similar octahedral or proximate positions with cations distributed among them in a variety of ways. The distribution of Fe^{2+} and Fe^{3+} between structural positions and the ratio Fe^{3+}/Fe^{2+} found from the Mössbauer spectra have become as important a feature characterizing these groups of minerals in petrographic investigations as is their iron content. X-ray structural valuations of this distribution require precise calculation of the structure. To do this with the content of impurity of cations in Mn^{2+}, Fe^{3+}, Cr^{3+} and others (see Sect 3.2), EPR is indispensible. Supplementary information on the distribution of cations is obtained from optical absorption and IR spectra.

Theoretical substantiation of the incorporation of cations in one or another octahedral position is given by calculating the energy of these positions, including that of crystal field stabilization energy. The computations of the lattice sums (coefficient A_2^0) that form part of the crystalline field gradient explain the correlation between the quadrupoles' splitting values, and the degree of distortion of the octahedral positions corresponding to them.

The degree of the cationic order of distribution is made use of in apprizing the physicochemical parameters of the mineral formation.

Methodologically, the interpretation of the spectra is done in three stages.

1. Since the nonequivalent positions in the above mineral groups are similar, the Fe spectra obtained in them closely resemble one another in many cases. First of all, one should obtain well-resolved spectra. To this end, frequent use is made of a Fe^{57} source in Pd, which yields narrower lines. In the event of superpositions and overlappings of the lines, the spectrum decomposes into its components. This is usually done by means of computers. In many instances it is only the computer-retrieved spectra that yield a sufficiently reliable decomposition [62, 86].

2. Next, one has to assign the observed peaks (usually quadrupole doublets) to definite structural positions. To do so, the intensity of the outer doublet, i.e., of that with a greater quadrupole splitting, and of the inner doublet, which corresponds to a lesser quadrupole splitting, are compared with the iron occupancy in two structural positions, this occupancy being known from the X-ray structural data. Inasmuch as the mean interatomic distances and the degree of distortion of the two positions in the silicates are usually the same within the entire range of

solid solutions, the assignment of the outer and inner doublets in them also remains the same.

The correlation of outer doublets with less distorted positions, and of the inner doublets with more distorted positions [11], is not based on an exact determination of the degree of distortion. Moreover, the calculation of the gradient of the crystalline field, which causes quadrupole splitting, requires consideration not only of the immediate environment, but of the whole of the crystal lattice, as well as diverse overlappings of the wave functions. Empirical observations have shown the outer doublets in the Mg-Fe-pyroxenes and amphiboles to correspond to more distorted, and in the Ca-Fe-pyroxenes to less distorted positions.

3. Quantitative assessment of the relation n_2/n_1, i.e., Fe^{2+} contents in M_2 and M_1 (as well as Fe^{3+}/Fe^{2+} relation) is made on the ground of the ratio A_2/A_1 of the peak areas plotted, if necessary, by using computers, or in conformity with the relation I_2/I_1 of the peaks intensity [14, 58, 121]:

$$A_2 A_1 = C n_2/n_1 ,$$

where constant C is determined for a given series of compounds from the spectra of samples with known distribution of iron, or from those of the end members. For olivines, pyroxenes and amphiboles $C = 0.9$. Sometimes C is taken to equal unity.

If the Fe content in the mineral is known, then by taking the ratio n_2/n_1, found by the Mössbauer technique, the occupancy X_1 and X_2 at both positions M_1 and M_2 can be defined. Then, from the X_1 and X_2 occupancies the disorder parameter $p(Fe^{2+})$ is calculated [60, 221]:

$$p(Fe^{2+}) = \frac{X_1(1 - X_2)}{(1 - X_1)X_2}.$$

Fe^{2+} in Olivines, Pyroxenes, Amphiboles and Micas. Let us first consider the distribution of Fe^{2+} only: in many species of these groups, Fe^{3+} is present in amounts which, given the prevalence of the Fe^{2+} content, do not show up in the Mössbauer spectra (2%–5% of Fe^{3+}), or display only a weak single peak.

When considering the distribution of Fe^{2+} in these mineral groups, one should first take into account the following data: (1) characteristics (Fig. 8) of the M_1 and M_2 positions (in amphiboles M_1, M_2, M_3 and M_4) according to the mean interatomic distance and the degree of distortion, (2) assignment of the inner and outer doublets to M_1 and M_2 positions, (3) evaluation of the dominant share of Fe^{2+} at M_1 or M_2, the occupancy of each site, and the relative order of distribution.

Table 1 lists the selected values of the Mössbauer spectrum parameters for these groups. The diagramatic illustrations of typical spectra are shown in Figure 9 and their parameters contrasted in Figure 10.

In *olivines* [153, 154, 222] the M_1 and M_2 positions are similar, and in the ordinary spectra only a single doublet is evident which is a superposition of peaks from two positions with a mean $\delta = 1.28$ and $\Delta = 3.02$ mm s^{-1}. A disordered Mg-Fe distribution causes the absence of even the asymmetry of the peaks. The taking of the spectra at elevated ($\sim$ 550 K) temperatures [155], or with high resolution [35]

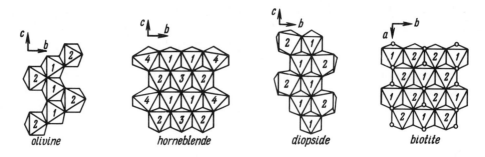

Fig. 8. Spectroscopic aspect of the crystal structures of rock-forming Ca-Mg-Fe silicates. Structural positions occupied by Ca, Mg, Fe are shown with octahedron and octuple coordination

allows it to discern two doublets. The inner doublet is assigned to M_2, and the outer to M_1. Fe^{2+} is distributed in a disorderly fashion, or with a slight enrichment of M_2 with Fe^{2+} in samples of different origin and composition. For $Fe_{26}Fa_{74}$, for instance, $[Mg_{0.13} Fe_{0.37} \rightarrow M_1] [Mg_{0.13} Fe_{0.37} \rightarrow M_2] SiO_4$.

In *pyroxenes* [28, 46, 57, 69, 91, 96, 131, 133, 137, 156, 163, 235] the two positions M_1 and M_2 differ strongly from each other. In all the pyroxenes, M_2 has greater interatomic distances and a more pronounced distortion; M_1 is a distorted octahedron and M_2 is a position with eightfold coordination. Two cases of the assignment of doublets to these positions are distinguished.

Pyroxenes	Outer doublet	Inner doublet
Diopside-hedenbergite	$M_2(Ca)$	M_1
Other Mg-Fe-Ca pyroxenes	M_1	M_2

Table 1. Isomer shift δ^a and quadrupole splitting Δ for Fe^{2+} in olivines, pyroxenes, amphiboles, mica at room temperature

Mineral	δ mm s^{-1}		Δ mm s^{-1}		Reference
	M_1	M_2	M_1	M_2	
Olivine $Fo_{26}Fa_{74}$	1.13	0.99	2.41	2.14	[35]
Olivine $Fo_{82}Fa_{18}$	1.11	1.00	2.47	2.25	[35]
Hypersthene 48% Fe^{2+}	1.24	1.22	2.39	2.01	[11]
Hedenbergite $Dy_{75} Ged_{25}$	1.25	1.26	2.15	2.71	[11]
Pigeonite	1.24	1.21	2.31	1.96	[11]
	M_{123}	M_4	M_{123}	M_4	
Anthophyllite 31.6% Fe^{2+}	1.22	1.18	2.58	1.80	[11]
Grünerite 35.4% Fe^{2+}	1.24	1.18	2.76	1.64	[11]
Actinolite 15% Fe^{2+}	1.24	1.22	2.82	1.89	[11]
	M_1	M_2	M_1	M_2	
Muscovite	1.25	1.20	3.04	2.14	[110]
Phlogopite	1.20	1.18	2.56	2.24	[110]
Biotite	1.22	1.21	2.61	2.19	[110]

^{a} Isomer shift values are re-calculated relative stainless steel (+0.161 relative Na nitroprusside).

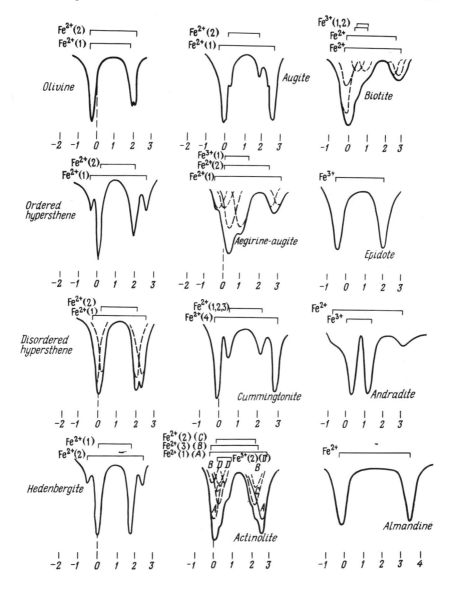

Fig. 9. Fe^{57} Mössbauer spectra in rock-forming silicates

The predominant occupancy of the sites corresponding to the outer doublet is Ca, then Fe and Mg, and to the inner doublet, Mg and Fe.

Thus first Ca enters M_2, and then Mg and Fe are distributed between M_1 and M_2, with Fe^{2+} entering predominantly M_2.

The occupancy of these positions in different Mg-Fe-Ca pyroxenes changes as follows:

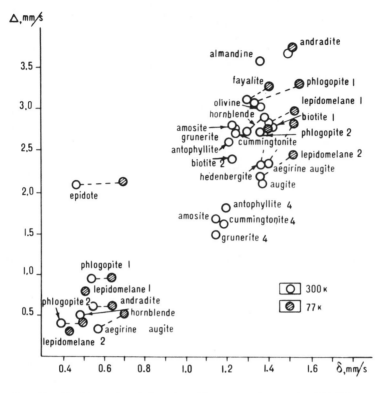

Fig. 10. Position of rock-forming silicates on the graph of Fe^{57} isomer shift (δ) versus quadrupole splitting ($\varDelta$)

M_2	M_1	
Ca	Mg, Fe	Diopside-Hedenbergite
Ca, Fe	Mg, Fe	Augite
Fe, Ca	Mg, Fe	Pigeonite
Fe, Mg	Mg, Fe	Bronzite-Hypersthene-Ferrosilite

Hence the salient features common to the spectra of these pyroxenes are understandable (see Fig. 9): hedenbergite has a strong inner doublet M_1, and a very feeble or no outer doublet M_2; augite has a more intensive outer doublet M_1 with varying M_1 and M_2 intensities, depending upon the composition (Ca-Mg-Fe relations) and the Mg-Fe order; pigeonite and orthorhombic pyroxenes again have a stronger inner doublet M_2.

Thus the occupancy of M_1 and M_2 positions in the monoclinic pyroxenes depends, above all, on the composition (Ca $\rightarrow M_2$, Mg $\rightarrow M_1$) and, even with Fe^{2+} entering both sites, it is essentially ordered. In the orthorhombic pyroxenes an essential degree of the Mg-Fe disorder is observed as it is temperature-dependent, and determined by the distribution coefficients.

Amphiboles [4, 16, 53, 54, 92, 102, 132, 145, 199] have four positions: (a) M_1, M_2, M_3 with very similar interatomic distances usually indiscernible by the Mössbauer technique, and (b) M_4 with greater, more distorted, interatomic distances. The occupancies of these sites in the Mg, Fe, Ca-amphiboles are:

M_4	$M_1M_2M_3$	
[Ca]	[Mg, Fe]	actinolite
[Fe, Mg]	[Mg, Fe]	cummingtonite-grünerite
[Fe, Mg]	[Mg, Fe]	anthophyllite

The outer doublet corresponds to positions $M_1M_2M_3$, and the inner to M_4.

In actinolite a single doublet Fe^{2+} ascribed to positions $M_1M_2M_3$ (and a small Fe^{3+} peak) is usually observed. In the Mg, Fe-amphiboles, both monoclinic and orthorhombic, the form of the spectrum depends largely on the Fe content: in magnesian members a strong inner doublet from Fe^{2+} in M_4 is in evidence with a very weak outer doublet from Fe^{2+} in M_1, M_2, M_3; in the magnesian-ferruginous members, both doublets have a similar intensity; in the ferruginous ones the outer doublet from Fe^{2+} in M_1, M_2, M_3 becomes more intensive (not equal to M_4, but still stronger, since the number of positions $M_1M_2M_3 : M_4 = 5:2$ in the ferruginous end members).

In high-resolution spectra, and by using computer technique, the Fe^{2+} doublets in M_1, M_2 and M_3 could be resolved (Fig. 11). The separation of Fe^{2+} in each M_1, M_2 and M_3 site is done by means of the IR spectra [31, 32]. A comparison of commungtonites from volcanic and metamorphic rocks showed the absence of any essential difference in the distribution of Fe^{2+} [12, 70].

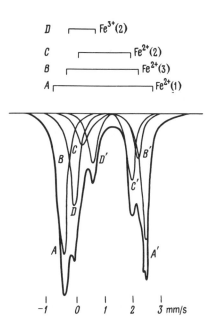

Fig. 11. Computer decomposition into components of the actinolite Fe^{57} spectra [11, 12]

In *micas* there are two types of octahedral positions $MO_4(OH)_2$. In M_1 the hydroxyl groups (OH) are in the opposite vertices of the octahedron, in M_2 in the adjoining ones. The number of the M_2 sites is twice that of M_1. In the dioctahedral mica (muscovite) it is only the M_2 positions (mainly Al^{3+}) that are occupied, while the M_1 ones are vacant. The M_2 dimensions are smaller that those of M_1 ($M_2 - O_{av} = 1.95$ Å, $M_1 - O_{av} = 2.20$ for muscovite). In the trioctahedral mica (phlogopite, biotite, and others) both positions are occupied (mainly by Mg^{2+}, Fe^{2+}) and their dimensions are practically the same ($M - O_{av} = 2.106$ Å).

In obtaining the spectra of micas and other sheet silicates [5, 8, 50, 80–82, 103, 120, 130, 165, 183] one has to eliminate the influence exercised by the orientation of the mica flakes, which manifests itself in the greater intensity of the low-velocity peaks. To do so, the specimens are broken up into fine fragments (but not triturated), and then mixed together with a powdered neutral substance.

In the spectra of micas are seen two broad, usually unsplit bands of Fe^{2+} in M_1, and M_2, together with Fe^{3+} peaks superposed on the low-velocity peak as a shoulder. The high-velocity band represents a superposition of two peaks; in appearance these are two Fe^{2+} doublets, the inner, weaker, from Fe^{2+} in M_2 and the outer that corresponds to Fe^{2+} in M_1. The Fe^{3+} peaks are less clearly discernible, being composed of one or two doublets with small quadrupole splittings.

Fe^{2+} Spectra in Alkali Pyroxenes and Amphiboles. The occupancy of the M_2 site in alkali pyroxenes or of the M_4 site in alkaline amphiboles by Na stipulates the entering of Fe^{3+} in M_1, or in M_1, M_2, M_3:

M_2	M_2	
[Na]	[Fe^{3+}]	aegirine
[Na, Ca]	[Fe^{3+}, Fe^{2+}, Mg^{2+}]	aegirine-augite
[NaCa]	[Mg^{2+}, Fe^{2+}, Fe^{3+}, Al^3]	omphacite

M_4	$M_1M_2M_3$	
[Na]	[Fe^{3+}, Al^{3+}, Fe^{2+}, Mg^{2+}]	riebeckite (crocidolite)-glaucophane
[NaCa]	[Fe^{2+}, Fe^{3+}, Mg^{2+}]	arfvedsonite

Accordingly, in the Mössbauer spectrum a Fe^{3+} doublet is observable at M_1 in pyroxenes, or at M_1, M_2, M_3 in amphiboles, with a weak Fe^{2+} doublet at the same site. A combination with the IR spectrum permitted determination of the prevalent incorporation of Fe^{3+} in crocidolite in M_2 [29].

Fe^{3+} in the Tetrahedral Position in Silicates. For minerals notoriously known to carry Fe^{3+} in silicon-oxygen tetrahedra, the parameters obtained are: for synthetic ferridiopside $CaMg_{0.74}$ $Fe^{3+}_{0.26}$ $Si_{1.74}$ $Fe^{3+}_{0.26}$ O_6): $\delta = 0.18$ mm s^{-1}, $\Delta = 1.49$ mm s^{-1} [94]; for ferruginous orthoclase: $\delta = 0.46$, $\Delta = 0.68$ mm s^{-1} [26]; for sapphirine: $\delta = 0.28$–0.31, $\Delta = 0.78$–1.37 mm s^{-1} [94]; (all values relative to Fe] and for synthetic Ca, Mg, Fe-pyroxenes with a $CaFeFeSiO_6$ molecule [178]. Through decomposition of composite bands, peaks assigned to Fe $^{3+}$ are selected at the tetrahedral position in ferriphlogopite [109, 110], and in some sheet silicates [20, 226].

Iron in feldspars. Besides the Fe^{3+} impurity spectrum in the ferruginous orthoclase [7, 96] the Fe^{2+} spectrum in amazonite feldspar was also observed. In basic plagioclases of lunar rocks the Mössbauer spectra (Fig. 12) allow determination of Fe^{2+} impurity ions at two sites of feldspar structure: in Ca position and in a position with a lower coordination [95, 197].

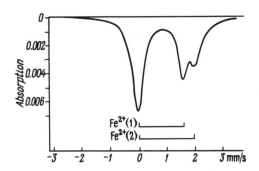

Fig. 12. Mössbauer spectra of Fe^{2+} in feldspars of lunar basalts [95]

In *garnets* Fe^{2+} is in a position with an eightfold coordination (one quadrupole doublet in almandine $Fe_3Al_2Si_3O_{12}$ with a very large splitting), Fe^{3+} in the octahedral position (in andradite $Ca_3Fe_2Si_3O_{12}$ a single Fe^{3+} doublet with a very weak Fe^{2+} doublet similar to almandine one). Y-Ga-, Y-Fe-, RE-garnets crystallizing in the garnet structural type find application in techniques because of their special magnetic properties.

In contrast to the general formula of the silicate garnets $[R_8^{2+}]_3 [R_6^{3+}]_2 [Si_4^{4+}]_3 O_{12}$, where R_8^{2+} is Ca, Mg, Fe^{2+} in the eightfold coordination, R_6^{3+} is Al, Fe^{3+} in the octahedron, the general formula for these garnets is written in the form $[R_8^{3+}]_3$ $[R_6^{3+}]_2 [R_4^{3+}]_3 O_{12}$, where R_8^{3+} is Y^{3+}, RE^{3+}; R_6^{3+} is Ga^{3+}, Fe^{3+}; R_4^{3+} is Ga^{3+}, Al^{3+}, Fe^{3+}. Mössbauer spectra of these have been studied quite extensively [2, 3, 57, 147, 147].

In *tourmalines* (Fig. 13) Fe^{3+} populates pre-eminently the Mg-octahedron but also enters the Al-octahedron (in black tourmaline) [22, 23, 83, 164, 215].

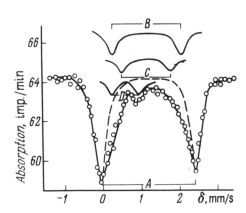

Fig. 13. Mössbauer spectra of Fe^{57} in tourmalines (after [22]). *A, B, C,* Fe^{2+}; *D,* Fe^{3+}

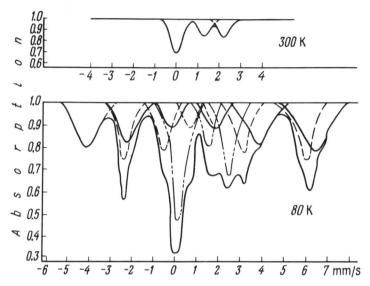

Fig. 14. Mössbauer spectra of ilvaites at 300 and 80 K [24]

In *ilvaites* (Fig. 14; Fe^{2+} and Fe^{3+} in octahedral positions) the $Fe^{2+} \rightarrow Mn^{2+}$ replacement raises the Curie point. At 80 K this determines the transformation of the simple spectrum of two doublets (Fe^{2+} and Fe^{3+}) into a complex spectrum with a magnetic hyperfine structure of both atoms Fe^{2+} and Fe^{3+}, when the Mn proportion is high enough [24].

In *stavrolite* Fe^{2+} spectrum was obtained in a tetrahedral coordination with a low Fe^{3+} content [11, 186].

In epidote a quadrupole splitting uncommonly great for Fe^{3+} has been observed with $\Delta = 2.15$ mm s^{-1} and the isomer shift δ of 0.47 [163].

In *danalite-helvine* with Fe^{2+} in an inordinate fourfold coordination (a trigonal pyramid with sulfur at its summit and three oxygen parts at the base) a quadrupole splitting extremely small for Fe^{2+} occurs (at room temperature): for danalite $\Delta = 0.24$ mm s^{-1}, $\delta = 1.07$, and for helvine $\Delta = 0.11$ mm s^{-1} $\delta = 0.81$ (190).

Mössbauer spectra have been investigated also for vesuvianites [160], beryl [180, 185], kyanite and cordierite [180].

Investigations of transformations made by means of the Mössbauer technique covered a number of slilicates. They included weathering of biotites and vermiculites [25, 187], and high-temperature reactions in the kaolinite group [149].

1.4.2 The Spectra of Sulfide Minerals

The specificity and significance of nuclear gamma resonance specira of sulfides are determined by a complex state of chemical bonding in them, complicated valence interrelations, possible existence of low-spin states alongside the high-spin ones, varying magnetic (para-, ferro- and antiferromagnetic), and electric (semiconducting and other) properties.

The parameters of the sulfide minerals' spectra are listed in Table 2 and are presented in Figure 15, typical spectra being shown in Figure 6.

Distinction is made (see Fig. 6) among: (1) spectra represented by a single line (bornite), i.e., without quadrupole splitting and with no hyperfine structure; (2) quadrupole doublets (pentlandite, sphalerite, stannite, berthierite, represented in Fig. 6, and pyrite, marcasite, arsenopyrite, and loellingite yielding the similar doublets); (3) spectra composed of six lines of magnetic hyperfine structure (chalcopyrite, cubanite, with troilite and pyrrotite displaying the spectra of the same type).

The parameters of the sulfide minerals spectra are characterized by the following values [76, 77, 84, 87, 118, 151, 152, 161, 174–176, 198, 205, 218].

1. Isomer shift. A highly marked covalence of sulfides by comparison with oxides results in a reduced density of the s-electrons on the Fe^{57} nucleus and, by virtue of this, a lesser isomer shift: in sulfides Fe^{2+} δ are 0.38–1.20 (in mm s^{-1} relative to

Table 2. Mössbauer spectra parameters in sulfide minerals. (After [161])

Mineral	Ion	T, K	δ	Δ	H_l
	System Fe – S				
Sphalerite ZnS:Fe	Fe_4^{2+}	300	0.75	0.80	—
		80	0.95	2.04	—
Wurtzite Zns:Fe	Fe_4^{2+}	300	0.78	0.80	—
		80	1.00	2.05	—
Pentlandite (Fe, Ni)S	Fe_4^{2+}	300	0.58	0.40	—
		80	0.61	0.57	—
Pyrite FeS$_2$	Fe_6^{2+}	137	0.42	0.63	—
			0.48	0.64	—
Marcasite FeS$_2$	Fe_6^{2+}	144	0.38	0.52	—
			0.45	0.53	—
Smithite Fe$_3$S$_4$	Fe_6^{2+}	300	0.56	0.35	
	System Fe – As				
Lõellingite FeAs$_2$	Fe_6^{2+}	293	0.39	1.68	—
		80	0.41	1.73	—
Arsenopyrite FeAsS	Fe_6^{3+}	293	0.39	1.68	—
		80	0.35	1.06	—
	System Cu – Fe – S and others				
Chalcopyrite CuFeS$_2$, tetr.	Fe_4^{3+}	80	0.36	0	365
Chalcopyrite (talnachite)	Fe_4^{3+}	80	0.50	0	370
Cubanite Cu$_2$Fe$_4$S$_5$	Fe_4^{2+}	300	0.51	0.88	320
		80	0.12	1.02	332
Bornite Cu$_5$FeS$_4$	Fe_4^{3+}	300	0.47	0	—
Stannite Cu$_2$FeSnS$_2$	Fe_4^{2+}	300	0.87	2.76	—
		80	0.93	3.20	—
Berthierite CuSb$_2$S$_4$	Fe_6^{2+}	300	1.20	2.67	—
		80	1.27	3.60	—

δ is isomer shift relative to Fe^{57} in stainless steel, mm s^{-1}; Δ is quadrupole splitting $(-1/2\ e^2qQ)$ mm s^{-1}; H_l is local magnetic field, kOe.

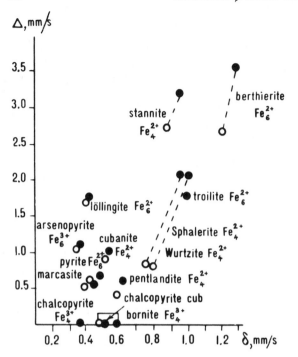

Fig. 15. Parameters of Mössbauer spectra of sulfide minerals

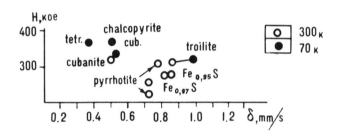

stainless steel), Fe^{3+} δ 0.34–0.50; in oxides, silicates and others Fe^{2+} δ 1.00–1.60, Fe^{3+} 0.35–0.80.

The δ variations range in Fe^{2+} sulfides is, however, quite extensive, and these variations cover values characteristic of both the most covalent compounds and the essentially ionic ones.

As in oxygen compounds, the δ-values are used to distinguish Fe^{2+} and Fe^{3+}, but in most covalent sulfides the values δ for Fe^{2+} and Fe^{3+} overlap. However, these δ values, small for Fe^{2+}, are observed only in low-spin compounds with complex covalent radicals $[S_2]^{2-}$ and $[As_2]^{2-}$.

In the case of simple sulfides (with no complex radicals) the isomer shift increases, depending on the valence and the coordination number: $Fe_6^{2+} > Fe_4^{2+} > Fe_4^{3+}$. The isomer shift values are rather sensitive to a change in the distance Fe- S, rising with the increase of this distance. This, however, is not the sole factor influencing the isomer shift.

As demonstrated by the Mössbauer spectra [161] a factor of great importance affecting the bonding state and, consequently, all the parameters of the spectra, is the formation of complex radicals in covalent sulfides: $[S_2]^{2-}$ in pyrite and marcasite, $[As_2]^{2-}$ in loellingite, $[AsS]^{3-}$ in arsenopyrite, and of more "ionic" ones $[SbS_2]^-$ in berthierite and $[SnS_4]^-$ in stannite.

For sulfides with complex radicals, the isomer shift values diverge sharply from values that are average for sulfides, either toward extremely covalent, low-spin (pyrite, marcasite, loellingite) or in the direction of highly "ionic" (for sulfides) values (stannite, berthierite).

2. The quadrupole splitting Δ occurs in all the Fe^{2+} sulfides, but in the case of simple sulfides it is small (as compared to Δ for Fe^{2+} in oxygen compounds), while for sulfides with complex radicals it runs very high, both as concerns the covalent low-spin Fe^{2+} in loellingite and the high-spin Fe^{2+} in stannite and berthierite. For Fe^{3+}, the quadrupole splitting in a measurable form is lacking for sulfides in which no complex radicals evolve, but reaches a very high Fe^{3+} level in arsenopyrite with the complex radical $[AsS]^{3-}$. A noticeably great temperature dependence of the quadrupole splitting in stannite and berthierite deserves attention and, especially in sphalerite and wurtzite (see Fig. 15).

3. Magnetic local fields H_l at the Fe^{57} nuclei are seen to occur in magnetically ordered sulfides. These are sulfides without complex radicals, and with adjacent Fe-S polyhedra. In the Fe-S system, the magnetic hyperfine structure is observed in the ferromagnetic trolites and pyrrhotites in whose structure the octahedra faces contact one another. In the Cu-Fe-S system subject to examination were the spectra of three compounds in which the Fe-S_4 tetrahedra alternate with the Cu-S_4 tetrahedra:

$$Fe:Cu = 2:1 - Cu_2Fe_4S_5, \text{ cubanite, } Fe_4^{2+}, \text{ ferrimagnetic}$$
$$1:1 - CuFeS_2, \text{ chalcopyrite } Fe_4^{3+}, \text{ antiferromagnetic}$$
$$1:5 - Cu_5FeS_4, \text{ bornite } Fe_4^{3+}, \text{ paramagnetic.}$$

In bornite, the Fe-S_4 tetrahedra are not adjacent, and it is not paramagnetic. In chalcopyrite, the Fe-S_4 tetrahedra are linked by their vertices (the Fe–Fe distance being 3.67 Å), in cubanite, by their faces (Fe–Fe = 2.81 Å). The former is antiferromagnetic and the latter ferrimagnetic [151, 161]. For all three of them modifications exist, ordered in regard to the Cu and Fe distribution (low-temperature tetragonal or orthorhombic–pseudotetragonal) and disordered ones (high-temperature, cubic). The disorder leads to a slight increase in the Fe-S distance and, hence, also to a somewhat increased isomer shift, and to a very insignificant enlargement of the magnetic field.

The magnetic field value at the Fe^{57} nuclei depends on the valence and coordination and is increasing in the Fe_6^{2+}-Fe_4^{2+}-Fe_4^{3+} series. As compared to oxides, H_l in sulfides decreases owing to the covalence bonding in sulfides and this makes it possible in valuating the chemical bonding to use H_l (kOe, at 300 K):

in sulfides		in oxides
Fe^{2+} 225 (pyrrhotite)		–
332 (cubanite)		–
Fe^{3+}	(chalcopyrite)	400–515.

The states of iron with spin $S = 3/2$, assigned to some compounds with a strong rhombic distortion that yields a scheme of energy levels showing a lower doublet with two paired electrons, and with three unpaired electrons on the upper levels, apparently do not take place in sulfide minerals due to small distortions of co-ordination polyhedron.

1.4.3 Spectra of Ferric Oxides and Hydroxides (Table 3)

To this group belong minerals of important systems Fe_2O_3–Fe_3O_4, Fe_2O_3–TiO_2, Fe_2O_3–H_2O and others entering into the composition of iron and chromite ores, bauxites, oxidation zones of ore deposits, mantle of waste, etc. Spinels, chromites, ferrites, and iron garnets find wide application in technique and electronics.

Of prime importance among the Mössbauer parameters for these minerals is the local magnetic field H_l at the Fe^{57} nuclei. Its intensity helps distinguish valency, Fe^{2+} and Fe^{3+}, and state of chemical bonding, coordination and distribution between octahedral and tetrahedral sites. The presence or absence of the magnetic hyperfine structure, and its behavior with changing temperature and superposition

Table 3.   Parameters of the Mössbauer spectra in ferric oxides and hydroxides (at 300 K)

Compound	δ	Δ	H_l	Reference
$Fe_{0.941}O$	1.15	0.30	—	[36]
$(Fe_{0.75}Mg_{0.25})O$	1.15	0.32	—	[36]
$(Fe_{0.10}Mg_{0.90})O_4$	1.15	0.34	—	[36]
α Fe_2O_3 (hematite)	0.47	0.20	520	[36]
$(Fe_{0.9}Al_{0.1})_2O_3$	0.53	0.21	504	[36]
$(Fe_{0.76}Al_{0.23})_2O_3$	0.47	0.20	495	[36]
$(Fe_{0.5}Cr_{0.75})_2O_3$	0.47	0.18	435	[36]
$(Fe_{0.5}V_{0.5})_2O_3$	0.51	0.22	430	[36]
γFe_2O_3 (maghemite)	0.50	−0.10	505	[36]
Fe_3O_4 (magnetite A)	0.45	0.00	500	[56]
(magnetite B)	0.70	0.00	450	[56]
$FeTiO_3$ (ilmenite)	1.20	0.80	—	[20]
Fe_2TiO_5 (pseudobrookite)	0.48	0.35	—	[20]
$\alpha FeOOH$ (goethite)	0.50	0.31	379	[79]
$\gamma FeOOH$ (lepidocrocite)	0.52	0.59	—	[79]
$\beta FeOOH$ (acaganeite)	0.43	0.62	—	[79]

δ is isomer shift relative to stainless steel, mm s^{-1}; Δ is quadrupole splitting, mm s^{-1}; H_l is local magnetic field, kOe.

of an outer magnetic field allow the discernment of ferro-, ferri-, antiferri-, para-, and supermagnetic states, the investigation of thermal phase and magnetic transformation, the tracing of impurities in the minerals, vacant sites, the effects of heat treatment, hydration, and changes in the size of particles.

In particular one can determine the states of iron in superparamagnetic ultrafine particles, which are of prime importance for the understanding of magnetic properties and other features of minerals, rocks, ores, and also of the states of iron existing in the form of isomorphous impurities in dimagnetic minerals.

In FeO (wustite), and also in the system FeO–MgO, in NiO:Fe, CoO:Fe, and MnO:Fe [140, 143, 202], the values for the isomer shift and quadrupole splitting at 300 K are the same within the limits of the measuring error. However, depending upon the Neel temperature T_N, which differs in these compounds, the quadrupole doublet or magnetic hyperfine structure of six lines is observed either at room temperature, or at low or elevated temperatures. The course of the temperature-dependence for the quadrupole splitting and magnetic structure is determined by the effect of the crystalline field, and depends on the composition. In these compounds with NaCl structure and cubic local symmetry, i.e., in positions where there should have been no crystal field gradient, quadrupole splitting occurs just the same. A similar situation exists in spinel-chromites and in sphalerite. This can be related to the Jahn-Teller effect, to defects subsequent to incorporation of impurities, and to the spin-orbital splitting of the lower term.

In hematite, the αFe_2O_3 spectrum consists of six magnetic structure lines. The Mössbauer spectra of the αFe_2O_3–Al_2O_3, αFe_2O_3–Cr_2O_3 and other systems have been investigated, and showed the dependence of H_i and T_N values on the changes of composition.

The spectrum of magnetite which has the structure of inverted spinel AB_2O_4 (A—tetrahedral, B—octahedral positions) and the distribution of cations $[Fe^{3+}]_A$ $[Fe^{3+} + Fe^{2+}]_B$, represents a superposition of two (but not three as one might have expected) spectra with six lines from Fe^{3+} at position A and Fe^{3+} + Fe^{2+} at position B, where the electron exchange Fe^{3+}⇌Fe^{2+} results in their being indistiguishable at all temperatures, save the helium ones, when the spectrum lines are seen to split into three components [56, 167, 189]. The presence of impurities and vacant sites has the effect of changing the H_i, T_N values and the relative intensities of peaks from the A and B positions.

In maghemite Fe_2O_3 [Fe^{3+}] [Fe$^{3+}_{13.38}\square_{2.67}$] O_{32} only a single six-line spectrum is observed, which indicates the presence of only a single kind of the Fe^{3+} ions.

In a large number of investigated ilmenites stemming from many deposits, the spectra parameters proved to be practically analogous [20, 73, 22], but many samples displayed changes (due to weathering), reflected in their spectra.

The Mössbauer spectra help to discern distinctly goethite $\alpha FeOOH$, lepidocrocite $\gamma FeOOH$, which are major natural ferric hydroxides. In the former, the spectrum is represented by a six-line magnetic structure, and in the latter by a quadrupole doublet (Fig. 16). All the ferric oxides have characteristic hydrated states: hydrohematite, hydromaghemite, hydrogoethite, and hydrolepidocrocite. These states are identified in the Mössbauer spectra [9, 44, 60, 67, 79, 85, 113, 124, 125, 143, 146, 223, 224].

In *spinels* MAl_2O_4–*chromites* MCr_2O_4–*ferrites* MFe_2O_4 the Mössbauer effect enables discernment of normal $[M^{2+}]_A[M^{3+}]_B$ and inverted $[M^{3+}]_A[M^{2+}]_B$ structures, and among the latter, ordered and disordered ones. From among numerous compounds of these types [58, 66, 93, 122, 123, 135, 196] the parameters for compounds close to the natural minerals are presented in Table 4.

So far the spectra of compounds with iron forming their basic component have been considered. By using more powerful Fe^{57} emission sources, it has now become possible to obtain Mössbauer spectra from impurity ions (down to $0.0n\%$). This is of special importance in clarifying the nature of magnetic properties of paramagnetic and dimagnetic minerals. With small concentrations and at low temperatures, they begin to display a hyperfine magnetic structure.

Superparamagnetic Ultrafine Particles of Ferric Oxides and Hydroxides. Ferrimagnetic and antiferrimagnetic compounds can have the form of ultrafine particles measuring less than 200–50Å. They then assume a superparamagnetic state, i.e., they turn into single-domain particles with a uniform spontaneous magnetization. In these particles the direction of the magnetic moment at temperatures below the Curie transition point, i.e., when the compound is ferrimagnetic, can become subject to thermal fluctuations. The probability of such fluctuations is determined by the correlation $kT \gtrless kV$, where k is Boltzmann constant ($\sim 10^{-16}$ erg K^{-1}), T is temperature (K), K is magnetic anisotropy constant of the material (~ 107–103 erg cm^{-3}), V is volume of the particle. For instance, with $K = 10^4$ erg cm^{-3} and $T = 100$ K ($kT \sim 10^{-14}$ erg), the particle volume $V = kT/K = -10^{-14}/10^4 = 10^{-18}$ cm^3, i.e., its linear dimension is 100 Å (10^{-6} cm). With the particles size exceeding 100 Å or at a temperature below 100 K, the thermal fluctuations do not suffice to overcome magnetic anisotropy; the substance behaves as a ferrimagnetic, and the Mössbauer spectra show it to have a six-component magnetic structure. Conversely, with the particles size less than 100 Å (with a given magnetic anisotropy constant), or at a temperature higher than 100 K, the substance behaves as a superparamagnetic and its Mössbauer spectrum does not display magnetic structure (see Fig. 16).

In the superparamagnetic state can be hematite, magnetite, goethite, magnesioferrite, ferrous stannates and others [139, 173, 209]. The forms of occurrence of the

Table 4. Parameters of the Mössbauer spectra in spinels, chromites, ferrites [100, 169, 192]

Compound	Distribution of cations	δ	$\varDelta$
$FeAl_2O_4$ *A*	$[Fe^{2+}, Al^{3+}]$	1.10	1.39
B	$[Al^{3+}, Fe^{2+}]$	1.52	2.76
$ZnCr_2O_4$: 3% Fe^{2+}	$[Zn^{2+}, Fe^{2+}] [Cr^{3+}]$	0.02	0
$ZnCr_2O_4$:2% Fe^{3+}	$[Zn^2][Cr^{3+}, Fe^{3+}]$	0.65	0
$MgCr_2O_4$:3% Fe^{2+}	$[Mg^{2+}, Fe^{2+}][Cr^{3+}]$	0.99	0
$MgCr_2O_4$:1% Fe^{3+}	$[Mg^{2+}][Cr^{3+}, Fe^{3+}]$	0.66	0
$ZnFe_2O_4$	$[Zn^{2+}][Fe^{3+}]$	0.77	0.36
$Mn_{0.43}Fe_{0.57}O_4$	$[Mn^{2+}][Fe^{3+}]$	0.50	—

δ is isomer shift relative to stainless steel, mm s^{-1}; $\varDelta$ is quadrupole splitting, mm s^{-1}.

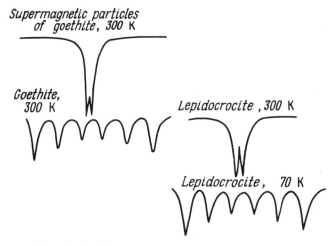

Fig. 16. Mössbauer spectra of iron hydroxides

ultrafine particles can vary greatly; the weak ferrimagnetism of cassiterite is associated with ultrafine particles of the type $Fe_2 SnO_4$ (ferrous stannate) [15, 89], the Mössbauer spectra of the ferro-manganese concretions from the bottom of the Indian and Pacific oceans are attributable to the presence of microcrystalline (less than 90 Å) particles of goethite and lepidocrocite [79, 129].

Bauxites stemming from various deposits contain in the supermagnetic state [126] aluminiforous hematite in the shape of flakes measuring less than 400 Å and aluminiferous goethite of a size about 50 Å (Fig. 17). With these is associated some of the alumina not extractable from the ore by usual procedures.

1.4.4 Spectra of Iron in Different Classes of Minerals, in Meteorites, Tektites and Lunar Samples

Each class and group of iron-containing minerals presents its own problems, determined both by the possibilities of the method and the mineralogical objectives.

In the case of *phosphate minerals*, Mössbauer spectra were used in investigating the process of the vivianite oxidation, and in defining from its example the method of the $Fe^{3+}:Fe^{2+}$ ratio determination [144, 181]. In pure vivianite one single quadrupole doublet only is observed from two Fe^{2+} sites, while upon its oxidation (attrition of the specimen) a peak of trivalent iron makes its appearance. Iron in the form of Fe^{2+} has been introduced by way of diffusion, into natural apatite substituting Ca^{2+} in two structural positions. Such a "probe" helped to describe in detail the crystalline fields in the both positions [99, 177].

In *borates* of the ludwigite–voncenite series, Fe^{2+} has been found to be distributed over three nonequivalent sites [52, 154, 159]. The Mössbauer technique proved helpful in investigating the ferroelectric properties of boracites [214].

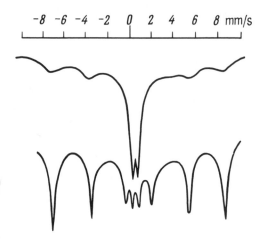

Fig. 17. Mössbauer spectra of goethite (doublet) and hematite (sextuplet) supermagnetic particles in bauxites [126]

In sulfates, iron spectra have been obtained for paramagnetic alums, and for numerous compounds of the jarosite group [112].

In *ferberites-wolframites*, the Fe^{2+} spectra, used in conjunction with neutronography and magnetic susceptibility measurements, enabled determination of the distribution of Mg-Fe and plotting of a magnetic phase diagram for this group [107].

Siderite was one of the carbonate minerals in which structural-magnetic transformations after heating have been studied by using the gamma-resonance iron spectra [141].

By implantation of the Co^{57} isotope into the *diamond* structure, the interstitial iron spectra with its tetrahedral positions in diamond has been obtained [18].

The availability of the Mössabuer characteristics for major ferruginous rock-forming and ore minerals makes it possible to put gamma-resonance spectra to direct use in investigating rocks. The information thus obtained includes an approximate evaluation of the total iron content, the Fe^{3+} Fe^{2+} ratio in the rock, the identification of the ferruginous minerals, the quantitative mineralogical analysis, the distribution of Fe between co-existing minerals, determination of accessory magnetite and hematite.

The importance of these observations for paleomagnetic research is quite obvious. However, of particular value are measurements of the rock spectra in the case of fine-grained sediments, clays, and soils [65].

Mössbauer spectroscopy has been used also in studies of *plant* and *animal fossils* [51], *ozocerite, asphalt, oil-shales, kukersite, shungite, graphite* [231], while the NGR lines of Fe^{57} in oil, bitumen, and coal are too weak to be measured.

High-pressure Mössbauer studies of minerals allow the tracing of changes of spin-multiplicity of Fe^{2+}, reduction of high-spin Fe^{3+} to Fe^{2+}, and electron and magnetic behavior of iron compounds under conditions modeling the earth's interior [47, 61, 116].

In *meteorites*, Mössbauer spectra help determine and characterize olivine, pyroxenes, and small amounts of metallic iron: nickel and troilite [48, 182, 206,

207]. They serve as a basis for compiling a classification of stone meteorites [104]. The products of the oxidation of meteorites, especially fully oxidized fragments, and also the recovered material of returned artificial satellites of the earth and re-entered spacecraft, where oxidation occurs during their re-entry into the atmosphere, and as a result of terrestrial processes of weathering, can be identified by means of the Mössbauer technique. The presence therein of α, β, and γ modifications of FeOOH, αFe_2O_3, Fe_3O_4 has been ascertained.

In *tektites*, a glass of extraterrestrial origin, carrying no microlites at all, and containing 1% to 5% of iron, the Mössbauer spectra display two broad lines of unequal intensity and width with an isomer shift of 1.80–1.20 mm s⁻¹ (relative to stainless steel) and a quadrupole splitting of 1.84–2.08 mm s⁻¹. This points to the existence of a short-range order, and to deviation of coordination from the cubic one.

Many Mössbauer studies have been made for the *minerals of lunar rocks*: olivines, pyroxenes, ilmenite, troilite, metallic iron, plagioclases (oxydation state of iron, site occupancy, order–disorder relations, cooling history of lunar basalts) [45, 59, 91, 95, 96, 97, 108, 131, 157–159, 197, 222]. In the fine-grained lunar regolith and "soil", the lines of the same minerals were separated through computer-retrieved spectral decomposition into individual components and by using the known parameters of the minerals (Fig. 18).

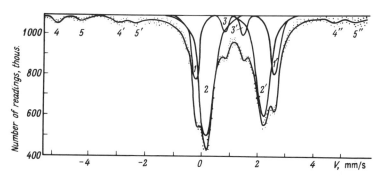

Fig. 18. Mössbauer spectra of lunar regolith [157]. 1, 1', Olivine peaks; 2,2', pyroxene peaks; 3,3', ilmenite peaks; 4,4'4'', metallic iron peaks; 5,5',5'', troilite peaks

The *Mössbauer spectra of Sn¹¹⁹* [36, 38] have been studied in cassiterite [37, 75, 89, 162, 204, 211], stannite [162], and biotite [19], and are used in tin ore prospecting In a number of sulfide and oxide *antimony minerals Sb¹²¹* Mössbauer spectra have also been investigated [208].

2. X-Ray and X-Ray Electron Spectroscopy

2.1 Basic Concepts

2.1.1 Development Stages

Geology has always been one of the principal fields for the application of X-ray spectral analysis. However, in the 60s, its position in geological sciences acquiried a qualitatively new significance, due to its definite place in the cycle of instrumental analytical methods, to the appearance of new devices and instruments offering unique possibilities (X-ray microanalyzers with an electron probe, quantometers, X-ray electron spectrometers) and to its application in investigating the electronic structure of the solids (chemical bonds) in crystals and molecules. The scope of the present book does not include a detailed description of all the applications of X-ray spectroscopy in quantitative analysis (242, 277, 324), the only investigations under consideration have being those carried out with its help into the electronic structure of solids, especially minerals.

In the history of X-ray and X-ray electron spectroscopy three stages can be distinguished.

1. The first stage lasted from the beginning of X-ray spectroscopy till the appearance and elaboration of the qualitative and quantitative X-ray spectroscopic analysis (1913–1950).

Observations of X-ray diffraction by crystals (Laue, 1912) marked the beginning of the X-ray structural analysis. W.H.and W.L.Braggs (1913) showed also that crystals can serve as analyzers of the X-ray spectral composition. X-ray spectra were then obtained for some elements and wavelengths of the X-ray lines measured.

The fundamental law of X-ray spectroscopy was established by H.Moseley back in 1914; he found the existence of a relation between the wavelength of X-rays emitted by a given element and its atomic number. The way to the discovery of this law was paved by the application of Bohr's quantum theory to the interpretation of optic emission spectra of the atoms and, in its turn, provided the most convincing support to the Bohr theory of atomic structure.

The development of techniques of exciting X-rays, analyzing their spectral composition, determining their intensity, establishing K-, L-, M-, N-, O-series, and measuring lines for all the elements that took place in the years 1920–1930 enabled X-ray spectra to be used in the qualitative and quantitative analysis.

Starting from the first years of its analytical application, X-ray spectral analysis has been used in geochemistry and mineralogy. It was helpful in discovering new chemical elements: hafnium in zircon minerals from Norway (Heveshi and Koster in 1922) and rhenium in the Mansfeld shale concentration products (Noddaks in 1925). Next came systematic determinations of the rare earths in minerals and ores, made by V.M.Goldshmidt with co-workers, of the relations of pairs of elements

Hf-Zr, Nb-Ta, V-Ti and others. By 1940 to 1950, X-ray spectral analysis had become one of the standard geochemical methods.

2. The second stage (1950–1960) is marked by the construction of unique types of devices and instruments, whose appearance led to the emergence of new trends in X-ray and X-ray electron spectroscopy. It is quite remarkable that in this seemingly well established branch of research whose theoretical principles and concepts had already been known quite well, the attainement of further, purely technical limits of resolution, focusing and registration led to the opening up of new prospects for geologists, chemists and metallurgists. The principles of these new devices are (1) the X-ray electron probe micro-analyzer, designed in 1950 by R.Castaing in France and by I.B.Borovsky and N.I.Iliin in the USSR, which has found wide application in mineralogy and geochemistry (2) the quantometer (3) the ironless high-resolution magnetic spectrometer, designed by K.Siegbahn with co-workers [308]. This spectrometer made it possible to analyze the energy of photoelectrons knocked out by the X-rays, and to determine with a high degree of accuracy the energy of the electron levels in the atom. The construction of the magnetic spectrometer resulted in the development of electron spectroscopy for chemical analysis (ESCA).

To this time belongs also the designing of the ultrasoft X-ray emission spectrometers that offered new possibilities for investigating the electronic structures of molecules and crystals.

3. The third stage (1960–1977) is the period of applying X-ray spectroscopy to research into the electronic structure of molecules and crystals. Here, a different aspect of the X-ray and electron spectroscopy comes to the fore, which distinguishes it from spectral analysis and makes it enter the domain of spectroscopy of solids with its complex problems and possibilities. Measurements of the X-ray line shifts in diverse compounds have begun to be interpreted in terms of the theories of effective charges, while measurements of the electron levels position by X-ray and electron spectra yielded the quantitative backgrounds for constructing schemes of the electronic structure of concrete crystals within the framework of the energy band theory and molecular orbital theory. It is just these aspects of X-ray and electron spectroscopy that are chiefly to be considered in this chapter.

2.1.2 Systematics of X-Ray and X-Ray Electron Spectroscopy Types

The X-ray spectra together with the optical spectra of the atoms make up the atomic spectroscopy (as distinct from spectroscopy of solids).

Optical spectra appear as a result of transitions between the levels of outer, valence electrons, transitions between ground level and excited levels, being empties in normal states. The X-ray spectra arise due to transitions between the levels of the inner atom shells.

Optical spectra are the spectra of free (unbound) atoms, i.e., of the atoms transferred into the gaseous state or into a solution. They are used for analytical purposes (spectral emission analysis, atomic-absorption analysis), but unlike the spectra of solids, they fail to yield information on the state of atoms in compounds.

In this respect, X-ray spectra play a dual role. They are obtained from atoms in solids, but since X-ray spectra are associated with transitions between inner shells, they change but little from one compound to another, and it is this on which their analytical application is based, (X-ray spectral analysis). Nonetheless, however, in different compounds, some changes in the spectra of one and the same atom are observed and, for this reason, X-ray spectra are used also to characterize features specific to atoms in solids.

In their ground state the inner atomic shells with which the X-ray spectra are associated are filled, and no transitions between them are possible. The first condition necessary for the appearance of the X-ray spectra is to knock out an electron from one of its inner shells.

Let us consider the spectroscopic characteristics of an atom with an electron removed from one of the inner shells and then the process itself involved in the removal of the electron, and the possible ways of restoring to its ground state the atom excited in this manner, i.e., the origination of the X-ray and electron spectra.

The Scheme of Energy Levels and the Terms of Atoms Excited Through Knocking Out of Electrons from the Inner Shells (Fig. 19). In optical spectroscopy each electronic configuration (s^1, s^2, p^1, p^2, ... d^1, d^2, d^3, ..., f^1, f^2 ...) leads to quite different schemes of energy levels. In X-ray spectroscopy, the atomic spectra of all the elements are described (qualitatively) by one and the same energy level scheme (see Fig. 19), corresponding to the order of the electron shells filling. Only the energy of these levels in passing from one element to another is subject to change.

The optic term that describes the state of the atom does not include the principal quantum number ($3d^1$ and $4d^1$ are described by the same terms). In contrast to these the X-ray transitions take place between shells that are characterized by the principal quantum number n. In X-ray spectroscopy it is customary to designate the shells with n from 1 to 7 with letters K, L, M, N, O, P, Q, respectively, and the levels in each shell with Roman numerals in indexes:

$1s$	$2s\ 2p$	$3s\ 3p\ 3d$	$4s\ 4p\ 4d$
K	$L_IL_{II}L_{III}$	$M_IM_{II}M_{III}M_{IV}M_V$	$N_IN_{II}N_{III}N_{IV}N_VN_{VI}N_{VII}$
K-shell	L-shell	M-shell	N-shell
$n=1$	$n=2$	$n=3$	$n=4$
$5s\ 5p\ 5d\ 5f$		$6s\ 6p\ 6d$	$7s\ 7p$
$O_IO_{II}O_{III}O_{IV}O_VO_{VI}O_{VII}$		$P_IP_{II}P_{III}P_{IV}P_V$	Q_IQ_{II}
O-shell		P-shell	Q-shell
$n=5$		$n=6$	$n=7$

It is useful to contrast mentally these X-ray designations K_I, L_I, L_{II} etc. with the denotations of shells $1s$, $2s$, $2p$ ..., so as to relate changes in the position of the X-ray bands and lines to those in the state of the chemical bond of the atom in the compound.

Optic terms of atoms with a single electron knocked out from the inner shell, that supplement and unfold these X-ray designations, are easy to obtain, when one bears in mind two conditions (1) an atomic term is determined by the state of the shell from which an electron has been removed (for the rest of the shells being completely filled have zero spin and orbital moments) (2) since only a single electron is being removed, all these terms are related to the single-electron states (the absence of one electron in a filled shell is equivalent to the presence of one electron in this shell: np^1 - np^5, nd^1 - nd^9, nf^1 - nf^{13}).

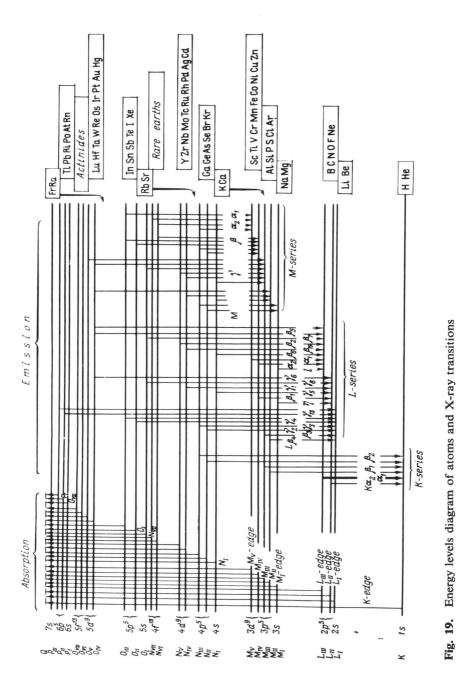

Fig. 19. Energy levels diagram of atoms and X-ray transitions

This simplifies the determination of terms. Spin S of the atom for all such states equals spin of the electron: $S = s = 1/2$; multiplicity $2S + 1 = 2$, i.e., all the terms are doublets. The orbital quantum number of the atom is equal to the orbital quantum number of the electron: $L = l = s(S), p(P), d(D), f(F)$. The total quantum number of the atom equals the total quantum number of the electron: $I = j = l \pm s(L \pm S)$. Hence, the terms of atoms with a single missing electron in filled shells will be as follows:

ns^1 - $^2S_{1/2}$ $(L = S = 0; I = L \pm S = 0 \pm 1/2 = 1/2)$
np^1 - $^2P_{1/2}, {}^2P_{3/2}$ $(L = P = 1; I = L \pm S = 1 \pm 1/2 = 3/2, 1/2)$
nd^1 - $^2D_{3/2}, {}^2D_{5/2}$ $(L = D = 2; I = L \pm S = 2 \pm 1/2 = 5/2, 3/2)$
nf^1 - $^2F_{5/2}, {}^2F_{7/2}$ $(L = F = 3; I = L \pm S = 3 \pm 1/2 = 7/2, 5/2)$.

The K-shell has a single level (1s-state), in the L-shell there are three levels (2s, $2p_{1/2}$, $2p_{3/2}$ = L_I, L_{II}, L_{III}), in the M-shell five levels (3s, $3p_{1/2}$, $3p_{3/2}$, $3d_{3/2}$, $3d_{5/2}$), N and O have seven levels (4s and two for 4p, 4d, 4f; the same applying to 5s, 5p, 5d, 5f), in P there are five levels (6f with no electrons in normal states of the atoms), and Q has two levels (7s and 7p).

In the case of the second-row elements (in oxygen with the electronic configuration of $1s^2 2s^2 2p^4$, for instance) the energy level scheme includes only two shells: K and $L_{I,II,III}$, whereas in case of the third row elements (for aluminum with $1s^2 2s^2 2p^6 3s^2 3p^3$, for example) there are already three shells in the energy level scheme: K, $L_{I,II,III}$, $M_{I,II,III}$; for transition metals of the iron group there exist K, $L_{I,II,III}$, $M_{I,II,III,IV,V}$, and so on.

In optical and X-ray spectroscopy the terms denote different states. In the former this is the ground and excited states that tend to a limit, which is the state of ionization (detachment of the valence electron). In X-ray spectroscopy, on the other hand, it is the state of an ionized atom (with removed inner electron). In X-ray spectroscopy, the diagrammatic presentation of the levels is often reversed in comparison with the energy level scheme in optical spectroscopy, so that the K level lies at the top. Inasmuch as in spectroscopy only the relative disposition of the levels is of significance, both ways of presenting them are equivalent.

The Origin of the X-Ray and Electron Spectra. In the ordinary spectroscopic pattern of "emission source—specimen—spectrometer" (Fig. 20), when inner electron levels are subject to spectroscopy, it is the X-ray tube that serves as a source of emission (radiation) and recording may be made of the following spectra (Fig. 21):

1. X-ray absorption spectra (X-ray absorption spectroscopy, see Fig. 21a) where transitions correspond to the energies necessary to knock out an electron from the inner shell beyond the bounds of the atom (from K-shell in Fig. 21a, or from L_I, L_{II}, L_{III}, M_I etc.).

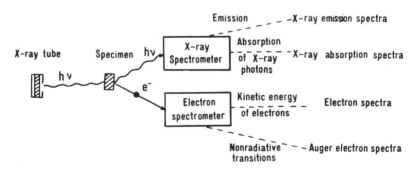

Fig. 20. Basic types of X-ray and X-ray electron spectroscopy. Schematic diagrams illustrating the ways of obtaining the spectra

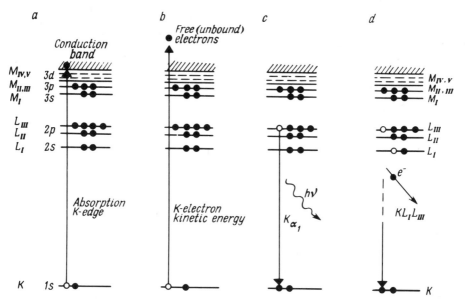

Fig. 21 a-d. Systematics of major types of X-ray and X-ray electron spectroscopy: types of transitions between inner electron energy levels (on an example of Al: $1s^2 2s^2 2p^6 3s^2 3p^3$). **a** X-ray absorption spectrum; **b** electron spectrum; **c** X-ray emission spectrum; **d** Auger-electron spectrum

2. X-ray emission spectra which are analogous to the optical luminiscence (fluorescence) spectra. These are secondary spectra emitted by a study substance. The primary emission of X-ray from the X-ray tube, by knocking out an electron from one of the inner shells, brings the atoms of the study substance into an excited state (see Fig. 21a), the return to the initial state being brought about through emission transitions from one of the upper shell to that with the missing electron (in Fig. 21c from the L_{III} shell to that of K from which the electron has been knocked out); in like manner the transition to the K shell from L_I, L_{II}, M_I etc. occurs, this being accompanied by emission of radiation corresponding to the energy difference of these shells, and also by transitions to the shell L_I, L_{II}, L_{III}, M_I and others, if an electron has been removed from them.

3. The electron spectra (the X-ray photoelectron spectra) (see Fig. 21b); upon knocking out electrons from the inner shells, one can register not the energy of the X-ray photons absorption, which is spent for this knocking out (as in X-ray absorption spectra), but the kinetic energy of the electrons ejected from the substance. The resulting electron spectrum indicates the relationship between the amount of such electrons and their energy, and thus permits determination of the position of the inner shell levels.

4. The Auger-electron spectra (see Fig. 21d); just as electron spectra are an alternative possibility to X-ray absorption spectra, so are also the Auger-electron spectra with respect to the X-ray emission spectra, i.e., deactivation of the excited

atom (with a knocked out electron) can be realized through emission transitions (X-ray emission spectra), or by way of emissionless transitions, whose energy is spent for knocking out one more, second electron from one of the overlying shells. This process (the Auger effect) is described as an autoionization of the excited atom, stemming from an internal redistribution of the excitation energy (inner conversion). The kinetic energy of the Auger-electrons allows the energy of the given levels to be determined.

In obtaining X-ray spectra of absorption, photo-and Auger-electrons, the primary X-ray emission from the X-ray tube acts as the source of radiation. In the case of the X-ray emission spectra the X-radiation of the specimen can be obtained by (1) electron bombardment of the specimen (primary excitation), the specimen being then rubbed into the anode of the X-ray tube, or serving itself as an anode, emitting X-rays due to the effect of a microprobe electron gun; (2) the action of the X-ray tube emission on the specimen (secondary excitation, X-ray fluorescence).

Let us now consider the general characteristics of the types of X-ray and electron spectroscopy.

2.1.3 X-Ray Emission Spectra

The energy levels scheme of inner electron shells (see Fig. 19) and the selection rules for the transitions between them are the same as in optical atomic spectroscopy: $\Delta l = \pm 1$ (and also $j = 0, \pm 1$). They make it possible to determine all the possible, transitions from the upper shells to the inner one from which an electron has been knocked out. The resultant lines make up the X-ray emission spectrum. The transition energies (the line frequency in the emission spectrum) equal the difference between the initial and final states.

All the lines appearing during transition to K level from all the upper ones form K-series, to L-level, L-series, and to those of M,M-series. Each series, except for K, consists of subseries: L of three subseries: L_I, L_{II}, L_{III} (transitions to $2s$, $2p_{1/2}$, $2p_{3/2}$ levels of L hsell with $n = 2$); M of five subseries: M_I, M_{II}, M_{III}, M_{IV}, M_V, etc. Individual lines are designated as α, β, γ . . . with numeral indexes. In the case of the K-series, for instance:

Transition	Designation
$L \rightarrow K$	K_α
$L_I (2s) \rightarrow K (1s)$	Forbidden (Δl should be ± 1)
$L_{III} (2p_{3/2}) \rightarrow K (1s)$	$K_{\alpha 1}$
$L_{II} (2p_{3/2}) \rightarrow K (1s)$	$K_{\alpha 2}$
$M \rightarrow K$	K_β
$M_I (3s) \rightarrow K (1s)$	Forbidden
$M_{IV,V} (3d) \rightarrow K (1s)$	Forbidden
$M_{II,III} (3p) \rightarrow K (1s)$	$K_{\beta 1}$
$N \rightarrow K$	
$N_{II,III} (4p) \rightarrow (1s)$	$K_{\beta 2}$

Table 5. Nomenclature of X-ray emission lines (selection rules: $\Delta l = \pm 1$, $\Delta j = 0, \pm 1$)

Final states (series)	Initial states of emission transitions					
	L_I $2s_{1/2}$	L_{II} L_{III} $2p_{1/2}$ $2p_{3/2}$	M_I $3s_{1/2}$	$M_{II}M_{III}$ $3p_{1/2}$ $2p_{3/2}$	$M_{IV}M_V$ $3d_{3/2}$ $3d_{5/2}$	N_I $4s_{1/2}$
$K(\rightarrow 1s, K)$	—	$K_{\alpha 2}K_{\alpha 1}$	—	$K_{\beta 3}K_{\beta 1}$	—	—
$L_I(\rightarrow 2s, L_I)$	—	—	—	$L_{\beta 4}L_{\beta 3}$	$(L_{\beta 10})$ $L_{\beta 9)}$	—
$L_{II}(\rightarrow 2p_{1/2}, L_{II})$	—	—	L_η	—	$L_{\beta 1}$	$L_{\gamma 5}$
$L_{III}(\rightarrow 2p_{3/2}, L_{III})$	—	—	L_l	—	$L_{\alpha 2}L_{\alpha 1}$	$L_{\beta 6}$

Final states (series)	$N_I N_{III}$ $4p_{1/2}$ $4p_{3/2}$	$N_{IV}N_V$ $4d_{3/2}$ $4d_{5/2}$	$N_{VI}N_{VII}$ $4f_{5/2}$ $4f_{7/2}$	O_I $5s_{1/2}$	$O_{II}O_{III}$ $5p_{1/2}$ $5p_{3/2}$	$O_{IV}O_V$ $5d_{3/2}$ $5d_{5/2}$
$K(\rightarrow 1s, K)$	$K_{\beta 2}$	—	—	—	$++$	—
$L_I(\rightarrow 2s, L_I)$	$L_{\beta 2}L_{\gamma 3}$	—	—	—	$L_{\gamma 6}$	—
$L_{II}(\rightarrow 2p_{1/2}, L_{II})$	—	$L_{\gamma 1}$	—	$L_{\gamma 8}$	—	$L_{\gamma 6}$
$L_{III}(\rightarrow 2p_{3/2}, L_{III})$	—	$L_{\beta 15}L_{\beta 2}$	—	$L_{\beta 7}$	—	$L_{\beta 5}$

The nomenclature of the X-ray emission lines is listed in Table 5 and shown in Figure 19.

Of all the X-ray emission lines doublet $K_{\alpha 1,2}$ displays the greatest intensity. It is precisely this doublet that is used in the X-ray structural investigation of crystals ($CuK_{\alpha 1,2}$, $MoK_{\alpha 1,2}$ and other emission) and also in the X-ray spectral analysis.

The emission of X-ray lines occurs as a result of transition from one excited state to another, i.e., both the initial and final states are excited. An emission of K_α line (K - L_{II}), for example, implies transition of an electron from level L_{II} to that of K, accompanied by emission of a photon whose energy equals the difference between these levels. Then, however, the atom passes from the initial state K (the state of the whole atom ionized in shell K) with a missing electron in the K-shell—$1s\,2s^2\,2p^6$ . . . into final state L_{II} (into a state of the atom ionized in the L_{II}-shell) also with a missing electron, but this time in the L_{II} shell: $1s^2\,2s^2\,2p^5$ . . . .

On the energy level diagrams (see Fig. 19) transitions attended by emission are denoted in a variety of ways: by an arrow pointing from level L_{II} to level K, indicating the direction of the electron transition with emission of an X-ray photon, or by an arrow directed from level K to that of L_{II}, thus indicating the transition of the atom from state K to that of L_{II}.

The lines related to transitions from valence to inner levels are called the last lines of the series (with the shortest wave length and the greatest energy in a given series). They are of special importance in investigating the chemical bond of atoms in compounds. One and the same line K_β ($M_{II,III} \rightarrow K$, i.e., $3p \rightarrow 2s$), for instance, is the last in compounds of the third group elements ($1s^2 2s^2 2p^6 3s^2 3p^n$), but in the case of heavier elements, it corresponds to a transition between the levels of the inner shells.

The lines of one series (even arising due to transitions to the level of a given shell from levels of various shells, for example, K_α, K_β, or $L_{\beta4}$, $L_{\beta5}$, $L_{\beta10}$, $L_{\beta9}$) lie much farther away from the lines of other series than from the lines within the given series. (Note is to be taken of the fact that in Fig. 19 the disposition of levels is shwon on a logarithmic scale). They lie in regions of different wavelengths and are obtained under dissimilar conditions of registration, or even by using different types of spectrometer (thus the lines of the light elements of the L-series come in the region of ultrasoft X-ray emission).

Apart from the transitions indicated in Figure 19 and listed in Table 5, allowed by selection rules for electric dipole transitions ($\Delta l = \pm 1$, $\Delta j = 0, \pm 1$) sometimes much weaker lines arise from transitions forbidden by these rules, but which nevertheless make their appearance as a result of electric quadrupole ($\Delta l = 0, \pm 2$, $\Delta j = 0, \pm 1, \pm 2$), or magnetic dipole ($\Delta l = 0$, $\Delta j = 0, \pm 1$) transitions.

The emission spectra lines of transition metals with unfilled d-shells are characterized by their asymmetric form (with smeared long-wave branch) and a considerable breadth. This feature is a result of a multiplet splitting of the transition element levels. In the electronic configurations of these atoms in excited states, not only is one electron knocked out from the inner shell (which leads to the appearance of doublet terms), but the d-shell also remains unfilled. For instance, $K_{\alpha1,2}$ Mn^{2+} doublet appears as a result of transition from the K-state ($1s \ldots 3d^5$) to $M_{II,III}^-$ state ($2p^5 \ldots 3d^5$), the terms of both states having not a doublet, but a multiplet structure, which in this case leads to the splitting of the line into four overlapped components, and thus causes broadening and asymmetry of these lines. Not the LS-interaction, but that of jj (see 1) is a point of departure for calculating the multiplet splitting of the transition element lines.

Dependence of Position of Emission Lines on the Atomic Number. The most characteristic feature of the X-ray spectra is the fact that the lines observed on them are the same for all the elements (within the set of lines possible for a given element), and that they shift continuously with the change in the element's number Z (Fig. 22). This naturally follows from the scheme of energy levels in the inner shells which is the same for all the elements (see Fig. 19), in which it is merely the energy of these levels that undergoes a change. The cause of this shift is the change of the atomic charge.

The quantitative dependence of the terms' energies and, consequently, of the emission lines transition frequencies as a difference of terms between which the transition occurs, as well as of the position of the absorption edges on the number of element Z are determined from Moseley's diagrams (see, e.g., [256]).

The electron E_{nl} energy with given principal n and orbital quantum numbers is

$$E_{nl} = \frac{2\pi^2 me^4}{h^2} \cdot \frac{Z^2}{n^2}$$

or, by substituting here the Rydberg constant $R = \dfrac{2\pi me^4}{ch^3}$ we find

$$E_{nl} = -hcR\frac{Z^{*2}}{n^2} = -hcR\frac{(Z - \sigma_{nl})^2}{n^2},$$

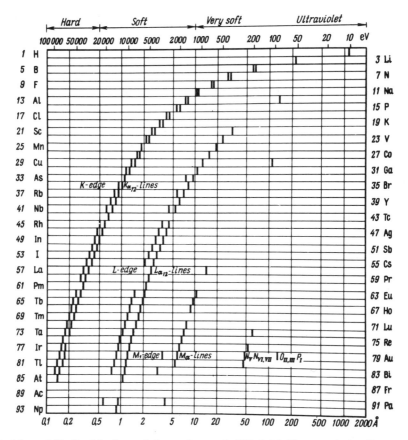

Fig. 22. Position of K-, L-, M-edges of absorption and of $K_a L_a M_a$ X-ray emission lines, depending upon the atomic numbers [256]

where Z^* is the effective charge of the nucleus less than Z by the value of the screening constant σ.

Let us take note of the dimensionality of h (erg s) $\times$ c(cm c^{-1}), R(cm^{-1}), i.e., E_{nl} is here expressed in ergs.

Term T (cm^{-1}) $= E$ (erg)$/hc$ (erg $\cdot$ c $\times$ cm $\cdot$ c^{-1}) equals

$$T = \frac{E_{nl}}{hc} = R\frac{(Z - \sigma_{nl})^2}{n^2} \text{ (cm}^{-1}\text{), or } \frac{T}{R} = \frac{Z^{*2}}{n^2}.$$

The square root of the absorption edge frequency (cm^{-1}), or, what is equivalent to it, of term T (cm^{-1}), i.e., $\sqrt{\nu}$, or $\sqrt{T}$ and terms to correct for screening σ is taken to be constant (for K-terms $\sigma \approx 1$, for L-terms $\sigma \approx 8$) and the values $\sqrt{\nu/R}$ increase almost exactly linearly with the atomic number. For outer M-, N-, O-shells, deviations from the linear dependence and the presence of breaks, accounted for by the shells undergoing completion, become noticeable.

For emission lines the frequency of transition ν_{1-2} is

$$\nu_{1-2} = T_2 - T_2 = R\left(\frac{Z^{*2}_1}{n^2_1} - \frac{Z^{*2}_2}{n^2_2}\right).$$

With σ approximately constant for initial and final states $Z^{*}_i \approx Z^{*}_k$. By denoting also $\sqrt{1/n^2_1 - 1/n^2_2} = a$, we obtain the Moseley expression for the X-ray spectra emission lines which corresponds to a simple linear dependence of the lines frequency ν on the atomic number Z:

$$\sqrt{\frac{\nu}{R}} = a(Z - \sigma),$$

where a is the same for the similar lines of all the elements (for $K_{\alpha 1}$ of all the elements, for instance); σ is approximately equal for all the lines in a given series (for all the K-lines of all the elements, or for all the L-lines of all the elements, etc).

For example, in the case of K_{α}-lines $n_1 = 1$ (1s-state); $n_2 = 2$ (2p-state); $a = \sqrt{1/1^2 - 1/2^2} = 0.874$; σ for K-series ≈ 1; $\sqrt{\nu/R} = 0.874\,(Z - 1)$ or ν (cm^{-1}) $= 0.248\,10^6\,(Z - 1)^2$.

X-Ray Emission (Doublet) Spectra. As concerns the nature of dependence of the distance between the doublets on atomic number, two types of doublet are distinguished.

Spin doublets (or regular doublets) are pairs of lines appearing when single-term levels differing in the value of j are combined with levels $^2P_{1/2}$ and $^2P_{3/2}$, i.e., of L_{II} and L_{III}, for instance. Spin doublets are lines $K_{\alpha 1}$ and $K_{\alpha 2}$ (1s $\rightarrow$ 2p$_{1/2}$ and 1s $-$ 2p$_{3/2}$), L_η and L_l (3s $\rightarrow$ 2p$_{1/2}$ and 3s $\rightarrow$ 2p$_{3/2}$), $L_{\beta 1}$ and $L_{\alpha 2}$ (3d $\rightarrow$ 2p$_{1/2}$ and 3d $\rightarrow$ 2p$_{3/2}$) (see Table 5). The difference in frequencies of these pairs of lines is the same for a given element, but increases with rising atomic number Z.

Screening doublets (irregular doublets) are pairs of energy levels arising from different terms with common n, j, but dissimilar l: L_I and L_{II}, M_I and M_{II}, M_{III} and M_{IV}, etc. The distances between these levels are approximately constant for all the elements, i.e., on the Moseley diagram the pairs of corresponding lines run parallel with the changes of l and the screening constant varies. In emission spectra commonly only one of the transitions, linked with a given screening doublet, becomes manifest, since the second transition is forbidden by the selection rules (see Table 5 and Fig. 19).

Satellites (Nondiagram Lines). Besides lines arising from the scheme of the inner electrons' energy levels (see Fig. 19) and called "diagram" lines (including weak quadrupole transitions) weak lines are observed, not included in this pattern and referred to as "nondiagram". They occur usually near one of the diagram lines (more often on the long-wave side) and, therefore, are also designated as satellites. For instance, next to the most intensive lines $K_{\alpha 1}$ and $K_{\alpha 2}$ can be seen fairly bright satellites $K_{\alpha 3}$ and $K_{\alpha 4}$. The same satellites are seen to occur in a group of neighboring elements, and their position varies depending on the atomic number after Moseley's law, or is expressed in the form of Moseley's "pseudodiagrams" which indicate the dependence on the atomic number of the difference in wave numbers between the generating (diagram) line and its satellite.

The origin of satellites is related to transitions between the states of double and multiple ionization (of the Auger-transition type).

Satellites display a strong dependence on the state of chemical bonding, but the possible interpretation of this dependence is solely confined to empirical correlations so far.

In addition to satellites, the X-ray spectra demonstrate lines that appear only in the spectra of chemical compounds, and are absent in those of pure elements. The emergence of these lines is due to transitions from molecular orbitals (see Fig. 22), and they are not included in the X-ray satellites group.

Continuous X-Ray Emission. When X-rays are excited by electrons (primary excitation) then, in addition to the line (characteristic) spectrum related to transitions between discreet levels of inner electron shells, a continuous (of a few angström) spectrum of a different nature arises.

The spectrum appears to follow slowed-down movement of an electron falling on the substance in the Coulomb field of the nucleus. On the short-wave side the continuous spectrum has a sharp limit. Its position is the same for all elements, and is determined by the electron's energy depending on the X-ray tube voltage. By changing the voltage, it is possible to obtain the continuous spectrum in any wave length region. The cut-off wave length (A) is linked with voltage U (V) by a relation: $(A) = 12\ 395/U(V)$ (Table 6). The higher the voltage, the shorter the wave length. With 1000 V the wave length is 12.4 Å, and with 10,000 V 1.24 Å.

The position of the characteristic spectrum lines is independent of the voltage and therefore, with rising voltage the limits of the continuous spectrum will shift towards short waves, and it will be overlapped by emission lines of other elements.

The continuous spectrum finds application in investigating X-ray absorption spectra and in X-ray microscopy. In X-ray spectral analysis with primary excitation, it forms a background that reduces the sensitivity and accuracy of the analysis. With secondary excitation (in X-ray fluorescent analysis) the continuous spectrum does not appear at all. Therefore, in modern analysis X-ray fluorescence (apart from electron probe microanalyzers) is used.

Units of Measurement. Appreciably higher energies and much smaller wave lengths of the X-ray emission, as against the optical region of the spectrum, determine the choice of the wave length, frequency, and energy units of measurements. The wave length is expressed in Ångströms (1 Å $= 10^{-8}$ cm) or in X units (10^3X $= 1$KX $= 1.00202$ Å; λ in Å $= 1.00202$ λ in kX).

The wave lengths of the heavy elements X-ray spectra less than 1 Å are often given in X units and those of light elements in Angströms. The conversion factor 1.00202 appears due to the fact that the determination of the wavelength in X-ray spectroscopy is performed from the Bragg relation $n\lambda = d\sin\theta$, where λ is the wavelength of the analyzed X-ray emission, d is interplanar spacing of the crystal analyzer, used as a diffraction grating in the X-ray spectrometers, θ is the reflection angle of X-ray emission, which is directly measured in the spectrometer, and n is the reflection order. Primarily, the wavelengths of the X-ray spectrum lines were determined relative to the magnitude of the interplanar spacing of the calcite plane [100] with $d = 3.029$ Å. Subsequent measurements showed the magnitude d to be inaccurate and, consequently, the true values of the wavelengths differ from those measured relative to d of calcite. Therefore, in 1947 it was decided to continue using the earlier measured values for the wavelengths, but consider them to be expressed in a new conventional unit kX, equalling 1.00202 Å.

Sometimes the positions of the X-ray spectral lines are indicated directly in terms of angle θ. Their conversion in wavelengths expressed in terms of Å is done by using the mentioned Bragg formula. For instance, the crystal-analyzer is quartz with reflecting plane of the prism [10$\bar{1}$0] $d = 4.246$ Å and then $\lambda = 8.492 \sin\theta$; if measured $\theta = 78°30'$, $\sin\theta = 0.9799$ and then $\lambda = 8.311$ Å, this corresponding to to line $K_{\alpha 1}$ of aluminum.

To express frequencies in terms of cm^{-1} in the X-ray region of the spectrum is inconvenient, since this yields too large figures. Therefore, frequencies are here expressed in terms of Rydbergs (1 Ry = 109 678 cm^{-1} = 13.60 eV). As in the optical region, emission energy is expressed not in terms of ergs, obtained in theoretical formulas which include Planck's constant h (erg. cm), but in electron-volts: 1 erg = 6.2419 10^{-11} eV.

The conversion of some of the units of measurement used in the X-ray spectroscopy into others may be conveniently done by referring to Table 6.

Table 6. Correlations of different units of measurement of the transition energy in X-ray and electron spectroscopy

	cm^{-1}	eV	Ry[a]	Å	kX
cm^{-1}		8066 × eV	109 678 × Ry	10^8/Å	1.00202 · 10^8/kX
eV	1.24 · 10^{-4} × cm^{-1}		13.60 × Ry	12 395/Å	12,372/kX
Ry	cm^{-1}/109 678	eV/13.60		912/Å	910.2/kX
A	10^8/cm^{-1}	12,395/eV	912/Ry		1.00202 × kX
kX	1.00202 · 10^8/cm^{-1}	12,372/eV	910.2/Ry	Å/1.00202	

[a] For example, to express the transition energy measured in Rydbergs (Ry) in terms of eV (second line of table) the value in Ry should be multiplied by 13.60.

2.1.4 X-Ray Absorption Spectra. The Quantum Yield Spectra. Reflection Spectra. Isochromatic Spectra

In this section diverse types of the X-ray spectra are considered, arising due to different processes, but yielding analogous spectral curves and similar information on the substance.

In the X-ray absorption spectra three elements are distinguished: (1) a continuously rising absorption coefficient with increasing wavelength, (2) absorption edge (the K-, L_I-, L_{II}, L_{III}- etc. absorption edge), (3) fine structure of the absorption edges.

To understand the specificity of the X-ray absorption spectra it is necessary to consider the mechanism of the interaction of X-ray emission with the substance (the process of absorption itself).

When a beam of X-rays passes through a specimen, the intensity of the rays becomes weaker, not only because of the absorption, but also due to scattering. The mechanism of X-ray absorption differs from that of optical absorption, in that the absorption of X-ray emission energy occurs as a result of a single process: knocking-out of the electrons of the inner shells beyond the boundaries of the atom, i.e., due to ionization of the atom at the expense of inner electrons. The energy of the absorbed emission is thus transformed into kinetic energy of these knocked out electrons (photoelectrons), plus the potential energy of the excited atom that equals the bonding energy of the knocked out electron.

The X-ray emission of the least energy (of the greatest wavelength) forces out electrons from the outer shells. With mounting emission energy, an ever smaller part of it is required to knock out electrons from a given shell. This is attended by reduced absorption. A continuous lessening of absorption continues for as long as the emission energy is sufficient to force an electron out of the next, deeper lying inner shell. This gives rise to a sharp increase of absorption that manifests itself in absorption edge. Scattering is yet another phenomenon that produces weakening of X-ray emission intensity during the passage through the substance, and occurs in two types. Scattering takes place as a result of collision of the X-ray photon of energy $h\nu$ with the atom's electrons fixed on their levels of energy E_{el}.

Should energy of the X-ray photons be inferior to that of the electron bonding ($h\nu < E_{el}$) the photons cannot then knock out the electron from a given inner shell. Following an elastic collision with fixed electrons, the photons merely change their direction (are scattered); their energy and, accordingly, the wavelength remain unchanged. Scattering where the wavelength does not change is called coherent (Thomson) scattering. It lies at the basis of the X-ray diffraction utilized in structural analysis.

On the other hand, if the energy of the X-ray photons is greater than that of the binding electrons ($h\nu > E_{el}$), the photons then force the electron out of the corresponding inner shell, but when colliding with electrons, they pass a part of their energy on to them. This results in the scattering photons displaying a lesser energy and a greater wavelength. This scattering, attended by a change in the length of the wave, is known as incoherent (Compton), scattering. Inasmuch as the knocking-out of an electron is the first and foremost condition for the emergence of all X-ray and electron spectra, it is precisely the incoherent scattering that accompanies their appearance. However, since the atom has at the same time more and less strongly bound electrons (more or less deep lying internal shells), one can observe two lines in the spectrum of scattered emission, those with unchanged and changed (increased) wave length.

The intensity of scattering grows parallel to the atomic number, i.e., the more electrons are in are atom, the greater their scattering effect, in other words the X-rays are scattered poorly by light atoms and strongly by heavy ones.

The correlation of the coherent and incoherent scattering depends on two factors, namely, upon the wavelengths of the incident radiation and the atomic number of the absorbing atom. The smaller the wavelength of the incident radiation, i.e., the higher its energy, and the lesser the atomic number, i.e., the smaller the energy of the electrons bonding in these atoms, the greater part of the emission scatters incoherently.

A change in the wavelength of the incoherently scattered emission, as compared to the incident or coherently scattered one, depends neither on the atomic number, nor upon the wavelength of the incident emission, but is determined only by the angle through which the scattered photons are deflected. With the angle increasing from 0° to 180° (thrown-back photons) the increasing of the wavelength goes from zero to $\Delta\lambda_{max} = 0.0486$ kX.

The quantitative estimation of the reduced intensity of the X-rays following their passage through the substance is effected by using the attenuation factor μ, which is the sum of the coefficient of pure (photoelectric) absorption τ and the scattering (dispersion) factor σ. The attenuation factor is often referred to as the coefficient of absorption. With wavelengths exceeding 0.5 Å and for elements with $Z > 26$, the attenuation is practically caused by absorption, and then $\mu \approx \tau$.

The linear attenuation (absorption) coefficient μ_l (in cm^{-1}) is deduced from Beer's law $I = I\,e^{-\mu_l d}$ which establishes an exponential relationship between intensity of any incident radiation and the thickness of the specimen: $\mu_l = \lg I/I_0$ $\cdot 1/d$. It is employed in assessing the transparency or opacity of the specimen with its given thickness and for a given radiation. Since it depends on the state of the substance (solid, liquid, gaseous), it is not a constant characterizing the absorption of a given element.

More frequently use is made of a mass absorption coefficient which is equal to $\mu = \mu_l/\rho$, where ρ is density in g cm^{-3}, i.e., μ is expressed in cm^2 g^{-1} (cm^{-1} g$^{-1} \times$ cm^{-3}). The mass absorption coefficient is independent of the state of the substance (of its density), and of the compound of which the given element is a part, and this coefficient has a definite significance for the given element and for the given wavelength. Therefore, it is common to indicate the wavelength for which the value of the mass absorption coefficient is cited, or to point up a characteristic emission line, whose radiation is used in measuring the absorption. For example, the mass absorption coefficient for Al in emission of Sr K_α ($\lambda = 0.876$Å) is denoted as $\mu_{0.876}$ or $\mu_{SrK\alpha}$.

The tables of μ values for the most important K_α, K_β, L_α, and other emission lines of the elements have been published [242]. The mass absorption coefficient is linked with the wavelength λ by an empirical equation $\mu = C\lambda n_0$, where C and n_0 are constants for regions between different absorption edges [276].

To characterize the absorption per unit of concentration, use is made of atomic absorption coefficient $\mu_a = \mu/n$, where n is the number of atoms in 1 cm^3, i.e., μ is expressed in square centimetres (cm^{-1}/at cm^{-3}) and, accordingly, the atomic absorption coefficient is equal to the absorption cross-section: $\mu_a = A/N$, where A is the atomic mass and N is Avogadro's number of the element.

The mass absorption coefficient for any compound is calculated exactly on the ground of the μ values of the elements and weight parts of the latter in the compound. For albite $NaAlSi_3O_8$ and anorthite $CaAl_2Si_2O_8$, for example, this is done as follows:

Albite		Anorthite	
Weight % $\times \mu_{Co}K_\alpha$		Weight % $\times \mu_{Co}K_\alpha$	
Na	8.8×45.9	Ca	14.4×238.6
Al	10.3×75.0	Al	19.4×75.0
Si	32.1×92.8	Si	20.1×92.8
O	48.9×19.6	O	46.1×19.6

$\mu_{Co}K_\alpha$ for albite $= 0.088 \times 45.9 + 0.103 \times 75.0 + 0.321 \times 92.8 + 0.488 \times 19.6 = 52$ cm^2 g^{-1}.

$\mu_{Co}K_\alpha$ for anorthite $= 0.144 \times 238.6 + 0.194 \times 75.0 + 0.201 \times 92.8 + 0.461 \times 19.6 = 76$ cm^3 g^{-1}.

The results of actual chemical analyses may be used, and the presence of impurities can be taken into account through multiplying their weight parts by the mass absorption coefficients for the same emission, obtained from the pertinent tables [242].

The measurement of the mass absorption coefficient allows determination of the end member relations in binary solid solutions, for example albite-anorthite, forsterite-fayalite, diopside-hedenbergite, the iron content in sphalerite, etc. [282, 297].

Since a change in the ratio of Compton scattering (with variation in the wavelength) to Thomson scattering (without any change in the wavelength) is linearly related to the mass absorption coefficient, this ratio may be used for determining the latter. To do so it suffices to contrast the relative intensities of the corresponding two lines, slightly displaced with respect to each other in the scattering spectrum [297].

It is only with a low resolution of the spectrometer and as a preliminary picture that one may imagine the X-ray absorption spectra as a continuously rising absorption proceeding in parallel with increasing wavelength and abrupt discontinuities superposed. A more detailed examination reveals the presence of a fine structure made up of three distinctive elements: (1) the position of the absorption edge itself; (2) the structure of the initial region of absorption (Kossel's), additional weaker maxima near the absorption edge over the length from a few electron-volts up to 20 to 30 eV from it towards predominantly higher energies (shorter wavelengths); (3) an extended fine structure (Kroning's), continuing from the absorption edge over a length of up to 400 eV towards higher energies in the form of weak absorption fluctuations.

The nature of these fine structure transitions is dissimilar in the case of atoms (gaseous), molecules, metals, inorganic crystals (minerals), this being due to the difference in their electronic structure.

In atoms the absorption edge corresponds to the first allowed transition to the empty level. In the case of argon, for example, the electron configuration is $1s^2 2s^2 2p^6 3s^2 3p^6$; next come unoccupied levels $4s$, $4p$, $3d$ . . .; the K-edge of absorption corresponds to the transition to the first vacant level: $1s \rightarrow 4p$ (transitions $s \rightarrow s$, $s \rightarrow d$ being forbidden by the selection rule $\Delta l = \pm 1$). The next allowed transitions from the $1s(K)$ level will be $1s \rightarrow 5p$, $1s \rightarrow 6p$, $1s \rightarrow 7p$ and so on, to the limit of the p-levels series. Superposition of these lines forms a fine structure of the argon absorption K-edge.

In inorganic crystals, characterized by the energy band structures, the absorption edge corresponds to the transition to the free (conduction) band, onto a level to which transition is allowed by the selection rules. (The conduction band is composed by quantum states that accord with excited molecular orbitals). The fine structure of the initial region arises as a result of transitions into empty (excited) "antibonding" states of the conduction band. In the case of molecules transitions occur onto similar molecular orbitals not broadened into bands. In metals and alloys the absorption edge and the fine structure match a transition onto the first vacant level behind the Fermi surface and onto the subsequent levels of the conduction bands.

Upon transition onto still more remote levels and, consequently, with yet higher energies, the electron knocked out from inner shells occurs in the periodic field of the crystalline lattice. The absorption fluctuations in this energy region are determined by the fact that the knocked out semi-free electrons can possess energies only within the limits of allowed intervals, defined by the Brillouin zones. This explains the existence of a continuous fine structure. Here, slight variations in absorption are proportional (for cubic crystals) to $(h^2 + k^2 + l^2)\, a^2$, where h, k, l are indices of crystallographic planes, and a is the lattice parameter. The continuous fine structures for crystals of the same structural type are analogous, while the distances between the maxima of this fine structure vary with a change in the lattice dimensions. (Hence, it can be used in estimating interatomic distances.)

Thus, on its being knocked out from the inner shell, the electron is not simply removed beyond the bounds of the atom, but reaches unoccupied levels, similar to the way in which optical transitions take place. The common selection rules (for optical and X-ray spectra) determine the possibility of transitions onto these vacant levels: the K-edge (transitions from the $1s$-level), L_I ($2s$), M_I ($3s$)-edges from consequent upon transitions onto the p-levels; $L_\mathrm{II,III}$, $M_\mathrm{II,III}$-edges (transitions from $2p$- and $3p$-levels arise with transitions onto s-and d-levels). All these transitions leading to the formation of fine structure, as well as transitions onto the levels determined by the periodic lattice field lie in the shortwave region relative to the absorption edge, and yield information on energy levels that are outside the energy level scheme of the atom, molecule, or crystal (these unoccupied levels correspond spectroscopically to excited states, and chemically to antibonding levels). In the case of $3d$ ions, whose levels find themselves in the forbidden band, a weak peak can appear near the absorption edge on its longwave side.

The basic information obtained from the absorption spectra includes the position of its edge. The absorption edge is commonly approximated by a curve in the shape of an arctangensoid. The energy difference measured at 1/4 and 3/4 of the absorption edge height gives the width of the inner level.

The Extrinsic Photoeffect Quantum Yield Spectra. With the X-ray emission wavelength going up, the linear absorption coefficient rapidly increases and the substance in plaques whose thickness is up to 0.1 mm becomes opaque for the X-rays. In the long and ultralong-wave X-ray emission region (over 10–15 Å), in which spectra of the light elements occur, investigation of the absorption spectra is possible only on thin films up to $\sim$ 500–3000 Å (50–300 nm) thick. Such films are obtained by vacuum vaporization of the matter, this being possible only in the case of compounds undergoing no decomposition before reaching the fusion temperature.

The information rendered by the absorption spectra can be obtained by employing the method of investigating the spectral dependence on the extrinsic photo-effect quantum yield of the X-rays [257, 270]. Under the effect of X-rays the inner shell electrons are being knocked out (photoelectrons) i.e., emission of electrons from the matter occurs, which depends upon the bonding energy of the electrons, and on the wavelength of the incident readiation. The energy of the bonding electrons can be found by measuring their kinetic energy. It is also possible, however, to use a plate of the study substance as a photocathode of a secondary emission electron multiplier, i.e., the X-rays falling on the specimen-photocathode emit photoelectrons, and with the positive potential at the anode, a flow of free electrons is generated, registered in the form of photocurrent. The quantum yield spectrum (the number of photoelectrons per one incident X-ray photon) enables determination of the K-, L_{II}-, L_{III} energy level position ("absorption edges") and the fine structure associated with these levels, just as this can be found from the absorption spectrum.

Reflection Spectra in the X-Ray Region. Measurements of the reflection spectra in the region of soft and ultrasoft X-ray radiation (from 7 to 400–500 Å) also make it possible to determine the position of the absorption edges and the fine structure maxima and, moreover, permit one to calculate the refraction indices and absorption coefficients for this region characterized by a very great absorption (μ_l up to 10^5–10^5 cm^{-1}) The reflection coefficients depend, as in the optical range, on the angle of incidence. In the X-ray region use is made of small glide angles of reflection (from 1° to 10°). The critical angles of reflection are determined by the wave length and composition of the matter. Curves of dependence of reflectivity R on the angle of reflection and the wavelength are described by one of the Fresnel formula variants that links R with the complex refraction index, absorption coefficient, wavelength, and critical angle of reflection. On the R curves the X-ray absorption edges are matched by abrupt maxima situated at the sites of the absorption edges bends. The position of the R maxima is independent of the angle of reflection.

Isochromatic Spectra. As distinct from all other types of X-ray spectroscopy, where measurements are made with a varying wavelength and constant voltage, isochromates are obtained with a constant wavelength and changing voltage (whence the name of isochromates, i.e., wavelengths of equal "coloration"). The ordinary X-ray spectra represent the relationship of intensity I (in emission spectra), or absorption coefficient (in absorption spectra with the wavelength), and are linked with transitions onto the inner levels (emission), or from the inner levels (absorption).

The isochromates show the dependence of the intensity of radiation I given off by the matter on the energy of the electron falling on the specimen, determined by the voltage at the X-ray tube. They are not related to the inner levels. Depending upon the imparted difference of potentials, electrons reach diverse unoccupied levels of the matter, and from these they go to the lower-lying vacant levels down to the Fermi level.

The energy of electrons falling onto the matter gradually increases within the range of 15 to 25 V, as compared to the energy of any characteristic emission, or within the limits of ~ 100 V, compared with the energy of continuous radiation. The isochromates of the continuous spectrum reflect the continuous fine structure associated with the periodic field of the lattice. Isochromates of the characteristic spectra repeat the form of the absorption spectra (sometimes with better resolution). In both cases the isochromatic spectroscopy is, together with the absorption spectra, an alternative way of investigating free allowed states of the matter, in particular those of antibonding molecular orbitals.

2.1.5 X-Ray Electron Spectroscopy (Electron Spectroscopy for Chemical Analysis). Auger-Spectroscopy

Decisive for the appearance of the X-ray and electron spectra is knocking out of an electron from one of the inner shells (all the inner shells are occupied, and transitions between them are infeasible without a place being made free even by a single electron). Then two possibilities are given in principle: (1) registration of the X-ray photons, spent in absorption arising during an emission deactivation of the excited atom: the X-ray spectroscopy, or (2) registration of kinetic energy of the knocked out photoelectrons themselves: electron spectroscopy, including Auger-spectroscopy.

Measurements of the photoelectron energy were made by using magnetic or electrostatic beta-spectrometers. However, because of their low resolving power, the electrostatic spectrometers have found no further application. The advent of electron spectroscopy as a powerful alternative method to X-ray spectroscopy is associated with construction by Siegbahn and co-workers [307, 308] of a noniron magnetic beta-spectrometer with double focusing of the electron beam by the magnetic field. The accuracy of the electron energy measurement attained with it is of the order of 0.001 % (i.e., with energies of the order of thousands of eV, this comprises just 0.0n eV, and resolution around 0.02%. The linear nature of the electronic spectrum obtained, high resolution, and precision are the most outstanding features of this method.

At the root of electron spectroscopy lie the same theories as in X-ray spectroscopy, it being governed by the same scheme of energy levels and the same character of information about the substance.

For electron spectra (Fig. 23) a double scale on the x-axis is often given; the kinetic energy of photoelectrons E_{kin} and the electron binding energy at a given level E_b are unambigiously linked by a relation:

$$E_b = h\nu - E_{kin} - \varphi_{sp},$$

where $h\nu$ is the energy of X-ray emission used, φ_{sp} is spectrometer correction (working function of the spectrometer's material) that reduces the kinetic energy to a

Fig. 23. X-ray electron spectra of Na and Cl in NaCl, excited by the CuK_a-emission (after [308])

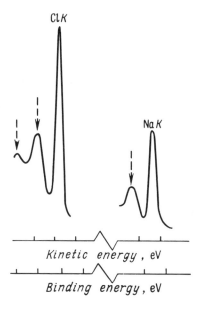

scale with zero reading at the Fermi level (equals $\approx 3 - 4$ eV). The correction for the recoil energy of the knocked out electron, with an appropriate choice of emission, is quite negligible.

For example, for Al peaks having kinetic energy of 1133 eV (for L_I) and 1178 eV (for $L_{II,III}$) obtained with MgK_a emission $h\nu = 1254$ eV, the binding energies are:

for L_I $E_b = 1254 - 1133 - 3 = 118$ eV,
$L_{II,III}$ $E_b = 1254 - 1178 - 3 = 73$ eV.

In order to measure the K-level energy for Al (with $E_b = 1560$ eV) the MgK_a emission ($h\nu = 1254$ eV) is insufficient, and a radiation of a higher energy should be used. With Mg emission, doublet $K_{a1,2}$ is not resolved, whereas with radiation of heavier elements ($CrK_{a1,2}$, $CuK_{a1,2}$), this doublet is resolved and, therefore, on the E_{kin} scale of the electron spectrum each line is doubled, but if one subtracts from each magnitude of $h\nu$ K_{a1} and $h\nu$ K_{a2} the corresponding E_{kin}^1 and E_{kin}^2 this will give a single value for the electron binding energy.

The electron binding energy corresponds to the energy of the electron levels, corresponding to positions of the absorption edges in the electron spectra, the selection rules are here nonexistent and, therefore, peaks of all levels are in evidence. Each level is matched by a single sharply marked line without fine structure, for, in contrast to the absorption edges, the electron spectra usually furnish no information on the free states. As distinct from the X-ray spectra, associated with transitions between two states, each one of which is shifted differently depending upon the compound in question, and from the emission spectra, where transitions occur between both excited states, in the electron spectra the position of each individual level is fixed independently.

The presence of one and the same element in different valence states and coordinations and in nonequivalent positions in different compounds results in shifting of the lines or in their splitting.

For this reacon, electron spectroscopy for chemical analysis (ESCA) includes not only analytical determinations, but also phase analysis, studies of chemical bonding states, valence, coordination, and the existence of nonequivalent positions. The method is universal: it is suitable for all elements and for all energy levels from the lowermost to the valence ones, and even for the excited states (by using uv emission); vibrational transitions are also susceptible to recording in the spectra, this being particularly important in studying surface states.

The distinctive feature of the ESCA method is the fact that, because the electrons penetrate the substance to only a shallow depth, the spectrum is obtained only from a thin superficial layer of the substance: up to 100 Å, and practically only to a depth of a few tens of angstroms. The condition of binding in this superficial layer can differ from that in bulk of the substance. Since the sensitivity of the mehtod is extremely great (10^{-7}–10^{-11} g), one should bear in mind the possible presence of atoms adsorbed on the surface of the substance, and in the case of metals and sulfides, the formation of an oxide film on the surface.

Electron spectroscopy is often subdivided into: (1) high-energy X-ray electron spectroscopy using soft X-rays of over 1000 eV, applied in the main for investigating solids and (2) low-energy photoelectron spectroscopy (vacuum ultraviolet up to 60 eV), frequently refered to as photoelectron spectroscopy.

In the electron spectra the energy of two types of electrons is subject to recording: directly knocked out photo-electrons, and Auger-electrons.

Nomenclature and Energy of Auger-Electrons [244, 294, 301]. If an electron is knocked out of any inner shell (from the K-shell, for instance) and an electron from a superjacent shell (from L_{II}, for example) passes on to it, the transition energy can then be spent in emitting an X-ray radiation, this leading to the appearance of emission spectra, or in knocking out of yet another electron from an overlying shell (from L_{III}, for example), in other words two holes remain. The primary hole in the K-shell is filled with an electron from the L_{II} level that is left with a hole, whereas the energy of this transition is spent (instead of emitting an X-ray quantum) in removing the electron from L_{III}. According to the designation of the shell with a primary hole (in the example under consideration: the K-shell) one distinguishes the Auger-electrons series; K-series, L-series, etc. According to the designation of the shells where, as a result of the Auger-effect, two holes have remained, one distingushes groups of the Auger-electrons: $K - LL$, $K - LM$, and $K - MN$ are the K-series group. Finally, each one of these groups consists of separate lines; KL_{II}-L_{III}, for instance, implies that primarily the electron was kncoked out of the K-shell and, due to the $L_{II} \rightarrow K$ transition, yet another electron was knocked out of the L_{III} shell, and the atom remained with two holes in two shells: L_{II} and L_{III}.

The kinetic energy of the Auger-electron, of $KL_{II}L_{III}$, for instance, is

$$E(KL_{II}L_{III}) = E(K) - E(L_{II}) - E(L_{III}) - \Delta E_{L_{II}L_{III}},$$

where $E(K)$, $E(L_{II})$, $E(L_{III})$ are electron binding energies on these shells or subshells; $\Delta E_{L_{II}L_{III}}$ is a value that takes care of an increase in the binding energy of the L_{III}-electron subsequent to ionization of the atom in the subshell L_{II} (similar to the way in which an increase of the atomic number influences the energy of the shell).

In published literature sources one find values for ΔE and tables listing the KLL-Auger-electron energies (308). For sulfur, for example: $E(K) = 2472$ eV, $E(L_{II}) = 165$ eV, $E(L_{III}) = 164$ eV, $\Delta E(L_{II}L_{III}) = 36$ eV, $E(KL_{II}L_{III}) = 2472 - 165 - 164 - 36 = 2107$ eV. A special case of the Auger effect are Koster-Kroning transitions; there the primary hole moves from one subshell to another (from L_{III} to L_{I}, for instance) remaining within the bounds of the same shell. In the group L-LX Koster-Kroning transitions $L_1L_3M_1$ (up to M_5), $L_2L_3M_1$ (up to M_5), L_1L_2N (up to N_7) are distinguished; in the group M - MX: $M_1M_2M_{4,5}$, $M_1M_2N_1$, etc. (In the K-shell the subshells are lacking and, therefore, there are no Koster-Kroning transitions associated with it.) During these transitions, electrons of very low energies are knocked out of the outer shells.

2.2 Application of X-Ray and X-Ray Electron Spectroscopy to Study of Chemical Bonding in Minerals

Two trends are distinguished: (1) investigation of transitions from molecular orbitals to inner energy levels, or from inner levels to molecular orbitals and (2) investigation of transitions between inner levels.

In the first case the spectra reflect changes in the schemes of molecular orbitals (or in energy band schemes) to which they conform completely, and in the second, shifting of atomic levels (for each of the atoms forming a compound) is registered without any direct connection with the molecular orbitals.

The results of the studies of chemical bonding are expressed in the first case by using the MO diagram (energy positions of the levels and c_i^2 coefficients in MO), and in the second, in the form of chemical shifts interpreted by employing the theory of effective charges.

The same types of X-ray transition (K_α, K_β, $L_{II,III}$ and others) can be related to transitions with participation of MO, or to transitions between inner levels, depending on the period to which the given element belongs (Figs. 24–27).

The most complete characteristic of chemical bonding is obtained by using all spectra possible for each given element: K_α, K_β, L, M emission and absorption spectra. The X-ray electron spectra yield directly a complete picture of inner levels and those of MO, but do not indicate positions of antibonding MO (empty) or any correlation of coefficients in MO. On establishing basic relationships between the X-ray spectra and characteristics of the chemical bonding for some individual classes of compounds, it becomes possible to trace changes of bonding in the series of compounds of the same type according to separate kinds of spectra, using for this purpose commercially manufactured apparatus for X-ray spectral analysis including electron micro-and macroprobes. Correlation of the X-ray transitions with the valence, coordination and effective charges enables it to combine X-ray chemical analysis with phase analysis.

2.2.1 Determination of Molecular Orbital and Energy Band Schemes from X-Ray and X-Ray Electron Spectra

It is only when applied to free atoms that emission lines K_α are described as a transition of $2p \to 1s$, K_β of $3p \to 1s$, etc. (see Fig. 19). In molecules and crystals the levels of valence electrons become split and form molecular orbitals. This results in that in place of a single line a more or less broad band appears that often represents superposition of several bands. At the same time the inner electron levels, though shifting, do not form molecular orbitals, continuing to retain the character of atomic orbitals. The X-ray and electron spectra of molecules and isolated groups in crystals (in dielectrics and semiconductors with narrow valence bands) may be conveniently described by drawing diagrams representing a mixture of the molecular orbitals scheme and schemes of the atomic orbital levels for each atom of the compound (see Figs. 24–27). The X-ray transitions are shown on such diagrams as transitions of MO to AO levels of the atoms (emission spectra), or as transitions from the AO level to the levels of empty MO (X-ray absorption spectra).

For semiconductors and metals the description of X-ray spectra should be made by a combination of an energy band diagram and the inner AO levels scheme.

Selection Rules. In order to assign the X-ray bands K_α, K_β, $L_{II,III}$, etc. to transitions from MO to inner levels, one should, first of all, take account of the selection rules. These rules are the same as in the case of dipole transitions in optical spectra: $\Delta l = \pm 1$. Transitions are possible from MO that correspond to the same symmetry as do the pertinent s-, p-, d-atomic orbitals. The relation between these AO and MO and selection rules for K_α, K_β and L spectra of AX_n compounds are as follows:

	AX_6	AX_4	K_α, K_β	$L_{II,III}$
A	$ns \to a_{1g}$	$ns \to a_1$	—	$\to 2p$
	$np \to t_{1u}$	$np \to t_2$	$\to 1s$	—
	$nd \to t_{2g}e_g$	$nd \to t_2e$	—	$\to 2p$
X_n	$ns \to a_{1g}e_gt_{1u}$	$ns \to a_1t_2$	—	$\to 2p$
	$np \to a_{1g}e_gt_{1u}$	$np \to a_1t_2$	$\to 1s$	—
	$t_{2g}t_{1g}t_{2u}t_{1u}$	t_1t_2e		

Transitions K_α and K_β will be the more intensive, the higher the degree by which MO t_{1u} or t_2, from where the transition proceeds, represents the p-orbital of the atom to whose inner $1s$-level the electron is made to pass.

All the bonding MO, mainly the ligand ones, are composed of $2p$ AO (in the case of O,F) or of $3p$ AO (S, Cl). K_α, K_β transitions in the spectra of these ligand atoms proceed from all bonding and nonbonding MO.

For the X-ray electron spectra, the selection rules are nonexistent, these spectra are obtainable for AO and MO of any symmetry.

Superposition of the Spectra. Diverse types of spectra: K_α, K_β, $L_{II,III}$ . . . both emission and absorption ones, are tied to the single MO scheme. Therefore, when considering the spectra they are often brought in line in such a way that the peaks correspond to the MO levels from which in emission spectra the transition is accomplished, or onto which the transition in absorption spectra takes place.

The combination is commonly made with reference to the $L_{II,III}$ ($2p_{1/2}$, $2p_{3/2}$) level. The position of this level is determined from calculation of the energy difference of K_β (MO $\to$ 1s) and K_α (2p $\to$ 1s).

Assignment of X-Ray Transitions in the Spectra of Octahedral and Tetrahedral Complexes. Let us consider the four complex types most important for us: transition metals in the octahedron and tetrahedron, and nontransition elements in the octahedron and tetrahedron. The methods employed in analyzing the X-ray spectra of such complexes have been considered in detail by Nefedov [287–290]. The X-ray and X-ray electron spectra of several molecules have been calculated minutely (CO_2, Cl_2, H_2S) [289, 309] as well as spectra of isolated groups in silicates [247, 250, 252, 257, 293, 321, 322].

An interpretation of the spectra of complexes includes that of the cation and anion spectra, and of the emission and absorption spectra; in the case of the emission spectra an analysis of K_α-, K_β and $L_{II,III}$-spectra, and in the case of the absorption of K- and L- spectra. All these can be conveniently matched on a single energy scale. Finally, all these spectra, as well as the X-ray electron spectra, can be naturally tied to a single scheme that represents a combination of the complex' molecular orbital scheme and the energy level diagrams of the cation and anion atomic orbitals.

The task of the direct interpretation of the X-ray spectra ends with the determination of the energy position of all the MO, deduced from the positions of lines in all these various spectra and with finding the correlation of the c_i coefficients in the MO from the intensity of these lines. Next comes the description of chemical bonding features, ensuing from the MO scheme.

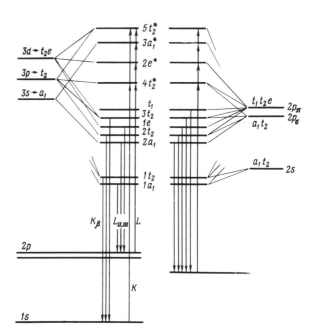

Fig. 24. Assignment of X-ray transitions in the spectra of the third row elements (Al, Si and others) in tetrahedral co-ordination

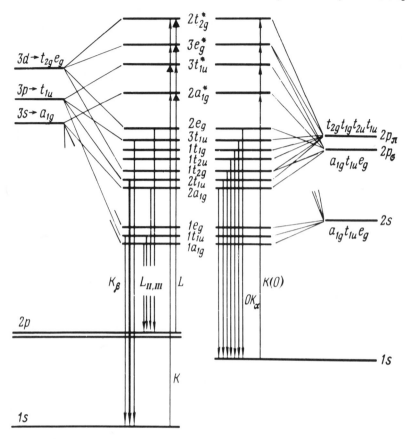

Fig. 25. Assignment of X-ray transitions in the spectra of the third row elements in octahedral coordination

Figures 24–27 illustrate typical MO diagrams for octahedral and tetrahedral complexes together with the AO of the central and AO of the ligand atoms, on which allowed emission X-ray transitions from MO to AO of the central atom or to AO of the ligands, as well as absorption transitions to empty antibondding MO, are shown.

Owing to different selection rules, the different transitions arise in the emission K_β spectra ($p \to s$), and in the $L_{II,III}$, as well as in K - and L absorption spectra. However, precisely by virtue of this, the bands of these spectra can be assigned to the transitions from concrete MO or to concrete MO, whereas in the X-ray electron spectra absolute values of all the AO and MO levels are obtained, but their assignment does not follow directly from the experiment and is effected on the ground of additional considerations, i.e., when the MO order is known from calculations or by analogy with other compounds.

The intensity correlation of individual transitions is determined largely by two factors: (1) to what extent a given MO in K_β-spectra is p-orbital, or in $L_{II,III}$ -

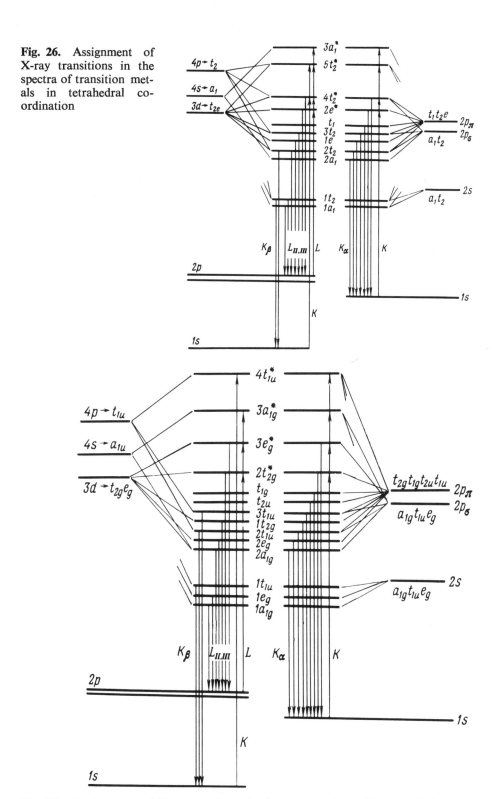

Fig. 26. Assignment of X-ray transitions in the spectra of transition metals in tetrahedral coordination

Fig. 27. Assignement of X-ray transitions in the spectra of transition metals in octahedral coordination

spectra s,d-orbital, and p- or s,d-orbitals being exactly of the central atom in the spectra of the latter and p- or s,d-orbitals of oxygen, fluorine, chlorine in the spectra of the ligand atom (2) the degree of MO degeneracy; all other conditions being equal, the intensities of the a:e:t molecular orbitals are related as the degree of their degeneracy, i.e., as 1:2:3.

The overall relative intensity of the entire emission spectrum of the central ion is the greater the more covalent the bond, i.e., greater the participation of the atomic orbitals of this ion in the bonding molecular orbitals.

Spectra of the Tetrahedrally Coordinated Third Row Elements (AlO_4^{5-}, SiO_4^{4-}, PO_4^{3-}, SO_4^{2-}, ClO_4^-) (see Fig. 24)

Emission Spectra. K_α spectra are related to transitions between the levels of inner electrons ($2p \rightarrow 1s$) and are considered below in connection with chemical shifts of the X-ray lines.

K_β spectra are related to the transitions from MO in which $3p$-atomic orbitals of the central ion take part; since in the tetrahedron p AO transforms as t_2 and there are three t_2 molecular orbitals, the K_β-spectra consist of three bands:

K_β-transition from $1t_2 = -0.30(3p\mathrm{Si}) - 0.26(3d\mathrm{Si}) + 1.06(2s\mathrm{O_4}) + 0.08 (2p\mathrm{O_4}) + 0.02 (2p\ \mathrm{O_4})$,

$$K_{\beta_1} \begin{cases} \text{-transition from } 2t_2 = -0.30(3p\mathrm{SI}) + 0.12(3d\mathrm{Si}) + -0.07 \times (2s\mathrm{O_4}) + \\ 0.51(2p\mathrm{O_4}) - 0.76(2p\ \mathrm{O_4}), \\ \text{-transition from } 3t_2 = -0.07(3p\mathrm{Si}) + 0.10(3d\mathrm{Si}) + -0.01 \times (2s\mathrm{O_4}) - \\ 0.78(2p\ \mathrm{O_4}) - 0.62(2p\ \mathrm{O_4}). \end{cases}$$

The c_i coefficients in the MOs (here and thereafter) are given for SiO_4^{4-} (see [1]); in spite of possible simplifications of the calculation they give an idea as to the character of each of these MOs, while squares of these coefficients indicate the relative intensity of the transitions from these MO.

The greatest amount of the $3p$ atomic orbital of Si are in the $2t_2$ molecular orbital, and it is exactly from there that the transition yielding the most intensive band in the K_β-spectrum occurs; onto this band a close and weak $3t_2$ band is superposed in the shape of a more or less clearly marked overthrust. The transition from the $1t_2$ band that consists largely of $2s$ group oxygen orbitals is pronounced rather weakly. The distance between $1t_2$ and $2t_2$ (14–18 eV) is almost the same as between the $2s$- and $2p$-orbitals of oxygen (17 eV).

K_β-spectra have been obtained for Al and Si in the tetrahedral coordination in a number of minerals [252, 253, 326] (see Figs. 28–30), and for PO_4^{3-}, SO_4^{2-} as well [289]. Initially, the interpretations of the K_β-spectra of minerals were not based on the MO diagram and, for this reason, they could not go beyond the attempts at establishing a correlation between the maximum of the K_{β_1}-band and coordination, as well as spacings of Al - O, Si - O. The K_β transitions in the tetrahedron and octahedron are related to different MO; in the tetrahedron the position of the K_β bands indicates the $1t_2$, $2t_2$ and $3t_2$ MO energies. Tetrahedrons Al - O_4 and Si - O_4

Fig. 28. SiK_β emission spectra (after [275]) (stishovite and Na_2SiF_6—octahedral coordination of Si; quartz, cristobalite, coesite—tetrahedral coordination)

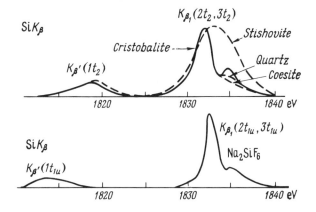

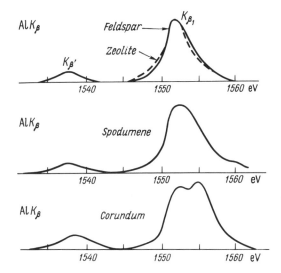

Fig. 29. AlK_β emission spectra (after [275]) (feldspar, zeolite-tetrahedral coordination of Al; spodumene, corundum octahedral coordination)

in aluminosilicates are almost always distorted, but no manifestation of this distortion in the splitting of the K_β-bands could be noted. It is advisable that the interpretation of the K_β-spectrum be made (in conjunction with other X-ray spectra) as interpretation of the MO scheme.

The $L_{II,III}$-spectra are the result of transitions from MO in which take part $3s$ and $3d$ AO of the central ion that transform in the tetrahedron as $s \rightarrow a_1$ and $d \rightarrow t_2$ and e. Therefore (see Fig. 30) the $L_{II,III}$-spectra are represented in the tetrahedral groupings largely by three bands:

$$1a_1 = [-0.36(3s\text{Si})] + [1.04(2s\text{O}_4) + 0.08(2p_\sigma\ \text{O}_4)],$$
$$2a_1 = [-0.83(3s\text{Si})] + [-0.22(2s\text{O}_4) + 0.66(2p_\sigma\ \text{O}_4)],$$
$$3t_2 = [-0.07(3p\text{Si})] + [0.10(3d\text{Si}) + 0.78(2p_\sigma\ \text{O}_4) - 0.62(2p_\pi\ \text{O}_4)],$$
$$1e\ = [-022(3d\text{Si})] + [1.02(2p\ \text{O}_4)].$$

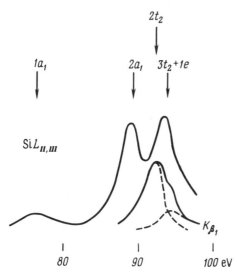

Fig. 30. Si$L_{II,III}$ emission spectrum of phenacite (after [290]) (tetrahedral coordination of Si)

Here the $1a_1$ and $2a_1$ bands represent $3s$Si but the former bonded with $2s$ of oxygens and the latter with $2p_\sigma$ of oxygens, while $3t_2 + 1e$ represents $3d$Si and may serve as a measure of the Si $3d$-orbitals (and also of other elements of the third period) participation in the chemical bonding. The two most closely lying and most intensive bands in the $L_{II,III}$-spectra of these elements represent transitions from $2a_1$ and $3t_2 + 1e$ MO. (see Fig. 30).

Absorption Spectra. K-absorption occurs consequent upon transitions from the $1s$-atomic orbital of Si to the antibonding t_2-molecular orbitals, where Si is represented by $3p$-orbitals: $4t_2$ and $5t_2$.

L-absorption is due to transitions from $2p$-atomic orbitals of Si to the antibonding $2e^*$ (with participation of $3d$Si) and $3a_1^*$ (with $3s$ Si).

The X-Ray Electron Spectra of Si. These spectra represent the energy of all AO Si and all MO, but to assign these energies and determine the relative amounts of different AO and MO, one has to resort to the X-ray spectra.

The Spectra of the Third Row Elements in Octahedral Coordination (AlO$_6$, SiO$_6$, SiF$_6^{2-}$, PF$_6^-$, SF$_6$) (see Figs. 25, 28, 29, 31, 32)

Emission Spectra. K_β-spectra for the case of octahedral coordination are similar to the tetrahedrally coordinated elements K_β-spectra; here too are in evidence three transitions, but this time from MO $1t_{1u}$, $2t_{1u}$, $3t_{1u}$, where there is no admixture of the $3d$-state.

$$K_\beta\text{-transition from } 1t_{1u} = 0.33(3p\text{Si}) + 0.93(2s\text{F}_6) + 0.15(2p_\sigma \text{ F}_6 + 0.10(2p_\pi \text{ F}_6),$$
$$= \text{transition from } 2t_{1u} = 0.60(3p\text{Si}) + -0.36(2s\text{F}_6 + 0.56 \times K_{\beta_1} \times (2p_\sigma \text{ F}_6) + 0.48 (2p_\pi \text{ F}_6)$$

Fig. 31. Emission spectrum (1) and absorption spectrum (2) of $AlL_{II,III}$ in corundum (octahedral coordination of Al) [263]

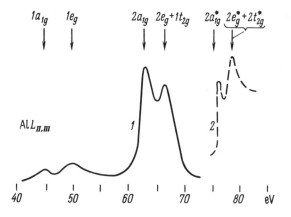

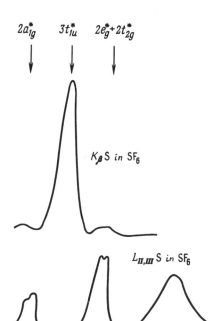

Fig. 32. Absorption spectra of sulfur in SF_4 (after [325]). Splitting of peaks in the L-spectrum is due to the spin-orbital interaction

$$= \text{transition from } 3t_{1u} = 0.26(3p\text{Si}) + -0.07(2s\text{F}_6) + + 0.45 \times (2p_\sigma \ \text{F}_6) - 0.86(2p_\pi \ \text{F}_6).$$

(Here the values of c_i are indicated for SiF_6^{2-} [203, 206].

In the Al - O_6 spectra of aluminosilicates and oxides and of Si - O_6 in stishovite (see Fig. 28), broadening of the $K_{\beta 1}$ band ($2t_{1u}$, $3t_{1u}$) is commonly in evidence as compared to $K_{\beta 1}$ in tetrahedral complexes. This may be attributed to the general broadening of the valence band for the octahedral coordination as against the tetrahedral one, to the $3t_{1u}$ intensity rising to that of $2t_{1u}$ [this being non-existent in Na_2SiF_6 (See Fig. 28) due to a greater coefficient at $2t_{1u}$ than at $3t_{1u}$, but in $Si-O_6$ and $Al-O_6$ these two bands are approximately of the same intensity]; it can also be attributed to a somewhat greater separation of $2t_{1u}$ and $3t_{1u}$, to the splitting of t_{1u} because of a usual distortion of the octahedral groupings in crystals.

The $L_{II,III}$-spectra usually represent two pairs of bands (see Figs. 31, 32): the first represent the transitions from $1a_{1g}$ and $1e_g$, consisting largely of the $2s$ ligand orbital and the second is related to the transitions from $2a_{1g}$, $2e_g$ and t_{2g}, made up mainly of $2p$ ligand orbitals:

$$1a_{1g} = [0.50(3sSi)] + [0.82(2sF_6) + 0.28(2p_\sigma F_6)],$$
$$1e_g = [0.37(3dSi)] + [-0.57(2sF_6) + 0.48(2p_\sigma F_6)],$$
$$2e_g = [0.64(3dSi)] + [0.33(2sF_6) + 0.69(2p_\sigma F_6)],$$
$$1t_{2g} = [0.30(3dSi)] + [c_i(2p_\pi F_6)].$$

Let us note that a_{1g} MO are formed without participation of $3d$ AO, and e_g MO without that of $3p$, both of them being σ orbitals.

From $L_{II,III}$-spectrum of octahedral complexes one can assess particularly clearly the participation of $3d -$ AO in the bond.

Absorption Spectra. K-spectra are related to the transitions from $1s -$ AO to antibonding $3t_{1u}$ MO, L-spectra from $2p$-AO to antibonding $2a_{1g}^*$ and $2e_{2g}^*$ MO (see Fig. 25).

Spectra of Transition Metals in Tetrahedral Coordination (see Fig. 26)

Emission Spectra. K_β-spectra of tetrahedrally coordinated transition metals and nontransition elements of the third period are similar, these are transitions from $3p$ t_2-orbitals. Unlike the third row elements, however, the transition from $1t_2$ in transition metals displays an intensity of the same order as from $2t_2$,, whereas $3t_2$ does not manifest itself in the spectra (see Fig. 36). Therefore two bands are evident:

K_β - transition from $1t_2$ and $K_{\beta_{2,5}}$ -transition from $2t_2$.

The transition metal K_β spectra are generally much weaker than those of the nontransition elements, for $t_2 -$ MO contains largely $3d -$ AO of the metal, $4p -$ AO appearing therein merely as an admixture state.

The $L_{II,III}$-spectra in the case of transition metals are much more intensive than those of K_α, because the amount of $3d$- and $4s$-AO in bonding MO of transition metals is greater than that of $4p$-AO. The following transitions are then possible: (in MnO_4, for example):

$$1a_1 = 0.12(4sMn) + 0.92(2sO_4),$$
$$1t_2 = 0.03(4pMn) - 0.16(3dMn) + -0.96(2sO_4),$$
$$1e = 0.74(3dMn) + 0.51(2p_\pi O_4),$$
$$2t_2 = 0.63(3dMn) - 0.17(4pMn) + -0.17(2sO_4) - 0.35(2p_\sigma O_4),$$
$$2a_1 = -0.03(4sMn) + 0.13(2sO_4) + 1.00(2p_\sigma O_4),$$

$3t_2 = 0.13(3d\text{Mn}) + 0.02(4p\text{Mn}) + -0.07(2s\text{O}_4) - 0.69 \times (2p_\sigma\text{O}_4) - 0.67$
$\quad (2p_\pi\text{O}_4),$
$2e^* = -0.72(3d\text{Mn}) + 0.90(2p_\pi\text{O}_4),$
$t_2^* = 0.80(3d\text{Mn}) - 0.14(4p\text{Mn}) + -0.30(2s\text{O}_4) - 0.67(2p_\sigma\text{O}_4) - 0.58$
$\quad (2p_\pi\text{O}_4).$

Here, in addition to transitions from bonding MO, are possible emission transitions from antiboding MO populated by $3d$-electrons.

From comparison of coefficients at $3d$- and $4s$-AO of a metal it follows that after two weak bands, $1a_1$ and $1t_2$, $1e$- and $2t_2$-transitions most intensive in the broad profile are, adjoined by other weak transitions.

Since in the case of transition metals the distance between $2p_{1/2}$ (L_{II}) and $2p_{3/2}$ (L_{III})-levels is great enough (from 6 eV for Ti to 20 eV for Cu), two spectra can be observed: L_{II} and a less intensive L_{III}, lying apart from it at a distance of $2p_{1/2} - 2p_{3/2}$, and this instead of a single $L_{II,III}$ spectrum of the third row elements, for which the spacing of $2p_{1/2} - 2p_{3/2}$ is about 1 eV.

Absorption Spectra. The K-spectra arise from the transition from $1s$ into $4t_2$ $(3d, 4p)$-MO, which yields a narrow strong peak and a more smeared $5t_2$; the L-spectra represent transitions to antibonding $2e^*$ and $4t_2^*$, whose spacing equals $10Dq$ (see Fig. 26).

The Spectra of Octahedrally Coordinated Transition Metals (see Fig. 27)

Emission Spectra. K_β has the same three bands as in the octahedral complexes of transition metals (with $1t_{1u}$, $2t_{1u}$, $3t_{1u}$), the most intensive transition being from $2t_{1u}$, but the distance between $2t_{1u}$ and $3t_{1u}$ is somewhat greater. These MO find themselves in different groups: (1) $1a_{1g}$, $1t_{1u}$, $1e_g$ (2) $2a_{1g}2t_{1u}2e_g$ and (3) $1t_{2g}3t_{1u}t_{1g}t_{2u}$.

K_β = transition from $1t_{1u} = [c_1(4p\text{Cr})] + [c_2(2p_\sigma\text{F}_6) + (2p_\pi\text{F}_6)],$
$K_{\beta5}$ = transition from $2t_{1u} = [0.1(4p\text{Cr})] + [0.87(2p_\sigma\text{F}_6) + 0.42(2p_\pi\text{F}_6)],$
$K_{\beta5}$ = transition from $3t_{1u} = [0.1(4p\text{Cr})] + [0.49(2p_\sigma\text{F}_6) - 0.87(2p_\pi\text{F}_6)],$

[Here values of c_i are indicated for CrF_6^{3-} (see 1)].
L_{II} and L_{III} spectra are represented by the following transitions:

$1a_{1g} = [c_1(4s\text{Cr})] + [c_2(2s\text{F}_6) + c_3(2p_\sigma\text{F}_6)],$
$1e_g = [c_4(3d\text{Cr})] + [c_5(2s\text{F}_6) + c_6(2p_\sigma\text{F}_6)],$
$2a_{1g} = [0.23(4s\text{Cr})] + [0.92(2p_\sigma\text{F}_6) + c_i(2s\text{F}_6)],$
$2e_g = [0.44(3d\text{Cr})] + [0.80(2p_\sigma\text{F}_6) + c_i(2s\text{F}_6)],$
$1t_{2g} = [0.39(3d\text{Cr})] + [0.86(2p_\pi\text{F}_6)],$
$2t_{2g}^* = [c_i(3d\text{Cr})] - [c_i(2p_\pi\text{F}_6)],$
$3e_g^* = [c_i(3d\text{Cr})] + [c_i(2p_\pi\text{F}_6) + c_i(2s\text{F}_6)].$

The last two MO are antibonding, populated by $3d$-electrons in the case of transition metal complexes.

In the case of ions with d^1–d^3 electron configuration, it is only $2t_{2g}^*$ — MO that is populated; with a greater number of electrons (in high-spin complexes) $3e_g^*$ is also populated and then one can see transitions from both these MO. The distance between them corresponds to $10Dq$.

Absorption Spectra. K-spectra are related to the transitions mainly from $1s$ to $4t_{1u}^*$ but also to weak transitions (forbidden, but occurring owing to distortions of the coordination polyhedra) to $3e_g^*$ and $3a_{1g}^*$; L-spectra represent transitions from $2p$ to $2t_{2g}^*$ (provided the latter is not completely populated), and to $3e_g^*$, $3a_{1g}^*$ (Fig. 33).

Spectra of Ligand Ions (O, F, Cl, S) Forming Octahedral and Tetrahedral Complexes

Only the K_α-emission spectra (MO → $1s$) are possible for oxygen and fluorine; for sulfur and chlorine K_α arise due to transitions between inner levels ($2p$ → $1s$), but K_β (MO → $1s$) arises from the same MO as do K_α of oxygen and fluorine.

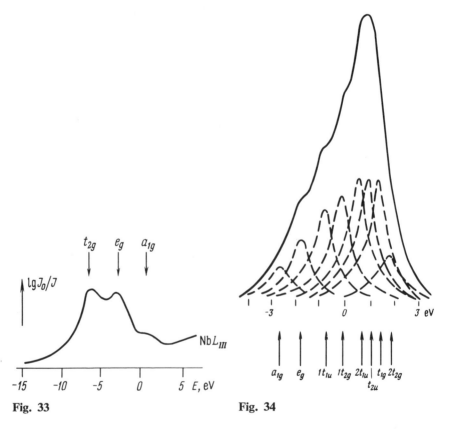

Fig. 33   **Fig. 34**

Fig. 33. L_{III} absorption spectrum of Nb in aeschynite [273]

Fig. 34. Decomposition of the K_β-band of ligand ions forming octahedral complexes around the transition metal ion (on an example of the Cl K_β-band in K_2PdCl_6[287])

Hence, K_α-spectra of O and F and K_β-spectra of S and Cl are analogous and depend in the main on the coordination, and upon the fact of whether the central ion is a transition metal or a nontransition element.

The common feature of all these ligand ions spectra is the fact that their $2p$ (or $3p$) atomic orbitals form group orbitals which make part of all the bonding, non-bonding and antibonding MOs. Therefore, in the O, F, Cl, and S emission spectra the transitions from all the MO manifest themselves with greater or lesser intensity; in the absorption spectra one observes that the transitions manifest themselves onto all the antibonding MOs. Because of the closeness of individual MOs the profile of the ligand ion spectra appears as a superposition of several bands according to the number of MOs that make a contribution conformable to the relation of coefficients at the group orbitals in different MO, and in compliance with the degree of degeneracy ($a:e:t_2 = 1:2:3$).

As an example in Figure 34 is shown the profile decomposition into eight components of the Cl K_β-spectrum in $PdCl_6^{2-}$ [287]. For the case of the octahedron with the central ion, which is a nontransition element, there are seven components in the emission spectrum of the ligand ion, for the octahedron with a transitional metal eight or nine components, this being accounted for by the occupancy of antibonding MO with d-electrons. In the tetrahedron with a nontransition element there are five components, and in that with a transition metal six or seven. The shape of the profile depends on the correlation of the transition intensities and on the relative spacing between individual MO.

Exhaustive data that mutually supplement and define one another are obtained by combination of all the transitions within a single MO scheme (Figs. 35–37).

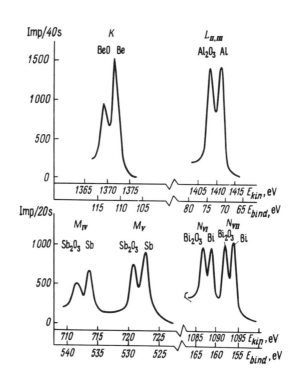

Fig. 35. K-, L-, M-, N-level shifts of light and heavie elements in the metal oxides as compared with the metals in the X-ray electron spectra [308]

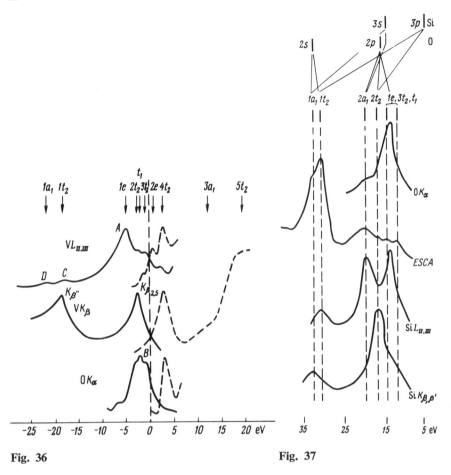

Fig. 36   **Fig. 37**

Fig. 36. Superposition of K-and L-emission and absorption spectra of tetrahedrally coordinated vanadium in Na_3VO_4 and correlating them to the MO levels [262]

Fig. 37. Superposition of SiK_β, $SiL_{II,III}$, OK_α and X-ray electron spectra (ESCA) of quartz and correlating them to the MO levels [290]

2.2.2 Chemical Shifts in the X-Ray and X-Ray Electron Spectra and Determination of Effective Charges

Changes caused by differences in the valence, coordination, and chemical binding become manifest not only in the X-ray spectra, related to transitions from the levels of outer electrons partaking in the formation of MO, but also in the spectra stemming from transitions between the levels of inner electrons that do not participate in the chemical binding.

In the X-ray spectra, the K_α spectra ($2p \rightarrow 1s$) more often subjected to measurements. The displacements of K_α-lines of the third row elements in compounds amount to up of $\sim$ 1 eV [210] relative to the spectrum of the elements (commonly about 0.2–0.5 V), the accuracy of the measurement being of the order of 0.02–0.005 eV. These K_α- shifts are noticeable even in the heaviest elements; for example K_α Mo in MO_3 shifts as against metallic Mo by $+0.192 \pm 0.007$ eV, W in WO_3 by $+0.110 \pm 0.033$ eV [313, 314].

It is expedient to separate in a clearcut fashion: (1) these shifts of the K_α-lines caused by transitions between inner levels, (2) changes in the spectra associated with transitions involving participation of MO. They differ both as concerns the nature of experimental data and the physical mechanism and chemical interpretation.

Transitions with participation of MO yield broad bands in lieu of individual lines, their number and intensity depending upon the number and the nature of MO. With changing coordination and electron configuration of the central ion, as well as of the AO ions energy, a change takes place in the entire MO scheme, the latter being then composed of the various MO types, their different number, and ordering of individual MO. For this reason, the shifts in the maxima of individual bands in such spectra are comparable only when they relate to one and the same MO. An interpretation of such a shift differs from that of the shifts of lines caused by transitions between inner levels. In consequence of an analysis of the spectra related to transitions from MO, an MO scheme results which yields the most complete description of the state of chemical binding.

Transitions between inner levels of an element yield one and the same line in all compounds, and the only parameter of such spectra is the chemical shift of the said lines. This single experimental parameter is associated with one parameter of chemical binding: the effective charge of the atom in a compound. Then the definition of the effective charge is given not within the framework of the MO method (by means of coefficients in MO), but within that of diverse physical models (see below).

Hence, changes in the K_β-spectra of the third row elements and of transition metals, as well as in the K_α-spectra of the second row elements (where $2p$-AO are valent) should be considered preferably within the framework of MO.

The displacements of K_α represent a difference of the $1s$- and $2p$-electron levels shifts. The X-ray electron spectra that fix the very energies of all the electron levels are helpful in directly determining chemical shifts of these levels.

These shifts become manifest most clearly in a simultaneous registration of an X-ray spectrum of a metal and of an oxide film developing on it (Fig. 35). The magnitude of shifts metal–metal oxide is around 2–3 eV. With a change in the valence (degree of oxidation) the shift comprises about 1eV; the ultimate changes: from Cl^- in NaCl to Cl^{7+} in $NaClO_4$ for $K(1s)$ and $L_{II,III}$ ($2p$), for example, amount to about 1.5 eV. For compounds of the same type, the shifts are of the order of 0.1–0.5 eV relative to NaF taken for zero; in the case of NaCl it is $+0.1$ eV, in that of NaBr $+0.5$ eV, in NaI $+0.3$ eV with a mean reproducibility of 0.2 eV[194].

Let us consider the physical meaning of the energy level shifts of inner electrons $1s$, $2s$, and $2p$.

Upon removal of the outer electron with a charge e and mean orbital radius r, the potential produced by the electrons screening the interaction of the inner electron with the nucleus diminishes by $V = e/r$. Accordingly, the binding energy of the inner electron with the nucleus increases with the removal of the outer electron. Conversely, with the electron joining the atom, the binding energy of the inner electron decreases.

The binding energy of the inner electron is an energy required for its separation from the atom, i.e., the energy of its ionization, or the potential of its ionization, since $E = eV$. This ionization energy of the inner electron depends on the charge of the ion (M^0, M^{1+}, M^{2+} . . .) and on its electron configuration (s^2p^2 or sp^3, etc.). In the case of inner electrons it is the same energy of ionization as that of various outer electrons with different valence states of the atom (VSIE—valence state ionization energy, see [1]). For these it is possible to construct the same dependences on the free ions charge (Fig. 38) as for the VSIE of outer electrons.

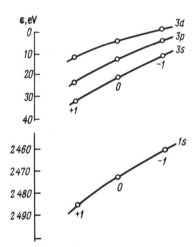

Fig. 38. Charge dependence of the levels energies of the outer $3s$-, $3p$-, $3d$-electron and inner $1s$-electron for sulfur [308]

However, for the outer electrons, the experimental VSIE values are taken as obtained from optical spectra of the atoms. As to the inner electrons, no data are available that would correspond to the X-ray spectra for free ions M^0, M^+, M^{2+} . . . and, for this reason, resort is taken to the inner electrons VSIE computations for different charges and diverse excited states of valence electrons.

In the case of ions in crystals, apart from VSIP, account has to be taken also of the potential created by all the lattice ions (Madelung's potential). It is just as necessary for inner as for outer electrons in calculations by the semi-empirical tight-binding method. The effect of the Madelung potential is opposite to that of VSIP, and it diminishes the chemical shift of the level.

Thus the dependence of the inner level chemical shift on the effective charge of the ion in a crystal appears as a result of a shift in the level due to a change of the electron binding energy in the free ion (VSIE), and an opposite shift of the level due to a change of interaction of this electron with the lattice field.

One inconvenience of this method of effective charge determination is that it requires, apart from estimating the dependence of the level shift on the free ion charge for each element, also calculating the dependence of the level shift on the Madelung potential for each compound.

Chemical shifts of K_α-lines appear as a difference of the 2p- and 1s-levels shifts, and it would seem that these values ought to be fully in line with each other. There is, however, a difference between them. This difference is due to the fact that the Madelung potential of the crystalline lattice produces an equal shift of both inner levels. Therefore, the K_α-lines shift (2p → 1s), unlike the 2p- and 1s-levels shifts, is independent of the Madelung potential. The ΔK_α versus effective charge curves do not have to be calculated for each individual compound by estimating the lattice potential for it. However, it is then obvious that the effective charge determined from K_α and from ΔE_{2p} or ΔE_{1s} assumes different magnitudes and these differ exactly by the contribution of the lattice potential.

This also explains a somewhat greater sensitivity of ΔK_α to a change in co-ordination. Octahedral and tetrahedral coordination produces not only different effective charges, which affects ΔK_α, but also different lattice potentials that compensate a change in the charges, thus causing slight shifts of inner levels [288].

Diverse ways exist of determining the relationship between the effective charge and the displacement of K_α-lines: (1) through estimating changes in the electron density in the atom sphere with a radius delimiting the sphere outside of which the electron density of the ligands is negligible [240, 306]; (2) by way of calculating changes in the screening of inner electrons [298, 313, 314]. The most direct method is to assess the shift of K_α-lines by calculating the wave functions:

$$\Delta K_\alpha = (E_{1s} - E_{2p})_{\mathrm{ion}} - (E_{1s} - E_{2p})_{\mathrm{atom}}.$$

The X-ray lines shift-effective charge curves (Fig. 39) plotted by Urusov [323] are in keeping with calculations made by Nefedov [286, 289]. In plotting them account was taken of semi-empirical corrections for a change in the wave functions of the atoms in crystals, as compared to free atoms.

Experimental values of ΔK_α were obtained for compounds of the third row elements and for those of iron [240, 323].

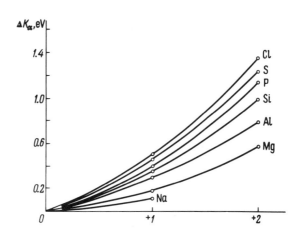

Fig. 39. Curves for determining an effective charge of the third row elements from chemical shifts of K_α X-ray lines [323]

3. Electron Paramagnetic Resonance

3.1 The Substance of the Electron Paramagnetic Resonance (EPR) Phenomenon

3.1.1 The Scheme of Obtaining EPR Spectra

This scheme can be compared with that used in obtaining optical absorption spectra (Fig. 40). The source of emission in the EPR spectrometers is klystron, i.e., an electron tube, operating within the range of a superhigh-frequency band, and yielding a strictly monochromatic emission with a wavelength in commercially constructed spectrometers equal to 3.2, 1.2, or 0.8 cm. Therefore no monochromator is required here, but a waveguide and resonator, holding the specimen to be studied, are needed.

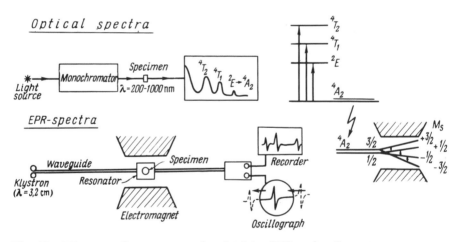

Fig. 40. Diagrams of arrangements for obtaining EPR and optic spectra

The magnetic field of the electromagnet causes splitting of spin sublevels with the energy difference which can be made equal to the energy emitted by the klystron. The resonance signal can be either seen on the screen of an oscillograph or registered with a recorder.

The condition necessary for obtaining the resonance is equality of the radio-frequency quantum $h\nu$ (h is Planck's constant, ν is frequency) and the energy difference between the spin sublevels, amounting to $g\beta H$ (g is a factor of spectroscopic splitting that takes account of the contribution of the orbital and spin moments to the state of the atom; β is Bohr magneton; H is intensity of the resonance magnetic field):

$$h\nu = g\beta H.$$

This expression establishes a triple dependence:

1. Magnetic field H brings forth the appearance of spin sublevels and determines the energy difference between them.

2. The radio-frequency energy quantum $h\nu$ causes transition from the lower spin sublevel to the upper one, attended by absorption of energy, and produces an absorption signal.

3. The g-factor defines the change in the position of the absorption line in the spectrum under given $h\nu$ and H, depending on the features particular to the state of the paramgnetic electron in the study specimen, thus presenting the characteristics of the substance in conditions of resonance. The resonance condition pre-supposes two technical variants for obtaining resonance absorption (Fig. 41): (a) on achieving the splitting of the spin sublevels to a definite extent by superposition of the magnetic field one can change the radiation frequency (as in optical spectrometers) and fix the equality of energies by means of the absorption curve (b) at a constant radiation frequency it is possbile to change the distance between the spin sublevels by varying the intensity of the magnetic field until the resonance signal is obtained. The second of these variants proved to be more convenient.

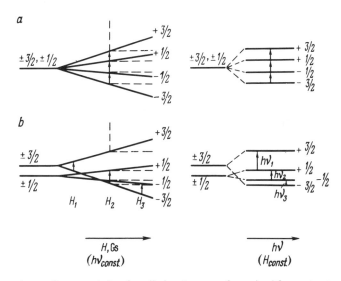

Fig. 41 a and b. Presentations of magnetic levels splitting (*upper schemes*) with constant frequency ($h\nu_{\text{const}}$) and varying magnetic field H and (*lower schemes*) with constant magnetic field (H_{const}) and varying frequency ($h\nu_1$, $h\nu_2$, $h\nu_3$). **a** Without initial splitting; **b** with initial splitting (on an example of a ion with spin $S = 3/2$, Cr^{3+}, for instance)

Transitions between spin sublevels in the magnetic field are considered in determining the concepts of the electron spin s and of the magnetic spin quantum number M_s. (see [1]). EPR is not only a part of spectroscopy, but it is also based on the theory of magnetism and, for this reason, we face here problems associated with magnetic properties of matter. Although magnetic properties lie at the root of EPR, we shall be concerned here largely with spectroscopic aspects, since it is with these that the application of the method is associated.

By using a vector model describing the behavior of the electron in a magnetic field (Fig. 42), let us obtain the values for the energy of the spin sublevels and the resonance frequency.

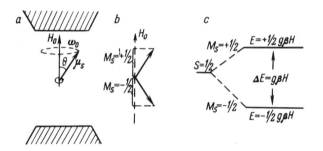

Fig. 42 a-c. Vector model of an electron in the magnetic field and splitting of spin sublevels. **a** precession of magnetic dipole μ_S (electron with spin S) about the orientation of the external magnetic field H_0; ω_0 is Larmor precession frequency (resonance frequency); **b** space quantization of the electron's magnetic moment: $M_s = +1/2$ and $M_s = -1/2$ are two possible magnitudes of the magnetic quantum number with the spin quantum number $S = 1/2$; **c** formation of two spin energy sublevels for the case of spin $S = 1/2$.

The angular moment p_s (mechanical moment or rotational moment) of the electron (as well as of the nucleus and atom) are measured in terms of the $\hbar$ units ($\hbar = h/2\pi$ where h is Planck's constant with the dimensionality of the moment of momentum: erg·s and it equals $p_s = s^*h$; $s^* = \sqrt{s(s+1)} = \sqrt{3}/2$; where s = 1/2 is the electron spin (the angular moment of the nucleus $I^*\hbar$ is written in a similar form, where I is nuclear spin, and angular moment of the atom $S^*\hbar$ with spin S).

The magnetic moment of the electron μ_s may be expressed in terms of two different factors that have the same meaning (conversion factors from the angular to the magnetic moment), but different dimensionality:

$$\mu_s = -g\gamma s^*\hbar,$$

where γ is gyromagnetic ratio (i.e., the ratio of the magnetic moment to the mechanical one equal to $\gamma = e/mc$ rad·s^{-1}. Gauss^{-1}, or

$$\mu_s = g\beta s^*,$$

where β is Bohr magneton[4] equalling $eh/4\pi mc$ erg Gauss^{-1}; $\beta = \gamma\, \hbar/2$.
The magnetic moment component in the direction of field H is

$$\mu_H = g\beta M_s.$$

The energy of magnetic dipole with moment μ_s in the field H_0 (see Fig. 42) is

$$E = \mu_s H_0 \cos\theta \text{ or } E = g\beta H_0 M_s.$$

Since the selection rules for magnetic dipole transitions require $\Delta M_s = \pm 1$, then
$E_2 - E_1 = g\beta H_0$.
The precession frequency ω_0 of magnetic dipole μ_s in field H_0 (Larmor frequency) is

$$\omega_0 = \gamma H_0.$$

When the radiation frequency equals the frequency of the electron dipole magnetic precession in field H_0, then the condition of the resonance absorption of the energy is realized, i.e., $(E_2 - E_1)/h = \omega_0$. Substituting in $\omega_0 = \gamma H_0$ in place of γ and ω their expressions $\gamma = g\beta/\hbar$ and $\omega = 2\pi\nu$, one obtains

$$\nu = \frac{g\beta H_0}{h} \text{ or } h\nu = g\beta H_0 .$$

Registration of the Spectrum: Absorption Curve and the Absorption Curve Derivative. Not the resonance absorption lines, but their derivatives are subject to registration in all modern spectrometers (Fig. 43). This is due, firstly, to a greater dis-

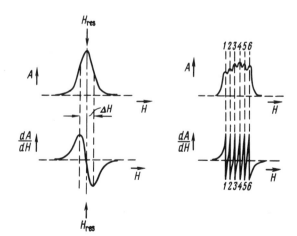

Fig. 43. Absorption lines (A) and absorption lines derivatives (dA/dH) in the EPR spectra

---

[4] Is also designated as μ_β, or μ_0.

tinction of the individual lines in complex spectra and, secondly, to the technical convenience of registering the first derivative [352]. Sometimes, a second derivative is recorded. The intersection of the first derivative with the zero line corresponds to the resonance value of the magnetic field, the line width being measured between inflection points. In complex spectra, the number of lines is counted with reference to the number of maxima or minima of the absorption line derivatives (see Fig. 43).

Units of Measurement. In order to estimate more concretely the spectral band in which EPR measurements are made and its specificity by comparison with the region of optical spectroscopy, and also for the purpose of converting various units of measurement encountered in literature, let us consider the numerical order of the values measured in different units and their conversion.

Direct measurements of the position of the lines in EPR spectra and their spacing are made in Gauss (Oersted[5]). But since the condition of resonance $H = g\beta H$ equates the energy magnitudes (in erg) in proportion to frequency ν (in s^{-1} and also in cm^{-1}), and to magnetic field H (in Gauss), all these three units: s^{-1}, cm^{-1}, Gauss (Tables 7 and 8) are used.

Table 7. Relations between units of measurement of energy used in radiospectroscopy

	Erg	Hz(s^{-1})	cm^{-1}	G
Erg	1	erg $\times$ $1/h$	erg $\times$ $1/hc$	erg $\times$ $1/g\beta$
Hz(s^{-1})	Hz $\times$ hc	1	Hz $\times$ $1/c$	Hz $\times$ $h/g\beta$
cm^{-1}	$cm^{-1} \times hc$	$cm^{-1} \times c$	1	$cm^{-1} \times hc/g\beta$
Gauss	G $\times$ $g\beta$	G $\times$ $g\beta/h$	G $\times$ $g\beta/hc$	1

Table 8. Relative order of values used in radiospectroscopy

	cm^{-1}	G	MHz
10^{-4} cm^{-1}	—	1.07	3
100 G	93.35 $\cdot$ 10^{-4}	—	280 (with $g = 2$)
100 MHz	33.3 $\cdot$ 10^{-4}	35.70(with $g = 2$)	—

All the relations are deduced from the condition of resonance:

$$h(\text{erg} \cdot \text{s}) \times \nu(s^{-1}) = g \times \beta(\text{erg} \cdot \text{G}^{-1}) \times H(\text{G}),$$

where $h = 6.62554 \cdot 10^{-27}$ erg$\cdot$s; $\beta = 9.27314 \cdot 10^{-21}$ erg G^{-1}; $\mu_{el} = 9.2837 \times 10^{-21}$ erg G^{-1} (the magnetic moment of the electron differs somewhat from the theoretical value of the Bohr magneton β; μ_{el}: $\beta = 1.0011596$; in calculations the value of

---

[5] The intensity of magnetic field H is measured in Oersteds, magnetic induction B in Gauss, but, since it is nonmagnetic substances that are investigated in EPR, here B and H represent one and the same field. In EPR, designations of the field in Gauss and Oersted are equally accepted. In EPR the Gauss designations were adopted from the technique of magnetic measurements, like those in Hz instead of s from radio technique.

μ_{el} is used precisely); $c = 2.997925 \cdot 10^{10}$ cm s$^{-1} \approx 3 \cdot 10^{10}$ cm s^{-1}; $hc = 1.98618 \cdot 10^{-16}$ erg·cm, $h/\beta = 0.7137 \cdot 10^{-6}$ s·G; $hc/\beta = 2.1419 \cdot 10^4$ cm G; $\beta/h = 1.40 \cdot 10^6$ s^{-1} G^{-1}; $\beta/hc = 0.4666 \cdot 10^{-4}$ cm^{-1} G^{-1}.

The energy of radiofrequency quantum $h\nu$ in ergs is found directly through multiplication of frequency ν in Hertz (s^{-1}) by Planck's constant h. However, as in optics, the EPR spectra parameters are almost never expressed in terms of ergs, but are given in MHz (GHz), in cm^{-1}, and in G.

cm^{-1} — Hz, MHz, GHz (or c/s, Mc/s, Gc/s); f (Hz) $= (s^{-1}) = \nu(\text{cm}^{-1}) \times c$ (cm s^{-1}) $= \nu(\text{cm}^{-1}) \times 3 \cdot 10^{10}$. For example: $f(\text{Hz}) = 0.3$ cm$^{-1} \times 3 \cdot 10^{10} = 0.9 \cdot 10^{10}$ Hz $= 9000$ MHz $= 9$ GHz.
1 GHz $= 0.033$ cm^{-1}; 1 MHz $= 0.33 \cdot 10^{-4}$ cm^{-1}; 100 MHz $= 33.3 \cdot 10^{-4}$ cm^{-1}
1 cm$^{-1} = 30$ GHz $= 30,000$ MHz $= 3 \ 10^{10}$ Hz.

cm-Hz, MHz, GHz; f (Hz) $= \nu(s^{-1}) = c(\text{cm s}^{-1})/\lambda(\text{cm}) = 3 \cdot 10^{10}/\lambda(\text{cm})$.
For example, the wavelength of 3.2 cm: corresponds to 9300 MHz.
For example, the wavelength of 1.2 cm: corresponds to 25,000 MHz.
For example, the wavelength of 0.8 cm: corresponds to 37,500 MHz.

G — Hz, MHz, GHz; H (G) $= h/\beta \cdot 1/g \ \nu$ (Hz) $= 0.7137 \ 1/g \ \nu$ (MHz); ν(Hz) $= \beta/h$ $gH = 1.40 \cdot 10^6 \ gH$; ν (MHz) $= 1.40 \ gH$ (G).
With $g = 2$, ν(Hz) $= 2.80 \cdot 10^6 H$; ν (MHz) $= 2.80 \ H$; with frequency of 9300 MHz the resonance magnetic field $H \approx 3300$ G.
With $g = 2$ 100 MHz $= 35.70$ G; 100 G $= 280$ MHz.

G — cm^{-1}; $H(\text{G}) = hc/\beta \cdot 1/g \ \nu(\text{cm}^{-1}) = 2.1419 \cdot 10^4 \ 1/g \ \nu(\text{cm}^{-1})$.
With $g = 2$; $H(\text{G}) = 1.0709 \cdot 10^4 \ \nu(\text{cm}^{-1})$; $100 \cdot 10^{-4}$ cm$^{-1} = 107.1$ Gν (cm^{-1}) $= \beta/hc \ gH = 0.4666 \cdot 10^{-4} \ gH$ (G).
With $g = 2$; ν (cm^{-1}) $= 0.9332 \cdot 10^{-4} \ H$ (G); 100 G $= 93.35 \ 10^{-4}$ cm^{-1}.

cm^{-1} — eV. In optical spectroscopy one often has to use the relation: 1 eV $= 8066$ cm^{-1}. In the EPR spectroscopy the order of energy is less than 1 cm^{-1}; 1 cm$^{-1} = 0.000124$ eV $= 1.24 \cdot 10^{-4}$ eV.

The X, K, Q Frequency Bands Used in the EPR Spectrometers. The commercially batch-manufactured EPR spectrometers commonly operate within band frequencies of:

Band frequency	cm	ν, MHz	H, G
X	3.2	9,300	3,300
K	1.25	24,000	8,500
Q	0.8	37,500	13,000

The equipment and technique of measuring EPR spectra are discussed in detail in a number of monographs [348, 441, 476].

3.1.2 The Energy Levels Scheme

Electron paramagnetic resonance (or electron spin resonance) is a part of spectroscopy which considers transitions between spin sublevels of the atoms, arising due to superposition of an external magnetic field; these transitions correspond to the microwave–radiofrequency region of the spectrum. As in optical spectra these are absorption spectra, but the EPR absorption bands find themselves not in the visible region, but in that of the mm-cm wavelengths, where the microwave (superhigh-frequency) region ends and the radio-frequency one begins. Both optical and EPR spectra are electron spectra (in contrast to the nuclear spectra NMR, NQR, and NGR), but in optical spectroscopy, transitions proceed between orbital levels, whereas in EPR spectroscopy they take place between spin sublevels emerging in the external magnetic field following splitting of the ground orbital state of the atom.

Thus, viewed from the standpoint of the electron structure of the atom, EPR spectroscopy is a natural sequel to optical spectroscopy. It comes as a concluding phase in the scheme of splitting the energy levels starting with terms of a free ion, passing on to energy levels that arise in the crystalline electric field and ending in the spin sublevels arising in the external magnetic field.

This whole sequence of the splitting of electron states is shown in Figure 44 in an example of the Cr^{3+} ion. In order to trace back this "geneological tree" of levels, and thus understand the essential substance of EPR as a part of spectroscopy, one has to be in command of the preliminary knowledge of basic theories in the field of atomic spectroscopy, the crystal field theory, and that of optical absorption spectra (see [1]). (The discussion of these problems constitutes a great part of the monographs on EPR.)

Basic Theories of Atomic and Optical Spectroscopy and Their Significance in EPR (see Fig. 44)

1. The number of electrons, i.e., the ion position in the periodic system, determines the electron configuration of the ion. For example, Cr^{3+} has 19 electrons, its electron configuration is $1s^2 2s^2 2p^6 3s^2 3p^6\ 3d^3$, or, without counting the electrons of the closed shells: $3d^3$, i.e., three d-electrons in the third electron shell.

From the electron configuration are determined the ground and excited terms of the free ion; for instance, the term of Cr^{3+} (see Fig. 44)^4F, i.e., orbital state $L = F = 3$; the multiplet $2S + 1 = 4$; spin $S = 3/2$.

Transitions between terms of the free atom (with due regard for the spin-orbital interaction in the free ion) correspond to spectra used in spectral and atomic absorption analysis.

2. In the cubic crystalline field, the terms of ions transform into states determined by the local symmetry of the ion's position. For example, in the case of Cr^{3+} (see Fig. 44) term ^4F splits in the octahedral field into $^4A_{2g} + {}^4T_{1g} + {}^4T_{2g}$. In the fields of lower symmetry a further splitting of the orbital states occurs. For instance, in the trigonal field $^4T_{1g}$ splits into $^4E + {}^4A_2$, $^4T_{2g}$ into $^4E + {}^4A_1$, and ^4A_g transforms into $^4A_2$. Here A, E, and T are designations of orbital states by the degree of

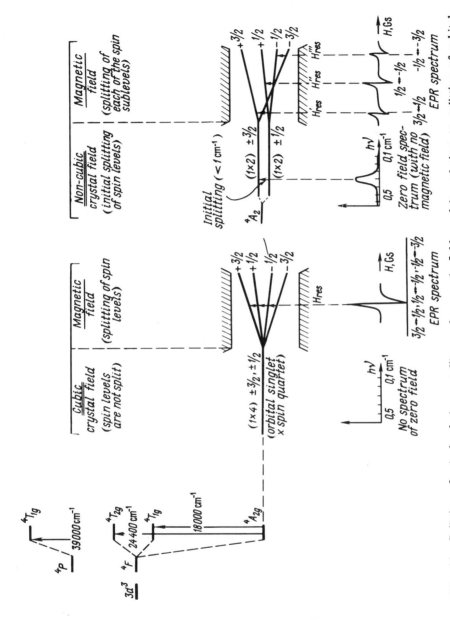

Fig. 44. Splitting of spin levels in crystalline and magnetic fields and its relation to splitting of orbital levels and to free ion terms

their degeneracy; A corresponds to nondegenerated (orbital singlet), E to twice degenerated (orbital doublet), and T to three times degenerated (orbital triplet) states (in crystal fields higher degrees of degeneracy are nonexistent).

The multiplicity of all these states originating from the single term ^4F is retained (^4A, ^4A, ^4T), i.e., the spin state is not subject to the action of the crystalline field in the first order of perturbation theory. The orbital singlet, doublet, and triplet represent spin quartets. Their total degeneracy is denoted as (1×4), (2×4), and (3×4).

Transitions between orbital states correspond to absorption bands in the optical spectra (crystal field spectra); the spin multiplicity determines 'the spin-allowed (between states of equal multiplicity) and spin-forbidden (different multiplicity) transitions.

3. Where the splitting of the levels of the optical transitions ends the splitting of levels of the EPR transitions begins, and is distinguished by the following features.

In EPR only the ground state comes under consideration, the excited orbital states separated from the ground one by energies of the order of 10,000–40,000 cm^{-1} being far beyond the limits of energies excited by the radio-frequency quanta (0.3 cm^{-1} at a frequency of 9000 MHz, 1 cm^{-1} at a frequency of 30,000 MHz, 3 cm^{-1} at '90,000 MHz). Therefore, even the levels that are more than 3–5 cm^{-1} away from the lower one lie far outside the EPR limits.

For EPR only the spin degeneracy of the lower orbital state is of interest; thus the $^4T_{2g}$ and ^4A states–the spin quartets—will yield the same pattern of spin sublevels.

The specificity of the action exerted by the crystal field on the spin state consists in the fact that the cubic field does not split the spin states, whereas noncubic fields lift the spin degeneracy (down to doublets and singlets); the noncubic fields then cause splitting of spin states even when the orbital state is a singlet. The order of value of this splitting of the spin levels by crystal fields is of the order of value of splittings by external magnetic field.

4. This splitting completes the scheme of energy levels of the ion in the crystal (i.e., the "last" corollary to the action of the crystal field are one or several spin levels with the degeneracy $2S + 1$, separated by the initial splitting).

Further splitting occurs due to the superposed external magnetic field (see Fig. 44), which completely removes the spin degeneracy. However, the scheme of magnetic splitting differs, depending upon the presence or absence of the initial splitting by the crystal field.

In a cubic crystalline field (without initial splitting) the spacing between the spin sublevels remains the same and, although the number of magnetic levels depends upon the spin multiplicity (i.e., is equal to $2S + 1$), the distance between the latter is equal and thus transitions between levels will occur at one and the same energy. In the cubic field there will be only a single line.

In noncubic fields, i.e., in the presence of an initial splitting, the magnetic field splits spin levels, thus leading to several lines in the EPR specta, referred to as fine-structure lines.

Comparison with Stark and Zeeman Effects of Atomic Spectroscopy. If a spectrum of an atom or ion is observed with superposition of an external electric field splitting of the atomic spectrum lines occurs and the Stark effect is in evidence. An analogous phenomenon is absorption band splitting of the ion in a crystal under the effect of a crystalline electric field. For this reason this is sometimes called Stark's splitting.

When observing atomic spectra in a superposed magnetic field splitting of spectral lines is also observed (but with a different number of lines and spacing between them), known as the Zeeman effect. Spin level splitting of an ion in a crystal, occurring under the influence of an external magnetic field that becomes manifest in the EPR, is often termed Zeeman splitting.

In the case of free atoms, however, the external electric and magnetic fields (Stark and Zeeman effects) split the atom states arising as a result of the spin-orbital interaction and characterized by quantum number I, while the levels forming in the external fields are defined by the magnetic quantum number M_J. In the case of ions in a crystal, the magnetic fields act on the spin states of the ion, these being characterized by spin quantum number S, while the spin (Zeeman's) sublevels that form in the magnetic field are defined by magnetic spin quantum numbers M_S, which takes values of $S, S - 1, \ldots -S$.

The selection rules for transitions between spin sublevels are determined by magnetic quantum numbers:

$$\Delta M_S = \pm 1.$$

(In some instances forbidden transitions with $\Delta M_S = \pm 2, \pm 3$ also become manifest.)

3.1.3 Substances Which Can Be Investigated by Using EPR

Two major domains of EPR application in mineralogy are determined by the existence in natural crystals of two paramagnetic center types: impurities of the paramagnetic ions and electron-hole centers.

Paramagnetic Ion Impurities. These are ions with unfilled inner d- and f- shells. It is precisely at the expense of unpaired d- and f-electrons that the magnetic moment of ions is produced. When forming chemical bonds in crystals (and also in glasses and solutions) ions with s- and p-valence electrons usually acquire electron configuration with filled shells in which the electron spins are paired, while the summated magnetic moment is zero.

The EPR spectra are observed in crystals for ions of the following transition groups:

$3d$ (iron group): Ti, V, Cr, Mn, Fe, Co, Ni, Cu;
$4d$ (palladium group): Zr, Nb, Mo, Tc, Ru, Rh, Pd, Ag;
$5d$ (platinum group): Hf, Ta, W, Re, Os, Ir, Pt, Au;
$4f$ (rare earths): Ce, Pr, Nd, Pm, Sm, Eu, Gd, Tb, Dy, Ho, Er, Tm, Yb;
$5f$ (actinides): Th, Pa, U, Np, Pu, Am, Cm, Bk, Cf . . .

These are the same ions which yield optical crystal field spectra.

However, not all the valence states of these elements are paramagnetic, for if all the inner d- or f-electrons become valent, the ion then ceases to be paramagnetic. Hereto belong the most common valence states in minerals such as Ti^{4+}, V^{5+}, Zr^{5+}, Mo^{6+}, Hf^{4+}, Ta^{5+}, W^{6+}, Th^{4+}, and U^{6+}. For this reason, for example, diamagnetics containing impurities, such as rutile (TiO_2), scheelite ($CaWO_4$), zircon ($Zr SiO_4$) etc. belong to this category.

Other limitations imposed on the EPR method are due to the defined ranges of the concentrations which can be investigated. Optimal concentrations of the paramagnetic ion vary from 0.1 to 0.001 %. In EPR are distinguished: (1) total number of unpaired electrons yielding a resonance absorption signal and (2) concentration of these electrons in the crystal. By increasing the size of the crystal (and this too restricts the number of natural objects to be investigated by this method) it becomes possible to raise the lower limit of sensitivity. In practice, crystals measuring up to 1 cm can be used, and in these contents of up to $0.000n$% are detected.

An upper limit to concentration also exists: O. n–1.0 %. This is connected with an exchange interaction between paramagnetic ions, owing to which a single extremely broad absorption line with an unresolved fine and hyperfine structure emerges, which usually makes it impossible to identify the ion and its structural position. A high concentration of any one of the paramagnetic ions may prevent observation of other paramagnetic ions that are present in an optimal concentration. In a number of instances, however, valuable information can be obtained for minerals with a high concentration of the paramagnetic ion as well.

The spectrum of such a character continues to be present in very small mineral samples with a high concentration of the paramagnetic ion. This can be made use of in distinguishing ion impurity incorporated in the mineral's lattice from micro-inclusions of the minerals; with a paramagnetic ion impurity concentration in a micro-inclusion of an order of less than 0.1 %, the total number of unpaired electrons in the micro-inclusion may well be below the sensitivity of the EPR spectrometers, while in concentrations surpassing 0.1 %–1 %, the totality of unpaired electrons may prove sufficient, but the exchange interaction will then yield a characteristic broad absorption curve.

Diamagnetism of many valence states common in minerals may, in some respect, be a favorable factor; had they been paramagnetic one would not have been able to observe impurities of other ions in such minerals as rutile, zircon, scheelite (with Ti^{4+}, Zr^{4+}, W^{6+}).

The skin effect, which reduces the absorption of superhigh-frequency emission, and complicates the absorption line, limits the possibilities of investigating some minerals, especially sulfides, sulfosalts, etc.

To have a complete picture of the EPR spectrum it is desirable to deal with crystals (not necessarily faceted) measuring about 2 to 9 mm; spectra of powders usually furnish much less complete information.

An additional difficulty in obtaining EPR spectra is the need for most of the ions to be observed at low temperatures, since at room temperature the spin-lattice relaxation time is too short, and no spectrum can be observed. Thus all the ions of rare earths, except for Eu^{2+} and Gd^{3+}, yield their spectra at the temperature of liquid hydrogen ($\sim$ 20 K) and below, while that of Fe^{2+} is usually observable at the temperature of liquid helium ($\sim$ 4 K).

The Cr^{3+}, Mn^{3+}, Ni^{2+} Cu^{2+}, Eu^{2+} and Gd^{3+} ions yield spectra at room temperature, and are most suitable for investigation by the EPR method. These impurity ions are utilized not only for detecting and investigating their position in the lattice, but also as "radioprobes" which transmit on centimeter waves the information relating to details of the electron and atomic structure of the crystal.

Electron-Hole Centers. While the object of EPR research was primarily the transition metal ions, the discovery of a wide distribution and diversity of electron-hole centers made the latter a major field for EPR application in mineralogy.

EPR would seem to be specially destined for investigating the centers: (1) a capture of an electron or hole attended by the formation of centers transforms the defects into the paramagnetic state; (2) these centers are present in an admixture in amounts that are optimal for being observed by means of EPR; (3) EPR furnishes comprehensive information on the structure of the anion group comprising the center. There is no doubt that it is only thanks to the use of EPR that this direction has achieved its present-day development. EPR parameters are also indispensable in describing the centers and the real structure of crystals, just as the optical and roentgenographic constants are in describing the crystals themselves and their atomic structure.

The centers are identified from EPR spectra in the same way as the transition metal ions are diagnosed by means of them, through the parameters of both the centers and ions vary within definite limits from one crystal to another.

The centers are widely spread in natural minerals. However, they can be created in these minerals by irradiation with X-rays, gamma-rays, neutrons and electrons and by means of irradiationn electron-hole centers can be obtained in practically all crystals. This broadens vastly the potentialities of the EPR.

The same parameters describe the EPR spectra of the transition metal ions and electron-hole centers. However, the description of the electron structure of the centers and its relation to EPR is made within the framework of the theory of molecular orbitals, and not within the crystal field theory.

Specific features of the centers in EPR spectra are considered in Chapter 7 of this book.

Among other objects of EPR investigations are: paramagnetic atoms and molecules (with unpaired electrons) in a gaseous state; liquids containing paramagnetic ions; free radicals in organic compounds that present an extensive domain of research in organic chemistry and biology, and which are of essential importance in geology (coal, bitumens, thucholites, fossil plants, etc), and metal conduction electrons.

3.1.4 EPR, Lasers[6], and Masers[7]

Both in the optical and EPR spectra of crystals absorption spectra arising as a result of the ion transition from a ground to an excited state are in evidence. An inverse process of returning to the ground state occurs by way of radiationless transitions (interaction with the lattice), or is attended by emission.

---

[6] Laser—light amplification by stimulated emission of radiation.

[7] Maser—microwave amplification by stimulated emission of radiation.

In the event of optical transitions, the emission manifests itself in the form of fluorescence (spontaneous radiation), or in that of stimulated radiation-laser emission (see Chap. 5.2). In the case of spin transitions one obtains stimulated emission of masers, as a result of transitions between the spin levels separated by initial splitting.

The choice of working medium for masers is made by means of determining the pattern of the spin levels with the help of EPR (these being the same levels which cause absorption in the EPR spectra).

A greater extent of initial splitting conducive to a higher radiation intensity and shifting the emission of the wavelength from the centimeter to the millimeter band was observed for a number of minerals (corundum, beryl, andalusite, cyanite, diopside, apatite, etc.). This draws attention to them as to potential maser crystals.

3.2 Physical Meaning of the EPR Spectra Parameters

3.2.1 The Order of Measurements; Experimental and Calculated Parameters

In addition to the resemblance of optical and EPR spectra, an essential difference also exists between them. In optical spectra the position of absorption bands directly determines the energy of the transitions corresponding to them between orbital levels of the ion in the crystal. In the EPR spectra, on the other hand, the position of the lines is determined by transitions between magnetic sublevels, whose energy is contingent not only on the ion's properties in the crystal, but also upon the emission frequency and the intensity of the external magnetic field ($h\nu = g\beta H$). From these one arrives by means of calculations at a value characterizing the splitting of the spin levels by the crystal field that supplements the characteristics of the splitting in electron states of the ion in the crystal (see Fig. 40); one finds also the value of the g-factor that already features magnetic splittings of the ion's spin states in the crystal.

The EPR spectrum itself registers only the H_{res} values of the magnetic field with which a resonance absorption signal is observed, commonly recorded in the form of an absorption line derivative. Such lines that correspond to the resonance values of the magnetic field, and constitute the EPR spectrum are indirectly associated with the paramagnetic ion properties in the crystal. To understand the substance of the EPR method means to trace the relations among the electron configurations of the ion, the pattern of levels, arising under the effect of the crystal field and the external magnetic field, the EPR spectrum resulting from this energy level pattern, the EPR parameters deduced from the spectrum then these parameters have to be related to the characteristics of features specific for the structural position and the electron state of the ion, established with the help of these parameters. Thus one has to trace back the way by which the EPR parameters were obtained from a spectrum representing a recording of resonance values of the magnetic field, establish just what they are and what kind of information can be derived from them.

For an ion in a cubic crystalline field (see Fig. 40) a single resonance line is in evidence, whose position is determined from the resonance condition $h\nu = g\beta H$, i.e., knowing frequency, and having measured the resonance magnetic field H, and substituting the values of the constants h and β, we obtain the value for the g-factor. This is the sole EPR parameter for an ion in the cubic symmetry field and in the absence of magnetic nuclei.

With lowering of the crystal field symmetry, several lines, known as the EPR spectrum fine structure, make their appearance in place of a single resonance line. This is due to the splitting of the spin levels by the crystal field (Figs. 45–47). The magnitude of this splitting by the crystal field ("initial splitting") is found from the position of lines in the magnetic field, and is described by means of parameters that depend on the symmetry of the crystal field. The position of the lines changes, depending upon the orientation of the crystal with respect to the magnetic field and, for this reason, instead of a single isotropic value of the g-factor, the spectrum is described by three principal values g_x, g_y, g_z, or, in the case of trigonal and tetragonal systems by two values : $g_{\parallel}$ and $g_{\perp}$.

In the presence of a magnetic moment of the nucleus of the paramagnetic ion (or of the nuclei of neighboring ions), each one of the fine structure lines experiences an additional splitting owing to the interaction with the ion of an unpaired electron, the extent of such a splitting being described by the hyperfine structure parameters.

Hence, the EPR spectrum is described by three groups of parameters that will be discussed below: (1) parameters of "initial splitting" (the effect of the crystal field); fine structure parameters; (2) g-factor (splitting by the applied magnetic field); (3) hyperfine structure parameters (interaction with the magnetic moment of the nucleus).

The interpretation of the EPR spectrum adds up to accurate measurements of the H_{res} of the spectrum lines, as well as to qualitative observations over the number of lines, their dependence on orientation and relative intensity.

A magnetic field is commonly measured through comparison of H_{res} of a given line against H_{res} of a standard with a known isotropic g-factor. The free radical of diphynylpicryl-hydrasil (DPPH) with $g = 2.0036 \pm 0.0003$ is often adopted as such a standard reference.

From the relation between two magnetic fields the g-factor of the resonance line is determined.

The frequency is habitually measured by using so-called proton probe (Gaussmeter); in the same magnetic fields as the EPR spectra a signal of nuclear (proton) megnetic resonance is observed.

As distinct from these exact measurements required for calculating EPR spectra parameters, a considerable body of information is obtained by a simple counting of a number of lines and by observing their behavior on the oscillograph screen.

1. The number of the fine structure lines defines the spin S of the paramagnetic ion (the number of lines equals $2S$, for Cr^{3+} with $S = 3/2$, for instance, three lines are seen, see Fig. 47). The larger the maximum spacing between fine structure lines, the greater the "initial splitting".

2. The number of hyperfine structure lines (equidistant and of equal intensity) determines the spin I of the nucleus (the number of lines is $2I + 1$, for Mn^{57} with

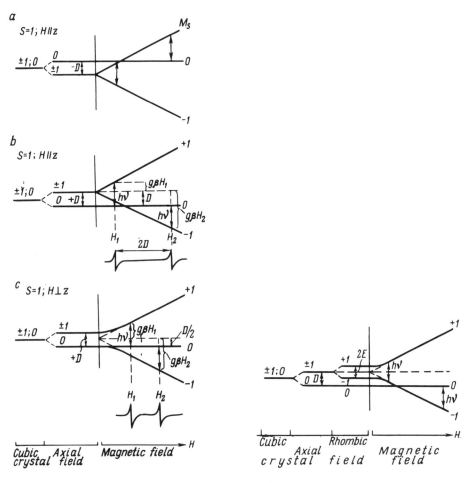

Fig. 45 a-c   **Fig. 46**

Fig. 45 a-c. EPR spectra fine structure for ions with spin = 1 in an axial crystalline field; for instance in V^{3+} (d^2, ^3F). In the cubic field a triple degenerated state ($\pm 1,0$) is observed, in the axial field splitting into two levels (± 1 and 0) occurs. Superposition of an external magnetic field results in splitting also of the level ± 1, the extent of splitting increasing with the intensification of the magnetic field. According to selection rules ($\Delta M_s = \pm 1$) two transitions are in evidence: $+1 \rightarrow 0$ and $0 \rightarrow -1$. **a** The magnetic field parallel to the crystalline field axis ($H \parallel z$) and D has a positive sign; **b** $H \parallel z$ but D has a negative sign; **c** magnetic field normal to the crystalline field axis ($H \perp z$). On the diagrams are shown relations from which the initial splitting D is determined

Fig. 46. Energy level diagram and initial splitting of the EPR spectra for ions with spin $S = 1$ in a rhombic crystalline field with $H \parallel z$

$I = 5/2$, for example, six hyperfine structure lines are in evidence). A direct measurement of the distance between these lines gives the hyperfine structure parameter (in Gauss).

Fig. 47. EPR spectra fine structure for ions with spin $S = 3/2$ in an axial crystalline field with $H \parallel z$; for instance Cr^{3+} (d^3, ^4F)

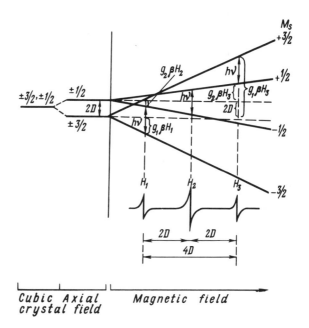

3. The spectrum symmetry found from the behavior of the lines during rotation of the crystal in a magnetic field reflects the symmetry of the ion position, enabling it to distinguish three cases: cubic symmetry (isotropic spectrum, unchanging during rotation of the crystal) trigonal and tetragonal systems (axial spectrum) and lower systems (rhombic symmetry of the spectrum).

The knowledge of the ion spin and that of the nucleus usually helps to identify the paramagnetic ion unambiguously. Observation of the orientational dependence of the spectrum points to the peculiarities of its entering the structure. Measurements of the EPR spectra parameters furnish the most comprehensive information as compared with all other methods of obtaining information on the properties of the ion in a crystal.

3.2.2 g-Factor and Splitting of the Spin Levels in a Magnetic Field

In the resonance condition $h\nu = g\beta H$, the g-factor determines the position of the line in the EPR spectrum and represents the sole value associated with the characteristics of the substance. Its physical meaning in actual crystals becomes apparent through a successive consideration of following interactions.

The g-factor of a free electron with no orbital moment is governed by its spin motion and the "purely spin value" of the g-factor is 2 and, by introducing the relativistic correction, $g_e = 2.0023$. (This is a correction because of which the magnetic moment of electron β differs from Bohr magneton μ_B; $\beta = 1.001146$. $\mu_B \pm 0.000012$; $g_e = 2 \cdot 1.001146 = 2.002292$).

The magnitude of g_e is very closely approached by the g-factors of atomic hydrogen in crystals, of F-centers, and of free radicals, as well as of ions in the S-state. The usual accuracy of the g-factor measurements is ± 0.001 and ± 0.0005, which permits distinguishing its slightest variations in a different environment, this being of particular importance in identifying the electron-hole centers.

The diphenyl-picryl-hydrazil (DPPH) free radical with $g = 2.0036$, often taken as an internal reference standard, separates magnetic fields into the ones that correspond to g-factors which are higher (weaker fields) and lower (stronger fields) than g_e.

The g-factor of a free paramagnetic ion (the spectroscopic splitting factor, or Lande factor) is conditioned not only by the spin, but also by the orbital movement of electrons, and it represents a hyromagnetic relation, determined by the Landé formula

$$g_L = 1 + \frac{I(I+1) + S(S+1) - L(L+1)}{2I(I+1)},$$

where L, S, I are quantum numbers of the orbital, spin, and total angular moments of the ion ($I = L \pm S$).

In the event of a purely spin state, when $L = 0$ (and, consequently $I = S$), $g = 2$, i.e., we obtain the g-factor of a free electron or of states with a "frozen" orbital motion. In the case of a purely orbital state ($S = 0$ and $I = L$) $g = 1$. For states intermediate between the purely spin and purely orbital states of the free ion, the g-factor can assume different magnitudes, including those below 1 and higher than 2. For example, for the free ion Ce^{3+} with ground state $^2F_{5/2}$, i.e., $L = 3$, $S = 1/2$, $I = 5/3$, the Landé factor $g_L = 6/7$. This Landé factor, however, does not correspond to the g-factor in the EPR spectra for reasons different in the cases of weak crystalline fields (realized for the rare earth ions) and medium or strong crystalline fields (iron and other transition group ions).

The g-factor of the rare earth ions in crystals is related to the spectroscopic splitting factor g_L by simple relations. Since the action of the crystalline field in the case of rare earths is weaker than the effect of the spin-orbital interaction, the quantum number of the total angular momentum I remains the same also for the ion in the crystalline field.

A crystalline field of noncubic symmetry splits the ground state of $^2F_{5/2}$, i.e., with $I = 5/2$, into three doublets: M_I equalling $\pm 5/2$, $\pm 3/2$, and $\pm 1/2$. Under dissimilar conditions some or other of these doublets may lie lower and, depending on the intensity of the crystalline field, the distance between them may vary from ~ 1–3 up to 100 cm^{-1} and greater, and therefore transitions can also occur between magnetic levels formed of different doublets. The energy of these magnetic levels with components $\pm I_z$ along the direction z of the magnetic field is $\pm g_I \beta H I_z$ (where $g_I = g_L$ and for Ce^{3+} $g_I = 6/7$). The energy difference between these levels is $(2g_I I_z \beta H)$ and, consequently, the g-factor along z axis ($g_{||}$) is $2g_I I_z$.

For Ce^{3+} g_z (or $g_{||}$) $= 2 \times 6/7 \times 5/2 = 30/7 = 4.286$ for doublet $M_I = \pm 5/2$; $g_{||} = 18/7 = 2.571$ for $M_I = \pm 3/2$; and $g_{||} = 6/7 = 0.857$ for $M_I = \pm 1/2$.

The actually observed magnitudes of $g_{||}$ and $g_\perp$ for rare earths differ from the calculated ones by nearer values of the $g_{||}$ and $g_\perp$. This is due to the mixing of different states by the crystalline field, i.e., to the contribution of states with different I values. Great differences in $g_{||}$ and $g_{\perp}$ cause a strong anisotropy of the rare earth EPR spectra. The g-factor of the iron group ions in crystals is not deduced from the Landé formula, for here the crystalline field is stronger than the spin-orbital interaction, there is no quantum number I, whereas the orbital states of the free ion, characterized by the quantum number L, transform into the symmetry types and are split by the crystalline field. Therefore, the interaction with excited orbital levels is described by the expression

$$g = g_e - \frac{8\lambda}{\Delta} = 2.0023 - \frac{8\lambda}{\Delta},$$

i.e., the g-factor is determined by the spin-orbital coupling constant λ, and by the distance to the nearest excited orbital levels Δ (for D-states this is 10 Dq, i.e., the strength of the crystalline field).

With positive values of λ (ions with less than half-filled d-shell) g-factor assumes magnitudes below $g = 2.0023$. With negative values of λ (ions with more than by half-filled-d-shell) $g > g_e$. In case of ions with the d^5-configuration $\lambda = 0$ and g-factor very closely approaches g_e (slight deviations are caused by an admixture of states with a different configuration).

Inasmuch as in the iron group for ions with $d^1 - d^7 -$ configuration λ is of the order of 100 cm^{-1} and Δ is of the order of 10,000–20,000 cm^{-1}, the departures of g-factor from 2 therein are not great (in contrast to the rare earth ions). The larger the orbital levels splitting by the crystalline field, the closer the g-factor is to 2. It is only in the case of Co^{2+} and Cu^{2+}, with considerable λ values, that the g-factor substantially exceeds this value.

Effective g-Factor and Effective Spin. Should the initial splitting of the spin levels be great (far greater than $g\beta H$ and, hence, than the radio-frequency quantum $h\nu$ used) the lower level then may be considered as an isolated one, described by effective spin S', which equals 1/2 in the case of a doublet lower level, and zero in case of a singlet lower level. In the latter case, no resonance is observed. An isolated doublet with $S = 1/2$, i.e., that separated from the higher-lying levels by a distance of $\gg h\nu$, is described by the effective g-factor: $g_{eff}\beta HS'$.

Anisotropy of the g-Factor. Yet another type of difference distinguishing the g-factor of the ion in a crystal from the g-factor of a free ion is its anisotropy, when ions find themselves in noncubic positions in a crystal structures. For this reason, with the crystal rotating in an external magnetic field, the position of lines in the EPR spectrum changes. In this the EPR spectra differ from optical absorption spectra, where the position of bands is independent of the orientation, since the position of levels in the crystalline field remains unchanged.

In EPR spectra the initial splitting, i.e., splitting by the crystalline field, is likewise independent of the orientation, but the splitting of the spin levels by the magnetic field shows an extremely strong dependence on the orientation of the crystal relative to the external magnetic field.

The g-factor anisotropy is explained by the fact that the crystalline field creates through the spin-orbital interaction an additional field of axial or rhombic symmetry. This field influences the spin moment of the electrons, determining its quantization along its axes. Therefore, upon superposition of the external magnetic field, the spin becomes oriented by the interaction with the crystalline field and magnetic splitting occurs depending on the existing orientation of the crystalline field axes relative to the magnetic field.

This allows it to utilize the orientation dependence of the EPR spectra in establishing local symmetry of the ion's position, in precisely defining the orientation of the crystalline field axes (coinciding with the principal values of the g-factor) and in revealing nonequivalent (differently oriented) positions of the ion in the crystal.

Like the axes of optical indicatrix (the refraction index indicatrix) the g-factor is a tensor determined by the symmetry not of the entire crystal, but by a local point symmetry at the site of the ion, i.e., by the symmetry of the crystalline field at the site of a given ion. For the latter, three types of local symmetry are distinguished: (1) cubic symmetry: isotropic g-factor ("indicatrix" is a sphere); (2) axial symmetry of the g-tensor (trigonal and tetragonal local symmetry): $g_{\parallel}$ and $g_{\perp}$ ("indicatrix" is ellipsoid of rotation); $g_{\parallel}$ is the axis of rotation and coincides with the sole axis of the higher order of symmetry, i.e., threefold or fourfold axis, while $g_{\perp}$ is normal to it; cases of $g_{\parallel} > g_{\perp}$ and $g_{\parallel} < g_{\perp}$ are possible; (3) rhombic symmetry of the g-tensor (lower systems: rhombic, monoclinic, triclinic local symmetry): g_x, g_y, g_z are mutually perpendicular and coincide with the crystalline field axes, which, in their turn, coincide with the twofold symmetry axes or with normals to the symmetry planes, if these exist (the "indicatrix" is triaxial ellipsoid).

The g-Factor Theories; Connection with Covalency. If the value of Δ (splitting by the cubic field), derived from the optical absorption spectrum, and the value of the g-factor, measured from the EPR spectrum, are substituted in the expression $g = 2.0023 - 8\lambda/\Delta$ and it then becomes possible to obtain a value for the constant of the spin-orbital interaction of the ion in a given crystal. Observations have shown that the values for λ obtained in this way are by 20% to 30% inferior to the values for free ions. This is explained chiefly by the contribution of covalency to the state of the ions bonding in crystals. However, a more detailed consideration by using molecular orbitals enables one to assess the part played by various mechanisms affecting the g-factor changes in crystals, and to relate it to the distribution of electrons among the metal and ligand orbitals during their overlapping, to bonding and antibonding orbitals, to the orbitals overlapping integral, to the mean orbital radius $< r^3 >$, etc.

3.2.3 Fine Structure of the EPR Spectra; Fine Structure B_n^m (or D, E) Parameters; Initial Splitting

Splitting of levels occurs in EPR under the effect of fields of two types; internal crystalline and applied magnetic fields, the latter being generated by the electromagnet of the EPR spectrometer (see Fig. 44). The number of EPR lines arising in the magnetic field depends upon the action of the crystalline field.

When interpreting the optical spectra, a splitting of the orbital states of ions is considered. Transitions between corresponding levels with an energy difference of the order of 10,000–40,000 cm^{-1} call forth optical transitions. The spin degeneracy here remains unchanged, i.e, the spin multiplicity experiences no change (the major term of the free ion ^5D for instance, splits in the cubic field into $^5T_{1g}$ and ^5E_g, in the trigonal into ^5E, ^5A and ^5E and in the rhombic into $^5B_1$, $^5B_2$, $^5B_3$ and ^5A, ^5A). The spin degeneracy is removed by the external magnetic field.

The spin states can, however, be split also by a crystalline field, i.e., in the absence of the magnetic field. This splitting is known as the splitting in a zero field (i.e., in a zero magnetic field), or as an initial splitting. The order of its magnitude (0.1–1.0 cm^{-1}) is quite negligible by comparison with optical transitions (10,000 cm^{-1}). In the case of EPR, however, this is in line with the energy height of the radio-frequency quantum (the wavelength used in spectrometers of 3.2 cm corresponds to ~ 0.3 cm^{-1} and of 0.8 cm to 1.25 cm^{-1}).

This magnitude, negligible from the standpoint of optical transitions, reflects the action of the very same crystalline field which splits orbital levels representing optical transitions, and spin levels that determine the EPR transitions. The initial splitting permits it to register with an extraordinary sensitivity and exactitude the specific distortion features of the coordination polyhedron around a given ion. It is described by a system of B_n^m parameters which expand the contribution made by components of different symmetry in this distortion.

The initial splitting determines maser transitions and manifests itself sometimes in linear fluorescence spectra, when transition from the upper orbital levels occurs to both spin sublevels separated by the initial splitting (see Chap. 5.3).

Direct measurements of the initial splitting can be made by using so-called zero-field spectrometers, where absorption is observed with changes in the wavelength of the micro-wave-radio-frequency emission (without magnetic field). However, it is usually estimated on the ground of the data derived from measurements of the resonance transitions position made with EPR spectrometers, for it is technically more convenient to change the magnetic field at a constant frequency and, moreover, this helps to obtain other EPR parameters simultaneously, such as the g-factor, hyperfine structure, and orientation dependence.

Whether the spin levels become split or not by the crystalline field, and what is the number of lines that appear in the EPR spectrum depends on the symmetry of the field, of which it suffices to distinguish mainly two cases: cubic and noncubic symmetry, depending on the number of unpaired electrons: odd (Kramers' ions) or even (non-Kramers' ions), and also on the spin S of the ion.

Kramers' and Non-Kramers' Ions. The Kramers theorem distinguishes the behavior of ions with odd and even numbers of electrons in the crystalline field: in the case of an odd number of electrons, no crystalline field can split the spin state below the doublet one, the spin doublets of these ions being split by the magnetic field.

A spin degeneracy of the level is determined by the states with possible projections of the spin quantum number S onto the axis of the electric or magnetic field: $M_S = S, S - 1, \ldots -S$. For example, in the case of Cr^{3+} ($3d^3$, ^4F) with spin $S = 3/2$, the ground state is four times spin-degenerated $M_S = +3/2, +1/2, -1/2,$

$-3/2$ (the spin degeneracy is equal to the spin multiplicity $2S + 1$), i.e., it represents a spin quartet irrespective of the orbitald egeneracy: $^4F \rightarrow {}^4T_{2g}$, ^4E_g, $^4A_{1g}$. According to the Kramers theorem a crystalline field of any low symmetry can remove this degeneracy only partially, leaving behind two doublets: $\pm 3/2$ and $\pm 1/2$. The magnetic field lifts degeneracy completely, splitting one doublet into levels $+3/2$ and $-3/2$, and the other into $+1/2$ and $-1/2$.

For V^{3+} (d^2, ^3F) with spin $S = 1$, the ground state is threefold ($2S + 1 = 3$) spin-degenerated: $M_S = +1, 0, 1$, i.e., it represents a spin triplet (see Fig. 45). Inasmuch as this is an ion with an even number of electrons, the Kramers theorem does not restrict the possibility of its states being split by the crystalline field and, therefore, in the field of a rhombic and a lower symmetry the spin degeneracy is fully removed with the formation of singlet levels ($+1$, 0, -1). Then no non-degenerated levels are left in the magnetic field. Should the singlet levels be separated from one another by a small initial splitting, an EPR signal can then be observed in the magnetic field; but if the distance between them exceeds the used radio-frequency quantum, no EPR spectrum is observed from such ions.

The doublets of ions with an odd number of electrons, whose degeneracy is not lifted by the crystalline field, are termed Kramers doublets and the ions themselves Kramers ions. For these it is always possible to obtain the EPR spectrum, whereas for the non-Kramers ions this is feasible only with small splittings in the zero field. The systematics of levels of these ions in noncubic fields differs too.

Splitting of Spin Levels in Cubic and Noncubic Fields; the Number of Fine Structure Lines. The systematics of the paramagnetic ions levels in the crystalline fields of different symmetry is convenient to consider separately for ions with odd number of electrons and a half-integral spin, called Kramers ions, and for ions with an even number of electrons and an integral spin, referred to as non-Kramers (Table 9).

For the Kramers ions in cubic fields degeneracy is not higher than 4 (quartets and doublets) and in all other fields there are doublets alone (Kramers doublets, whose degeneracy is removed only by the magnetic field). For non-Kramers ions in cubic fields the degeneracy is not over 3 (triplet, doublets); in fields of trigonal and tetragonal systems there are doublets and singlets, in rhombic ones the degeneracy is fully removed by the crystalline field (singlets).

In place of the spin quantum number S (for the iron group ions) the total quantum number I (for rare earth ions) may be substituted in Table 9. This same table is also valid for the orbital quantum number L (that takes on integral values only).

The number of lines in the EPR spectra is determined, above all, by the spin S and the symmetry, cubic or noncubic. With $S = 1/2$ there is a single doublet (Kramers); with any symmetry of the crystalline field, splitting in the magnetic field always leads to the appearance of a single line in the EPR spectrum. With $S = 1$ and $S = 3/2$ no splitting takes place in the cubic field of the triplet or quartet, respectively (see Table 9) and, for this reason, the magnetic field leads to an equidistant splitting, and there is only a single line in evidence. In all other instances, i.e., with $S = 1$ and $S = 3/2$ in noncubic fields, and with $S = 2$ and $5/2$ and more, both in cubic and noncubic fields, splitting of levels by the crystalline field is observed.

Table 9. Splitting of spin levels of paramagnetic ions in crystalline fields of different symmetry

			Kramers ions		
Number of electrons	S or I	Free ion; $(2S+1)$ or $(2I+1)$	Multiplicity and number of levels		
			Cubic field	Non-cubic fields Kramers doublets	
1	1/2	2	2		
3	3/2	4	4	2,2	
5	5/2	6	4,2	2,2,2	
7	7/2	8	4,2,2	2,2,2,2	

			Non-Kramers ions		
Number of electrons	S or I	Free ion; $(2S+1)$ or $(2I+1)$	Multiplicity and number of levels		
			Cubic field	Tetragonal and trigonal fields	Rhombic and lower fields
0	0	1	1	1	1
2	1	3	3	2,1	1,1,1
4	2	5	3,2	2,2$_a$,1	1,1,1,1,1
6	3	7	3,2,2	2,2,2,1	1,1,1,1,1,1

^{a} In the tetragonal field $2 \rightarrow 1,1$.

The levels forming in the magnetic field upon splitting of two and more spin levels shifted relative to one another cause the appearance of resonance transitions at dissimilar strengths of the magnetic field (see Figs. 44–47).

The appearance of several lines in the EPR spectrum of one and the same ion owing to splitting in the zero field (in the zero magnetic field, i.e., in the crystalline field) is called the fine structure of the EPR spectra. The number of the fine structure lines is either 1 (for $S = 1/2$ in any field and for $S < 2$ in the cubic field), or $2S$ (with $S < 2$ in a non-cubic field and with $S \geqslant 2$ in any field). In the case of Cr^{3+} ($S = 3/2$), for instance, a single fine structure line is seen in the cubic field and three lines in a noncubic one. For Fe^{3+} ($S = 5/2$) five lines in the field of any symmetry are observed.

The relation of the intensity of the fine structure lines is proportional to

$$S(S + 1) - M(M - 1),$$

where S is the spin of the ion, and M is a greater value of the M_S for the levels between which the resonance transition is accomplished. For instance, with $S = 5/2$ in case of transitions $5/2 \rightleftarrows 3/2$ and $-5/2 \rightleftarrows -3/2$ one has $[S(S + 1) - M(M - 1)] = 5/2(5/2 + 1) - 5/2(5/2 - 1) = 20/4$; for $\pm 3/2 \rightleftarrows \pm 1/2$ we obtain $32/4$ and for $\pm 1/2 \rightleftarrows \pm 1/2$ $36/4$; whence the intensities relation of five lines with $S = 5/2$ is $20:32:36:32:20 = 5:8:9:8:5$.

The fine structure lines intensity relations are:

$S = 1/2$ 1

$\quad\quad$ 1 $\quad$ 2:2 = 1:1

3/2   3:4:3
2   4:6:6:4 = 2:3:3:2
5/2   5:8:9:8:5
3   6:10:12:12:10:6 = 3:5:6:6:5:3
7/2   7:12:15:16:15:12:7,

i.e., the central lines are most intensive.

The distance between fine structure lines depends on the orientation and changes proportional to $(3\cos^2\theta - 1)$, where θ is the angle between the direction of the magnetic field H and the axis z in the crystal. The maximum distance between the lines is with $H \parallel z (\theta = 0)$. With $\theta = 54°44'$ the lines coincide since $3\cos^2\theta - 1 = 0$. With $H \perp z (\theta = 90)$ the distance between the lines is twice as short as in the case of the parallel orientation, and the transitions change places relative to the central line.

Splitting by the Crystalline Field in the Cases of $S = 1$ and $3/2$; Parameters of the Initial Splitting by Fields of Axial Symmetry (D) and Rhombic Symmetry (E). With spin $S = 1$ and $S = 3/2$, the effect of the crystalline field of tetragonal or trigonal symmetry is described by a single value, parameter D, and the action of the rhombic symmetry field by two parameters, D and E. In as much as the cubic field has no effect on the spin levels of ions with these S values, parameter D ("axiality") reflects directly the distortion of the coordination polyhedron along one of the symmetry axes L_3 or L_4, while parameter E ("rhombicity") describes an additional distortion normally to the former L_3 or L_4.

It can be readily shown (Figs. 45–47) how these initial splittings influence the splitting of the spin levels in the magnetic field, how the D value is determined from the resonance transitions for the cases of parallel and perpendicular orientations of the crystal field z axis relative to the magnetic field direction H. In the case of a purely axial symmetry (tetragonal or trigonal) with $S = 1$ and $3/2$, the distances between lines are $2D$ with $H \parallel z$ and D with $H \perp z$.

Splitting by the Crystalline Field in the Cases of $S = 3$ and $5/2$. Unlike in the preceding case, with $S \geqslant 2$ the splitting of spin states takes place in a cubic crystalline field: with $S = 2$ into triplet and doublet $(2S + 1 = 5)$ and with $S = 5/2$ into quartet and doublet $(2S + 1 = 6$; see Table 9). Therefore, for these ions the fine structure already appears in the cubic crystalline fields.

In the event of a lower symmetry, several parameters that take account of the contribution of the cubic, axial, rhombic, and lower fields are necessary to describe the splitting.

Description by Means of B_l^m Parameters of the Components of the Crystalline Field Acting upon the Spin Levels. Viewed from the standpoint of structural crystallography the position of an ion is defined unambiguously by the point group symmetry at the ion site without any additional complications. For instance, rhombic symmetry D_{2d} is rhombic symmetry, and there is no ambiguity about it. From the standpoint of spectroscopy a crystalline field, whose symmetry invariably coincides with the point symmetry of the ion position, is described by considering fields exhibiting several types of symmetry.

In optical spectroscopy, in addition to the parameter of the cubic field Dq, consideration is given simultaneously to parameters D_τ and D_s of the tetragonal field or V of the trigonal. The major contribution them comes from the cubic field, built up by charges lying on the vertices of octahedron, tetrahedron, or cube, while additional splittings occur owing to distortions along different axes of these polyhedrons.

The complexity of the coordination polyhedron distortions (which are usually not too great) is particularly manifest in the silicate structures, where octahedral positions can simultaneously experience the trigonal, tetragonal, and a lower distortion, while possessing a true monoclinic or triclinic symmetry.

In contrast to the optical spectra parameters, those of EPR embody with unique accuracy, sensitivity and completeness the contribution made by components of fields of diverse symmetry. The systematics of these contributions coming from different fields is based on the parameters entering the spherical function expansion of the crystalline field potential V:

$$V = \sum A_l^m <r^l> Y_l^m, \text{ or } V = \sum B_l^m Y_l^m,$$

where A_l^m are constants defined by all characteristics of the charge distribution ("lattice sums"); $<r^l>$ is the mean value of the l-power of the electron wave function radius; Y_l^m are spherical functions that take account of the angular dependence on the potential.

The B_l^m parameters retain their values as being products of $A_l^m <r^l>$, that enter the crystalline field potential, and reflect the charge distribution (symmetry and magnitude). In EPR they figure as experimentally determined values, i.e., they are of the same empiric nature as parameters Dq of the optical spectra, which are also linked with A_l^m.

Thus parameter B_2^0 corresponds to the above considered parameter D, i.e., the splitting by the axial field, and numerically it equals $B_2^0 = 1/3\ D$ and, for this reason, the distance between the spin levels split by the crystalline field is denoted in Figure 45 as $D = 3B_2^0$. It may be held as representing a different designation of an experimentally measured value describing the action of the crystalline field of tetragonal or trigonal symmetry. In like manner $B_2^2 = E$ corresponds to the splitting by the rhombic field (see Figs. 45–47).

The systematics of the B_l^m parameters, depending upon the symmetry of the crystalline field, the type of unpaired electrons (d in iron group ions or f in rare earth ions group), and spin S of the ion, is listed in Table 10.

In parameters B_l^m l and m can assume the following values: in iron group ions (d electrons) $l \leqslant 4$, in the group of rare earths (f electrons) $l \leqslant 6$.

$l = 0$ gives an additive value that does not affect spectroscopic calculations, but which makes part of the Madelung constant; odd l yields zero values.

With spin $S = 1$ and $3/2$, $l = 2$,
       $S = 2$ and $5/2$, $l = 2,4$,
    with $S = 3$ and above $l = 2,4,6$.

Table 10. Possible values of B_l parameters for crystalline fields of different symmetry

Header brace groupings (from widest to narrowest):
f = electrons (rare earths group)
d = electrons (iron group)
$I = 3;\ 7/2$ and higher
$S = 2;\ 5/2$
$S = 1;\ 3/2$

Symmetry	B_2^0	B_2^1	B_2^2	B_4^0	B_4^1	B_4^2	B_4^3	B_4^4	B_6^0	B_6^1	B_6^2	B_6^3	B_6^4	B_6^5	B_6^6
Cubic				+				+	+				+		
Trigonal	+			+			+		+			+			+
Tetragonal	+			+				+	+				+		
Rhombic and monoclinic	+		+	+		+		+	+		+		+		+
Monoclinic, C_s	+	+	+	+		+	+	+	+		+	+	+	+	+
Triclinic, C_t	+		+	+		+		+	+	+	+				+
Triclinic, C_1	+	+	+	+	+	+	+	+	+	+	+	+	+	+	+

Rhombic symmetry is matched by $m = 2$ and multiples (4,6), trigonal by $m = 3(6)$, tetragonal by $m = 4$, and axial by $m = 0$.

In the fields of cubic symmetry the crystalline field in the case of $S \geqslant 2$ is described by parameters B_4^0 and B_4^4 (for rare earths also by B_6^0 and B_6^4).

In the fields of trigonal symmetry:
with $S = 1,3/2$ by B_2^0,
with $S = 2,5/2$ by B_2^0, B_4^0 (for RE^{3+} also B_6^0, B_6^3, B_6^5)
In the fields of tetragonal symmetry:
with $S = 1,\ 3/2$ by B_{24}^0
with $S = 2,\ 5/2$ by B_2^0, B_4^0, B_4^4 (for RE^{3+} also B_6^0, B_6^4)
In the fields of rhombic symmetry:
with $S = 1,3/2$ by B_2^0, B_2^2,
with $S = 2,5/2$ by B_2^0, B, B_4^2, B_6^4 (For RE^{3+} also B_4^0, B_6^2, B_6^4, B_6^6).

Mention should be made that in the iron group, parameters B_2^0 and B_2^2 are usually by an order higher than other B_l^m.

Relations Between Designations B_l^m, b_l^m and D, E, a, F, c, d. In EPR, in addition to parameters B_l^m, parameters b_l^m, which have the same meaning, but differ only by

$D = b_2^0 = 3B_2^0$	Axial fields	
$E = 1/3\ b_2^2 = B_2^2$	Rhombic field	Iron
$a = 2b_4^0 = 2.60B_4^0$		group
$F = 3b_4^0 = 3.60B_4^0$	Cubic field	
$c = 2a = 4b_4^0 = 4.60B_4^0$		REE
$d = 4b_6^0 = 4.1260B_b^0$		group

constant common coefficients are of frequent use; $b_2^m = 3B_2^m$, $b_4^m = 60B_4^m$ II, $b_6^m = 1260 B_6^m$. Besides, one comes across designations, formerly traditional, that are linked with b_l^m and B_l^m by the following relations:

Large Initial Splitting. Simple pictures of the spectra with the fine structure lines numbering $2S$ are exhibited on condition that the initial splittings are not too great, not larger than the radio-frequency quantum $h\nu$ used, i.e., in operating an X-band spectrometer with wavelength $\lambda = 3.2$ cm the initial split must be not more than ~ 0.3 cm^{-1}, a K-band one with $\lambda = 1.25$ cm not more than 0.8 cm^{-1}, and a Q-band with $\lambda = 8$ mm not more than 1.25 cm^{-1}.

Should the initial splittings exceed the energy of the emission used ($\sum B_l^m > h\nu$), the situation becomes more complicated. Transitions $3/2 \rightleftarrows 1/2$ (and likewise $5/2 \rightleftarrows 3/2$) in weak magnetic fields cannot be accomplished because the $h\nu$ value is less than the distance between these levels (in a magnetic field these levels lie still farther asunder), whereas transitions $-3/2 - 1/2$ (and also $-5/2 - 3/2$) require the application of too strong a magnetic field (Fig. 48). Therefore not allowed ($\Delta M_S = \pm 1$), but forbidden transitions with $\Delta M_S = \pm 2, 3, 4$ and more comes into play.

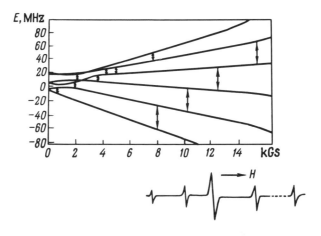

Fig. 48. Energy levels diagram for Fe^{3+} in oligoclase [419]. A case of large initial splittings. Shown are the transitions with a relatively low-frequency quantum (9.5 MHz) in lower fields and transitions with a high-frequency quantum (35 MHz) in higher fields

These transitions fall in two regions in the energy levels diagram (see Fig. 48): (1) in the region of intersection of levels which presents a rather complex picture of the spectrum and lends itself to interpretation only with difficulty, and (2) the region of weak magnetic fields (before intersection of the levels), where only transitions inside the doublet levels are in evidence, i.e., forbidden transitions 5/2–5/2, 3/2–3/2 and an allowed transition 1/2–1/2. In the majority of noncubic crystals with a Fe^{3+} ion, and in many with Cr^{3+}, large initial splittings are observed. In such cases in lieu of an X-band spectrometer (3.2 cm) one should use spectrometers of the K (1.25) cm or Q(8 mm) bands in measuring the spectrum. In passing to observations at higher frequencies the spectrum lines are "arranged" in conformity with the picture common to each value of the spin S. The parameters of the initial splitting or their relations may, however, be appraised also at frequencies not exceeding the initial splitting.

In the case of an axial field the parameter $B_2^0 = 1/3D$ is determined for $S = 1$ and 3/2 proceeding from anisotropy of the effective g-factors:

$$g_\perp = 3g\left[1 - 1/2\left(\frac{g\beta H_\perp}{D}\right)^2\right].$$

An approximate evaluation of $B_2^0 = 1/3D$ may be obtained from this expression also for ions with $S = 2$ and 5/2, provided the B_4^0 parameters are small enough (as often happens). In the event of a field with a stronger symmetry, the ratio E/D ($B_2^2/3B_2^0$), considered later, may be made use of.

In case of non-Kramers ions (with an even number of electrons) with increasing initial splitting doublet levels separated in axial fields from the singlet levels over a distance exceeding the radio-frequency quantum, only forbidden transitions inside doublets (2–2, 1–1) can be seen. In rhombic and lower fields there remain singlet levels only and, therefore, the distance between them exceeds the quantum in use, and no EPR signal is observed.

Determining the Ratio "Rhombicity/Axiality" E/D ($B_2^2/3B_2^0$). For ions finding themselves in low-symmetrical (rhombic and lower) crystalline fields with large initial splittings, measurements at frequencies that are below the splitting value enable one to gather the following information on the ion: site symmetry, the degree of distortion, orientation of the crystalline field axes, the presence of nonequivalent positions. These data are made available by studying the relation of the E/D ($B_2^2/3B_2^0$) parameters. Experimental findings for ions of the iron group demonstrated that of all the B_l^0 parameters characterizing the axial field, B_2^0 (equal to $1/3D$) exhibits the greatest value, and of the B_l^2 parameters describing the rhombic and lower fields, B_2^2 (equal to E) has the highest value. Therefore, the ratio of these two parameters may be taken to characterize the position of the ion with a sufficient degree of approximation. To do so the axes are chosen in such a way as to have B_2^0 exceed B_2^2. The resulting ratio is analogous and proportional to the parameter of asymmetry in the spectra of nuclear gluadrupole and nuclear magnetic resonance (see Chap. 4).

The E/D ratio is found [463] by constructing a diagram (Fig. 49). On this are shown the magnitudes of effective g-factors (and, consequently, the position of lines in the magnetic field), g_x, g_y, g_z for three transitions (between the doublet levels), observed in weak magnetic fields: 5/2–5/2, 3/2–3/2 and 1/2–1/2. Having determined the effective g-factor for any of these transitions, we obtain, by referring to the diagram, the value E/D. However, the transition 5/2–5/2 is least sensitive to the change of E/D, is of low intensity, and often cannot be observed at all. For the transition 3/2–3/2 the g-factor varies from 0 to 6, taking on isotropic value 4.3 with $E/D = 1/3$. Most convenient, however, is transition 1/2–1/2 and it is exactly from the difference g_y-g_x for this transition that the value E/D is determined most accurately. This determination is the more exact, the smaller the other B_l^m parameters and $D > g\beta H$.

Fig. 49. Determination of E/D (rhombicity axiality ratio for the cases of large initial splittings from effective lower-field values of the g-factor for the three lower doublets [463]

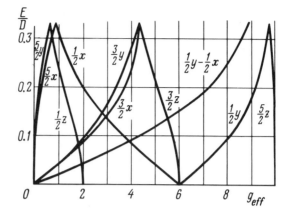

Mechanism of Splitting of the Mn^{2+} and Fe^{3+} Ions ^6S-State by Crystalline Fields. The splitting of the spin states in a zero magnetic field occurs in the general case under the effect of a crystalline field acting through a spin-orbital interaction.

With ions in the ^6S ground-state (of Mn^{2+} and Fe^{3+}, in particular) and a zero orbital moment, the crystalline field should not have split the ^6S-term, but had merely to transform it into ^6A_I. Experimental observations of the Mn^{2+} and Fe^{3+} EPR spectra show, however, that the splitting actually occurs. The mechanism of this splitting attracts considerable attention, for it is exactly for these ions that the greatest body of the EPR data has been obtained and, in order to establish the dependence of the data on the state of the chemical bonding, and to compare them with the parameters of optical and Mössbauer spectra, one has to comprehend the interactions leading up to the initial splitting of these ions in the ground ^6S-state.

A general cause accounting for this splitting is the admixture to the ground ^6S-state ($3d^5$ configuration) of excited states with different multiplicity (first of all of the nearest of levels arising from the ^4P-term) and of the excited $3d^44s(^6D)$ configuration states. Moreover, the model of the crystalline field should then include the effects of the bond covalency.

The predominant mechanism of the ^6S-state splitting are spin–spin interactions taking place owing to the admixture of the excited configuration (Price mechanism). Furthermore, spin-orbital interactions (Vatanabe mechanism) come into play here, causing splitting by the field through a complex interaction that includes a spin-orbital admixture of the excited ^4P-state to the ground ^6S-state.

The Hyperfine Structure (hfs) of the EPR Spectra for ions of the iron group is specified for each ion by the presence of magnetic and nonmagnetic isotopes, by the nuclear spin of each isotope, the magnitude of its magnetic moment and natural content of the isotope. Even isotopes have a zero spin and are nonmagnetic, whereas the odd ones have a spin that differs from zero, and are magnetic; an interaction with the latter, provided they occur in sufficient abundance, manifests itself in the appearance of hfs.

3.2.4 Parameters of Hyperfine Structure. Interaction with Magnetic Nuclei of Paramagnetic Ions

We have so far considered the emergence of resonance lines owing to the splitting of the electron spin states in the crystalline and applied magnetic fields and the fine structure of the EPR spectra.

However, many nuclei of paramagnetic ions or neighboring (ligand) ions have a spin, i.e., they possess a rotational moment of their own and, because of the presence of a charge in them, they also display a magnetic moment.

An interaction of the magnetic moment of an unpaired electron with the magnetic moment of the nucleus produces an additional weak splitting of the electron's spin levels that finds its expression in the emergence of hfs in the EPR spectra.

The hyperfine structure is a highly characteristic one, and presents a most convenient means for identification of paramagnetic ions by the number of lines observed on the oscillograph screen of a spectrometer, furnishing invaluable information for interpreting the electron-hole centers.

Moreover, it is of prime importance in determining the electron density at different points of the pramagnetic complex, and helps quantify the amounts of the s- and p-states and the sp-hybridization; it also furnishes direct data on the characteristics of chemical bonding, especially with the aid of molecular orbitals. The interaction of an unpaired electron with the nuclei of ligand ions leaves in the hyperfine structure a "marker", a trace of the direct "contact" of this electron with the nuclei of the adjacent ions.

Energy Level Diagrams with Due Regard for Hyperfine Interaction. In the presence of magnetic nuclei the splitting of spin levels proceeds not only in the applied magnetic field, but also in weak additional ones, created by different orientation of the nuclei (Fig. 50).

For nuclei with the spin $I = 1/2$, two orientations are possible (Fig. 51b): along the direction of the external magnetic field ($M_I = +1/2$) and in the one opposite to it ($M_I = -1/2$). Both orientations create a small additional field at the site of electron localization. With parallel orientation this local field of the nucleus ΔH_{nuc} is added to the external field H, and with an antiparallel orientation substracted

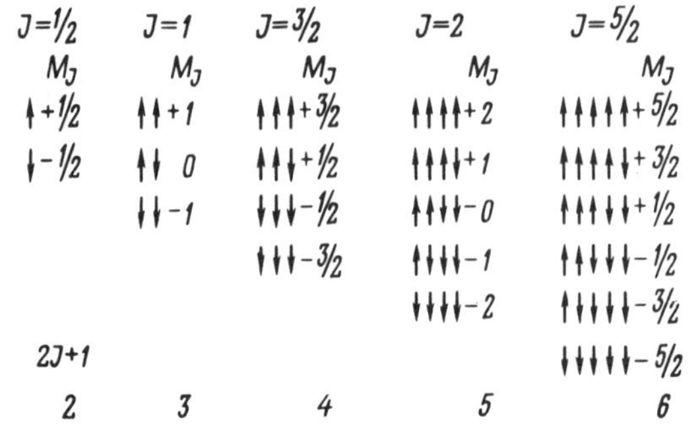

Fig. 50. Possible orientations of the nuclear spin in a magnetic field. The number of possible orientatioñs is $2I + 1$. M_I assumes values of $I, I - 1 \ldots -I$

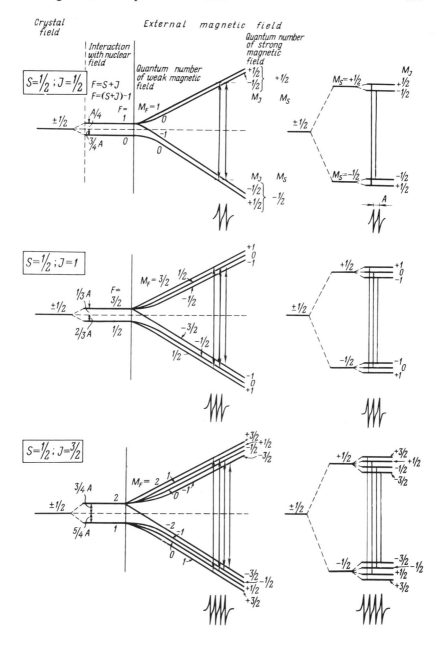

Fig. 51. Arising of EPR spectra hyperfine structure for ions with spin $S = 1/2$ and nuclear spin $I = 1/2$, 1, 3/2 **a**, energy level diagram with constant $h\nu$ and increasing external magnetic field H; **b**, energy level diagram with constant magnetic field H and varying radio-frequency quantum $h\nu$. Diagram **a** is in keeping with actual conditions of the experiment, but diagram **b** is simpler and more convenient for use in cases with high values of spin S

from it. Therefore, the unpaired electron will be in the field $H + \Delta H_{nuc}$ or in that of $H - \Delta H_{nuc}$, and the resonance condition may be written

$$h\nu = g\beta(H \pm \Delta H_{nuc}).$$

Hence, each one of the spin levels will experience an additional splitting into two sublevels (for $I = 1/2$), and each fine structure line will then be split into two hyper-fine structure lines, equidistant from the resonance position without interaction with the nucleus. The same conclusion may be arrived at in a different way (see Fig. 51a). Spin levels split by the crystalline field (only states with the spin $S = 1/2$ are not liable to splitting), when they interact with a magnetic nucleus with spin I in the absence of an external magnetic field, become split into hyperfine levels characterized by the resultant quantum number $F = (S + I), (S + I) - 1$.

On superposition of a weak external magnetic field, each one of these hyperfine structure levels with quantum numbers F splits into $(2F + 1)$ sublevels, characterized by magnetic quantum numbers of the weak field M_F.

In a strong magnetic field (counting more than several hundred Gauss, i.e., practically almost always when investigating EPR spectra) the linkage S and I is broken, and each one of these numbers is quantized in the direction of the external magnetic field separately, i.e., the levels are determined not by M_F, but by two magnetic quantum numbers M_S and M_I. Then $h\nu = g\beta H + \sum A M_I$, where A is the hyperfine structure constant equalling $g\beta \Delta H / M_I$.

As distinct from the fine structure lines number, defined by the ion spin and equalling $2S$, that of the hyperfine structure lines is determined by the nuclear spin I and equals $(2I + 1)$, according to the number of possible nuclear spin orientations (see Figs 50–52). This difference is due to the distinction between the selection rules: $\Delta M_S = \pm 1$, since in transition between the spin states, the electron spin undergoes a change (for instance, with $S = 1/2$ its orientation "reverses" or "flips" from $+ 1/2$ to $-1/2$), but $\Delta M_I = 0$, for in the electron transitions the orientation of nuclear spins remains unchanged. The quantization of nuclear levels with opposite values of M_S proceeds in an inverse order. This results in the emergence of $(2I + 1) -$ transitions, as shown in Figures 51 and 52.

Features Specific for the Hyperfine Structure Lines. With strong enough fields the hfs lines are equidistant, and are independent of the fields (see Figs. 51, 52). Because of a short distance between hyperfine interaction sublevels, at normal and not too low temperatures these are equally populated and, therefore the hfs lines display an equal intensity. The dependence of the hfs parameter on local symmetry is similar to that on the g-factor, in the case of a cubic local symmetry the hfs is characterized by one value of the hfs parameter (A), in that of axial symmetry by two ($A_{\parallel}$ and $A_{\perp}$, or A and B), in rhombic and a lower one by three (A_x, A_y, A_z).

In the presence of several isotopes of a single element the EPR spectrum demonstrates superposition of individual hfs from each isotope.

Two Basic Mechanisms of Hyperfine Interaction. With all the diversity of chemical types of the system where one sees hyperfine interactions, i.e., interactions of the electronic and nucleus magnetic moments, there are only two basic types of such interaction, i.e., the isotropic, or contact, and anisotropic, or dipole.

From the chemical viewpoint the hyperfine structure is interpreted in connection with the *sp*-hybridization, with changes in the ionicity-covalence, the type and occupancy of the molecular orbitals. Distinction is made between the interactions of a paramagnetic electron with the nucleus of the intrinsic ion and with the nuclei of neighboring anions (ligands), or even of adjacent cations, the interaction of an unpaired electron in free radicals with the nuclei entering the complex ions comprising this radical and so on. At all events, the physical mechanism of the interaction comes to the two basic types mentioned above. Expressions defining the components of the hyperfine interaction are deduced for these, and from the properties of the nucleus and electron of the free atom the constants of isotropic and anisotropic hyperfine structure for the free atom are calculated. By contrasting with them the observed values for isotropic and anisotropic hyperfine structure of these atoms in crystals, the features for the state of their chemical bonding are determined.

The consideration of this interaction has two aspects. The effect of the magnetic field of the nucleus on the state of the electron in the crystal produces splitting of the

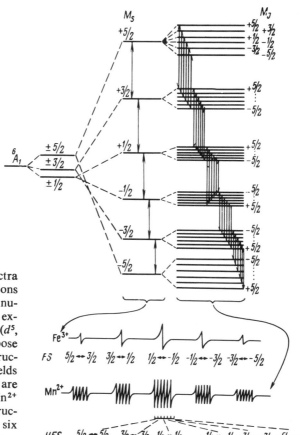

Fig. 52. EPR spectra hyperfine structure for ions with spin $S = 5/2$ and nuclear spin $I = 5/2$, for example in cases of Mn^{2+} (d^5, ^6S), Mn^{57}. For the purpose of comparison five fine structure lines of Fe^{3+} that yields no hyperfine structure are shown. In the case of Mn^{2+} each one of the five fine structure lines is split into six hyperfine structure lines

electron levels that describes the hyperfine structure of the EPR spectra. On the other hand, the action of the magnetic field of the electrons on the nuclear levels results in their splitting, which manifests itself in the emergence of the hyperfine structure of nuclear gamma-resonance (Mössbauer) spectra. The same mechanisms—isotropic and anisotropic—determine these interactions as well. In considering these phenomena the nucleus and electron are regarded as magnetic dipoles, the interaction between which is:

1. Proportional to the magnitude of their magnetic moments $\mu_{el} = g\beta S$ (g is g-factor of electron equalling 2, β is Bohr magneton, S is atom's spin); $\mu_{nuc} = g_N \beta_N I$ (g_N is nuclear g-factor, B is nuclear magneton, I is nuclear spin).

2. Inversely proportional to the cube of the distance between them, i.e., $< 1/r^3 >$ or $< r^{-3} >$ (here the sign $>$ denotes the mean value).

3. Dependent on the angle between the line connecting these dipoles and the direction of the external magnetic field: $(3\cos^2\theta - 1)$, i.e.,

$$H = \mu_{el}\mu_{nuc} < r^{-3} > \cdot < 3\cos^2\theta - 1 > = g\beta g_N \beta_N \frac{< 3\cos^2\theta - 1 >}{< r^3 >} IS.$$

Isotropic interaction is determined by the expression

$$A^*_{iso} = \frac{8\pi}{3} g\beta g_N \beta_N |\psi(0)|^2 \text{ (erg)}, \text{ or } A^*_{iso} = \frac{8\pi}{3} g_N \beta_N |\psi(0)|^2 \text{ (G)},$$

i.e., it will be the more intensive, the greater the magnetic moment of the nucleus, and the higher the density of the s-electrons on the nucleus. For the atomic hydrogen, for example, the isotropic hyperfine splitting is 500 G, but in interaction with remote protons in some free radicals it diminishes down to a few tenths of a Gauss.

The anisotropic interaction is determined by the expression

$$B^* = g\beta g_N \beta_N < r^{-3} > < 3\cos^2\theta - 1 > \text{ erg, or}$$
$$B^* = 2/5 g_N \beta_N < r^{-3} > < 3\cos^2\theta - 1 > \text{ (G)}$$

and it is precisely by using it that the position of the nucleus relative to the paramagnetic electron is ascertained.

In the above expressions for A^*_{iso} and B^*, all values are known, including the electron densities on the nucleus $|\psi(0)|^2$ and mean electron radii $<r>$, calculated in estimating wave functions of atoms, and for A and B the theoretical values have been obtained with which experimental findings can be compared.

The parameters are determined directly from measurements: A, in case of isotropic local symmetry (for the s-electrons only), $A_{||}$ and $A_\perp$ for axial symmetry and A_x, A_y, A_z for rhombic symmetry (as in the case of the g-factors).

These experimental values may be presented in the form of a sum of the isotropic term A_{iso} and the anisotropic term B: for axial symmetry $A_{||} = A_{iso} + 2B$ and $A_\perp = A_{iso} - B$; for rhombic one $A_{||} = A_z$ and $A_\perp = (A_x + A_y)/2$, if A_x and A_y stand close to each other.

In systems where molecular orbitals are derived from the s- and p- atomic orbitals, it is possible to appraise the contribution of the s-state; $c_s^2 = A_{iso}/A^*_{iso}$, i.e., as a ratio between an experimental isotropic component of the hyperfine structure and the theoretical one, indicated in tables of the book by Atkins and Symons (859), and also as a contribution of the p-state: $c_p^2 = B/B^*$. Hence the degree of hybridization is defined as

$$\lambda = c_p^2/c_s^2,$$

which permits one, in its turn, to find the valence angles, whose vertices coincide with the magnetic nucleus.

Hyperfine interactions can be observed in various systems. First and foremost, in the paramagnetic ions proper, provided their nuclei display a magnetic moment,

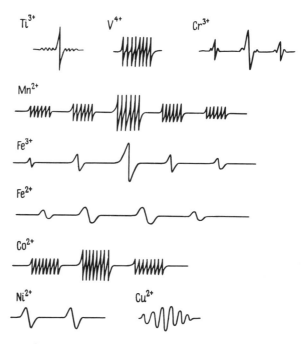

Fig. 53. Patterns of the iron group ions EPR spectra.
Ti^{3+}: $S = 1/2; 2S = 1$—a single fine structure line (fs); Ti$^{48}(I = 0)$; Ti^{47} (7.75%) $I = 5/2$; Ti^{49}(5.5%) $I = 7/2$; V^{4+} (d^1): $S = 1/2$; V^{51} (100%) $I = 7/2$; $2I + 1 = 8$—eight lines of hyperfine structure (hfs). Cr^{3+} (d^3): $S = 3/2; 2S = 3$—three fine structure lines (fs); Cr^{53} (9.5%) $I = 3/2$—apart from the even isotope central line each one of three fs-lines is split into four weak hfs lines, of which two are overlapped by the central line and two are seen in the form of weak satellites. Mn^{2+} (d^5): $S = 5/2; 2S = 5$—five fs lines; Mn^{57} (100%) $I = 5/2$; $2I + 1 = 6$—six hfs lines. Fe^{3+} (d^5): $S = 5/2$; $2S = 5$—five fs lines, Fe^{2+} (d^6): $S = 2; 2S = 4$—four fs lines. Co^{2+} (d^7): $S = 3/2; 2S = 3$—three fs lines; Co^{59}(100%) $I = 7/2$—eight hfs lines. Ni^{2+} (d^8): $S = 1$; two fs lines. Cu^{2+} (d^9): $S = 1/2$—one fs line; Cu^{63} (68.9%) $I = 3/2$; Cu^{65} (31.3%) $I = 3/2$—two series with four partially overlapping hfs lines in each one of them

for instance, in the ions of the iron group (Fig. 53) and in that of the rare earths. In these cases the configurations with unpaired d- and f-electrons, which cannot interact directly with the nucleus, either polarize the inner filled s-shells, or carry an admixture of s-configuration into their ground state.

Moreover, the same ions, which exhibit spectra with hyperfine structure owing to the interaction with their proper nucleus, can also interact with the nuclei of the near-neighboring atoms, if these nuclei are magnetic. In such cases one can observe an additional splitting of each hyperfine structure line, caused by the superhyperfine structure (shfs) or hyperfine structure of the ligand ions nuclei.

The shfs lines number $2mI + 1$, where m is the number of equivalent nuclei of the ligand ions. For example, in CaF_2: Mn^{2+} $I(F^{19}) = 1/2$, $m = 8$, $2mI + 1 = 9$, i.e., each one of the six hfs lines of Mn^{2+} splits into nine shfs lines (Fig. 54).

In the electron-hole centers, where the unpaired electron is distributed in the space embracing several nuclei with magnetic moments, each successive weaker interaction produces an additional splitting of every earlier emerged hfs line. For example, in apatite the unpaired electron of the O^--center (or of the F-center)

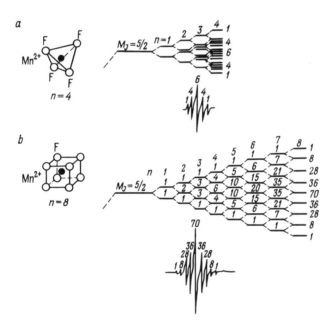

Fig. 54 a and b. Superhyperfine structure arising from interaction of Mn^{2+} with neighboring F^{19} nuclei. Each one of the fine structure lines (Mn^{2+} ion spin $S = 5/2$, number of lines $2S = 5$) is split into six hyperfine structure lines (Mn^{57} nucleus spin $I = 5/2$); the number of hfs lines is $2I + 1 = 6$, each one of which is split into $(2nI + 1)$ superhyperfine structure lines (n is number of neighboring F^{19} nuclei. I is F^{19} nucleus spin being 1/2). **a** In scheelite with $WO_4^{2-} \rightarrow MnF_4^{2-}$ five lines ($2nI + 1 = 2 \cdot 4 \cdot 1/2 + 1$); **b** in fluorite; nine lines ($2nI + 1 = 2 \cdot 8 \cdot 1/2 + 1$), (the Mn^{2+} in CaF_2 spectrum has altogether $5 \times 6 \times 9 = 270$ lines)

interacts not only with two F^{19} nuclei of the nearest fluorine ions, but also with more remote P^{31} nuclei of the phosphorus ions.

In these cases the possibility and the degree of the interaction is determined, above all, by the nature of the molecular orbital, i.e., its bonding, antibonding, or nonbonding type.

3.2.5 Spin-Hamiltonian; Order of Calculation the Paramagnetic Ions Spectra

Earlier we considered the interaction of an unpaired electron spin with the magnetic field, crystalline field (initial splitting) and with the magnetic moment of the nucleus (hyperfine splitting). Generally EPR results from a simultaneous manifestation of all these interactions determining the energy of the levels and the resonance values of the magnetic field.

All of them form part of the spin-Hamiltonian, which fully describes the behavior of the electron spin in the paramagnetic ion of a crystal in a magnetic field:

$$H = b_l^m O_l^m \qquad + \qquad g\beta H \qquad + \qquad ASI$$

initial splitting	splitting by	interaction with the magnetic
(fine structure)	magnetic field	moment of the nucleus
		(hyperfine structure)

A concrete form of the spin-Hamiltonian depends, above all, on the local symmetry that determines b_l^m and O_l^m (see Table 10), on the g-factor and the hyperfine structure constant A:

cubic symmetry: $g\beta H$	A
axial symmetry: $(g_{\shortparallel} + g_\perp)\beta H$	$A_{\shortparallel}$ and $A_\perp$
rhombic symmetry: $(g_x + g_y + g_z)\beta H$	A_x, A_y, A_z.

But what is the meaning of the spin-Hamiltonian? It describes all the interactions that determine the EPR spectrum. From it are derived concrete expressions for calculationg the spin-Hamiltonian parameters that describe the behavior of the paramagnetic ion in the crystal.

The sequence of calculating the spectrum is as follows: The primary data for the calculation are already obtained during observation of the spectrum which is made by: (1) varying the intensity of the electromagnet magnetic field, thus causing a successive appearance of lines with different values of H_{res} (resonance values of the magnetic field, i.e., of those satisfying the resonance condition) on the screen of the EPR radiospectrometer oscillograph, (2) changing orientation of the crystal relative to the direction of the magnetic field generated by the electromagnet.

The dependence of H_{res} on the orientation of the crystal-line field axes relative to the magnetic field enables each system of lines corresponding to a given paramagnetic ion in a definite structural position to be moved into the so-called "double extreme position". To achieve this the crystal is rotated in the resonator in two mutually perpendicular directions by using an arrangement resembling a two-circle goniometer, or the universal stage (of crystaloptics) by trying out each orientation through making changes in the magnetic field. When rotation of the crystal about both axes shows the extreme magnitude of H_{res} to have been achieved, this then indicates that one of the g-factor axes in fact coincides with the direction of the magnetic field.

Observations on the number of the fine structure lines and those of the hyperfine structure, as well as on orientation dependence allow determination of the spin of ion S, that of the nucleus I and the local symmetry and, consequently, the type of the spin-Hamiltonian incident to the given case. This usually already helps identify the paramagnetic ion.

Having ascertained the type of the spin-Hamiltonian that describes the observed spectrum, it is possible to take advantage of the already available expressions for the energy levels and for H_{res} deduced from it [333, 335, 410, 447].

Thus for the electron spin $S = 3/2$ (Cr^{3+}, d^3, for example) and for the rhombic symmetry (without hyperfine structure) the spin-Hamiltonian is

$$H = \frac{1}{3} b_2^0 O_2^0 + \frac{1}{3} b_2^2 O_2^2 + g_x \beta H_x S_x + g_y \beta H_y S_y + g_z \beta H_z S_z.$$

In this case, after orienting the magnetic field H parallel to the z axis the energy levels will be

$$\varepsilon_{1,2} = g_z \beta H/2 \pm \sqrt{(b_2^0 + g_z \beta H) + (b_2^2)^2},$$
$$\varepsilon_{3,4} = -g_z \beta H/2 \pm \sqrt{(b_2^0 - g_z \beta H)^2 + (b_2^2)^2}.$$

From this the resonance values for the magnetic fields H_{res} for three transitions observed in the case of Cr^{3+} are obtained.

By substituting the measured values for H_{res} of three observed lines we obtain the required parameters of the spin-Hamiltonian. In crystals geometrically non-equivalent (or magnetically nonequivalent) positions of one and the same ion (or center) are commonly seen, i.e, the position characterized by the same spin-Hamiltonian parameters, but with differently oriented axes g_x, g_y, g_z (or $g_{\parallel}$ and $g_{\perp}$). These positions arise owing to the action of the elements of the space group symmetry on the coordination polyhedron, being bound by these symmetry elements and, therefore, physically fully identical. The number and orientation of nonequivalent positions (the mutual position and the relation to crystallographic direction orientation) are determined by the types of the system of points in the space group, and usually permit it to establish unambiguously the structural position of the paramagnetic ion (Figs. 55–57).

Information on EPR spectra of powder (polycrystalline) specimens (Fig. 58) is much less, but in most cases it is still possible to determine the major parameters of the spectrum [404], and to identify ions or centers whose parameters have been earlier measured in monocrystals.

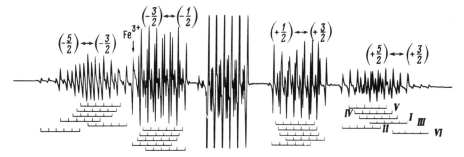

Fig. 55. EPR spectra of Mn^{2+} in six nonequivalent positions in spodumene, $LiAl(SiO_3)_2$ [387]

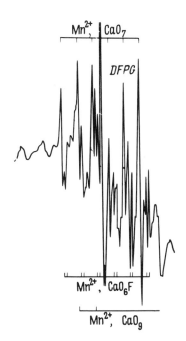

Fig. 56. Central transition $1/2 \rightarrow 1/2$ in the EPR spectrum of Mn^{2+} in hydroxyl-containing apatite. Mn^{2+} lines are distinguished from three complexes: CaO_7; CaO_6, F, CaO_9 [377]

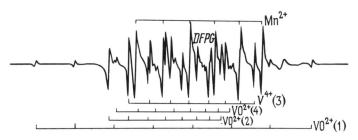

Fig. 57. Spectra of V^{4+}, VO^{2+}, and Mn^{2+} in Zoisite [395]

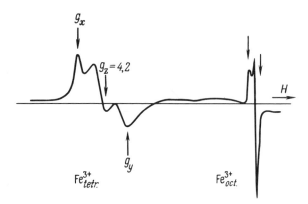

Fig. 58. Spectrum of Fe^{3+} in polycrstalline kaolinite in tetrahedral and octahedral positions [353]. Typical picture of a powder spectrum

3.3 Investigation of Minerals by EPR Spectra

3.3.1 Principal Applications of EPR Spectroscopy in Mineralogy and Geochemistry

According to the nature of substances that can be observed by EPR spectra, the following fields for the application of the method are distinguished.

EPR Spectra and the Principal Problems in the Chemical Bond Theory. The significance of EPR in the formation of principal theories describing the electronic structure of compounds is comparable only to that of X-ray spectroscopy (see Chap. 2), but the possibilities of EPR in this respect are unique: it is precisely with its help that differences between the concept of integral valence and continually changing charge are clearly defined and delocalization of the electron in a complex is ascertained unambiguously. The parameters of hyperfine and superhyperfine structure enable one to ascertain precisely the distribution of the electron density and *sp*-hybridization. All the EPR parameters are linked with the patterns of molecular orbitals and are used in assessing the ionicity-covalency. It is precisely in EPR that the most intimate relations between the spectroscopic and structural characteristics of the ion in a crystal exist. A set of the spin-Hamiltonian B_l^m parameters represents the most detailed and sensitive characteristics of the crystalline field at the site of the paramagnetic ion localization, with this field's axes being most precisely tied to the crystallographic orientations.

Paramagnetic Impurity Ions in Minerals; Objectives Facing Mineralogy and Geochemistry. EPR is capable of unambiguously differentiating impurity ions isomorphously entering the lattice from mineral micro-inclusions of microscopic or electron-microscopic dimensions. At the same time, it furnishes comprehensive information on a given ion in the crystal: its valence, coordination, local symmetry, the structural position it occupies, whether it assumes one or two non equivalent positions, and the orientation of the crystalline field axes at the site of this ion complete the characterization of the cyrstalline field and detailed information on the chemical bonding.

This is, however, not only the characteristic of the ion in a crystal, it is also the characteristic of the crystal itself, of features peculiar to the distribution of electron density, of the crystalline field, ionicity-covalency in the crystal and, finally, simply descriptive (diagnostic) characteristic of the mineral, for each ion in each mineral has parameters of its own, differing from the others. Here, the paramagnetic ion represents, as it were, a probe that furnishes spectroscopic characterization of the structure and the state of chemical bonding.

The EPR also yields information on the distribution of the paramagnetic ion impurities over structurally nonequivalent positions (in diopside, tremolite, apatite, etc., for example) i.e., the information on the intra- and intercrystalline distributions, which the Mössbauer and optical spectra provide on the basic components of the composition.

The contents of the impurities, the form of their incorporation and distribution are used as typomorphic characteristics of the minerals.

Electron-Hole Centers. These have become the most interesting field of EPR application in mineralogy and they exist in natural minerals or develop as a result of artificial irradiation (see Chap. 7). To these belong also complex investigations involving the use both of EPR and thermoluminiscence.

A promising trend is the application of EPR in investigating organic matter of sedimentary rocks, coal, tucholites, bitumens, petroleum, and fossil plants [364, 443].

Further broadening of EPR possibilities in mineralogy and geochemistry may be linked with: (1) carrying out measurements at low (helium) temperatures, which will increase the number of observed paramagnetic ions (of rare earths, in particular), (2) research not merely into the monocrystalline grains, but also into powders and rock fragments, which will drastically reduce the demands on the materials for investigation, (3) artificial irradiation of minerals, transforming the impurity ions into the paramagnetic state.

3.3.2 Survey of EPR Data for Paramagnetic Impurity Ions in Minerals

At the root of the whole of the EPR spectroscopy covering transition metal ions in crystals lie measurements of the spectra in each compound with a complete calculation of the spin-Hamiltonian parameters, representing the constants of a given compound.

The material made thus available embraces predominantly the spectra of the transition metal ions [407, 416] (see in [467] over 100 references till 1966). For some crystal types measurements for many ions have been made. Reviews of EPR for transition metal ions and rare earths have been compiled for CaF_2, $CaWO_4$ and other structural types (corundum, spinel, garnet, perovskite, etc.) [406, 407, 411, 416].

This material is systematized according to electron configurations (d^1: Ti $^{3+}$, V^{4+}, Cr^{5+}, Mn^{6+} etc.), to valence states of ions (for instance, in the case of manganese the EPR spectra of the Mn^0, Mn^+, Mn^{2+}, Mn^{3+}, Mn^{4+}, Mn^{5+}, and Mn^{6+} ions), and for each ion, the types of the spectra differing in coordination and local symmetry.

A good deal of material has been collected on natural minerals and their synthetic analogs. Particularly prominent are investigations of the following ions:

(1) Mn^{2+} and Fe^{3+}, which are most convenient for measuring by the EPR method and, at the same time, are most wide-spread geochemically, occurring in the form of impurities in many major mineral groups; (2) Cr^{3+}, Co^{2+}, Ni^{2+}, Cu^{2+}, which are also successfully measured, but geochemically are found in the form of impurities in a much more restricted number of minerals; (3) Ti^{3+}, V^{4+}, Nb^{4+}, Mo^{5+}, W^{5+}, i.e., ions with the d^1 configuration, which may be regarded as electron centers $M^{n+} + e^- = Mn^{n-1}$; Ti^{3+} being the commonest of these; (4) rare earths (Fig. 59), of which Gd^{3+} and Eu^{2+} can be observed at room temperature and are easily definable by their characteristic hyperfine structure, whereas for the majority of the other RE ions cooling down to the temperature of liquid hydrogen or helium is necessary.

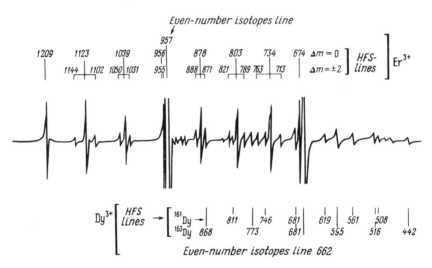

Fig. 59. Spectra of Er^{3+} and Dy^{3+} in zircon—$ZrSiO_4$, at 4.2 K and a frepuency of 9.5 mHz (magnetic field values in G) [340]

Below we list the EPR spectra of ions and minerals (EPR of the electron-hole centers, see Chap. 7) measured in detail and calculated. In brackets are indicated the Fe^{3+} ions established by means of E/D measurements only (see Fig. 49) and ions of Mn^{2+} observed in this mineral, but not measured in detail.

Silicates

Framework Silicates. In feldspars, a detailed study showed the spectra of Fe^{3+} (see Fig. 48) substituting Al^{3+} [372, 419, 473]: in microcline, orthoclase, adular, sanidin, albite, oligoclase, bytownite, and anorthite; they demonstrate moreover the presence of Ti^{3+} and Mn^{2+}, as well as the hole centers (see Chap. 7). In quartz this was carried out for Ti^{3+} [870] and Fe^{3+} (amethyst [392, 427]), in cristobalite for V^{4+} [380]; in petalite for Ti^{3+} [870] and Mn^{2+} [399]; in natrolite for Ti^{3+} [870] and Fe^{3+} [332], and in benitoite for Fe^{3+} [355].

Ring Silicates. In beryl: Cr^{3+} [375], Fe^{3+} [368], Ti^{3+} [870], Mn^{2+} [469]. In cordierite: Mn^{2+} [470].

Insular Silicates. In zircon the studies covered Ti^{3+} [870], Fe^{3+} [434], Nb^{4+} [468], Eu^{3+}, Dy^{3+} [340, 464], Tb^{4+}, Tm^{2+}, Y^{3+}, and Gd^{3+} [393, 468]; in forsterite they embraced Mn^{2+} [359, 427], in monticellite Mn^{2+} [361]; in andalusite Fe^{3+} [386, 467] and Cr^{3+} [471]; in cyantie Fe^{3+} [463]' Cr^{3+} [394] and Ti^{3+} [870]; in topaz Fe^{3+} [342, 447, 460] and Ti^{3+} [870]; in euclase (Fe^{3+}); in sphene V^{4+} [345], (Fe^{3+}); in clinohumite (Fe^{3+}); in eucryptite Fe^{3+} [362]; in zoisite VO^{2+}, Cr^{3+}, Fe^{3+} [395]; in epidote (Fe^{3+}), and in vesuvianite (Fe^{3+}).

Chain Silicates. In diopside Mn^{2+} [376], (Fe^{3+}); in spodumene Mn^{2+} [387, 388]; in enstatite Ti^{3+} [870]. (Fe^{3+}); and in tremolite Mn^{2+} [348, 378, 413], (Fe^{3+}).

Sheet Silicates. In apophyllite V^{4+} [345], Mn^{2+} [347], Ti^{3+} [870]; in datolite Mn^{2+} [349], (Fe^{3+}); in kaolinite Fe^{3+} [353]; in muscovite Fe^{3+} [402] and phlogopite Ti^{3+} [436] and Fe^{3+} [401]; and in allophane Cu^{2+} [350].

Oxides and Hydroxides

In CaO Mn^{2+} [453], Fe^{3+} [453], Gd^{3+}, Eu^{2+} [454]; in periclase Mn^{2+}, Fe^{3+}, Cr^{3+} and others [407, 474]; in zincite Mn^{2+}, Fe^{3+} [450]; in corundum Fe^{3+}, Mn^{2+}, Cr^{3+}, and others [407, 474]; in rutile Fe^{3+} [358]; in anatase Fe^{3+} [371, 389]; in brookite Fe^{3+} [446]; in cassiterite Fe^{3+} [336, 444] and Cr^{3+} [370]; in chrysoberyl Fe^{3+} [341, 465]; in spinel Mn^{2+}, Fe^{3+}, Cr^{3+}, and others [354]; in brucite Mn^{2+} [438] and in diaspore Fe^{3+} [374].

Phosphates

In apatite Mn^{2+} [377, 440]; in augelite Ti^{3+}, V^{4+} [870]; in wavellite V^{4+} [870]; in amblygonite Ti^{3+}, V^{4+} [870] (Fe^{3+}); in morinite Mn^{2+} [419]; in xenotime Yb^{3+}, Nd^{3+} [385], Er^{3+} [369], and Gd^{3+} [367]; in berlinite Cr^{3+} [384]; in turquoise Cu^{2+} [365]; in newberyite Mn^{2+} [415].

Tungstates

In scheelite Mn^{2+}, Fe^{3+} [346, 400], Gd^{3+} [431, 456], and RE^{3+} [406].

Sulfates

In astrakhanite Mn^{2+} [403].

Carbonates

In calcite Mn^{2+} [475], Fe^{3+} [338], Cu^{2+}, Ag^{2+} [351], and Gd^{3+} [338, 420]; in aragonite Mn^{2+} [412]; in magnestie, dolomite, ankerite Mn^{2+} [446]; and in smithsonite Cu^{2+} [350].

Arsenates

In adamite Fe^{3+} [396].

Halides

In fluorite Mn^{2+} [445], Gd^{3+} [468], Eu^{2+} [339], Tr^{3+} [468], and U^{4+} [461].

Sulfides

In sphalerite Mn^{2+} [451], Fe^{3+} [442], RE^{3+} [462], Co^{2+} [1]; and in galena, altaite Mn^{2+} [439].

4. Nuclear Magnetic Resonance (NMR) and Nuclear Quadrupole Resonance (NQR)

4.1 The Principle of the Phenomenon and Types of Interaction in NMR

The pattern of obtaining NMR spectra (Fig. 60) is analogous tor the EPR patten, but with differences conditioned by the nature of the electron and nucleus spin moments, and by the value of the electron and nuclear magnetic moments. The energy levels between which the transition is accomplished (Fig. 61). resulting in the appearance of the NMR signal, are nuclear levels, as in Mössbauer spectroscopy (NGR), but in NMR the transitions take place between sublevels of the ground state arising in the applied magnetic field, whereas in NGR, transitions occur between the ground and excited levels of the nucleus (see Fig. 7).

In NMR the source of emission is a 10–60 MHz radio-frequency generator (wavelength 30–50 m). A specimen in the form of a crystal or powder is placed be-

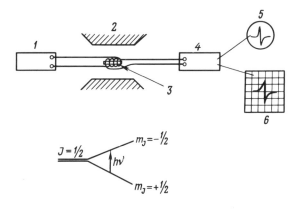

Fig. 60. Obtaining NMR spectra and the energy level diagram. I is nuclear spin; $h\nu$ is radiofrequency quantum. *1*, Radio-frequency generator; *2*, magnet; *3*, specimen; *4*, detector; *5*, oscillograph; *6*, pen recorder

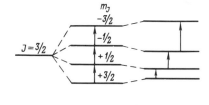

Fig. 61. NMR energy level diagram for the case of nuclear spin $I = 3/2$ without quadrupole splitting (*single NMR line*) and with quadrupole splitting (*three lines*)

tween the poles of the magnet. The spectrum is registered either visually on the oscillograph screen, or by means of a pen recorder.

As in the case of an electron or an atom, the nuclei can display a mechanical moment differing from zero, determined by the nuclear spin I and measured in term of the $\hbar$ units ($\hbar = h/2\pi$) i.e., $I^*\hbar = \sqrt{I(I+1)}\hbar$

The external magnetic field H generated by the magnet produces splitting of the spin sublevels, whose separation depends on the intensity of the magnetic field. If the energy of the radio-frequency emission becomes equal to the energy difference between the nuclear spin sublevels, a resonance absorption takes place recorded in the form of a NMR signal.

In NMR the resonance conditions are written in a manner analogous to the EPR resonance condition (see Chap. 3.1):

$$h\nu = g_N\beta_N H,$$

where g_N is nuclear g-factor, β_N is nuclear magneton. Nuclear g-factor equals

$$g_N = \mu I,$$

where μ is the magnetic moment of the nucleus in nuclear magnetons; I is nuclear spin.

The nuclear magneton β_N (μ_{nuc}) is an elementary value of the magnetic moment of the nuclei. It equals

$$\beta_N = \frac{e\hbar}{2Mc} = \frac{eh}{4\pi Mc} = 5.05038 \cdot 10^{-24} \text{ erg } G^{-1} = \frac{1}{1836}\mu_\beta,$$

where M is the proton mass; e is the magnitude of the charge; c is the velocity of light. The expression for β_N is analogous to the expression for β (or for μ_β, the Bohr magneton) $e\hbar/2mc$, only instead of the electron mass m here is the proton mass M, which is 1836 times the mass of the electron. Therefore, β_N is 1836 times less than is β (μ_β). Hence, with the electron g factor and the nuclear g_N factor being close to each other in a similar magnetic field H, the resonance in the NMR spectrum is observed at frequencies $\sim$ 1800 times inferior to those in the EPR spectrum (in the EPR usually at about 9000–38,000 MHz and in the NMR, accordingly, at about 5–20 MHz).

The magnetic moment of the proton does not equal the nuclear magneton, but

$$\mu_{proton} = 2.793\beta_N = 2.793 \cdot 5.05038 \cdot 10^{24} \text{ erg } \cdot G^{-1} = 1.41044 \cdot 10^{-23} \text{ erg } \cdot G^{-1}.$$

This is due to the fact that the proton and neutron are themselves of a complex nature.

The resonance condition is often written differently. Since

$$\beta_N = \frac{\mu(\text{erg} \cdot G^{-1})}{\mu(\text{nuclear magneton})} \text{ and } g_N = \frac{\mu(\text{nuclear magneton})}{I}$$

then

$$g_N \beta_N = \frac{\mu(\text{erg } G^{-1})}{I}$$

wherefrom we get

$$h\nu = g_N \beta_N H = \frac{\mu}{I} H.$$

The resonance condition written in this form proceeds from the quantum-mechanical description of NMR. In all works on NMR a vector model of the nucleus with precession of the nuclear magnetic moment about the magnetic field direction with the Larmor frequency of precession ω_0 is given. Inasmuch as the results thus obtained are fully identical, we will confine ourselves here to the quantum-mechanical description, but shall also state the formulation of the resonance condition ensuing from the classical model.

The cyclic frequency ω_0 (rad·c^{-1}) is linked with the frequency ν (s^{-1}, i.e., Hz) by the expression: $\omega = 2\pi\nu$. The gyromagnetic ratio is expressed in the form

$$\gamma = \frac{\mu(\text{erg } G^{-1})2\pi(\text{rad})}{I \cdot h \text{ (erg s)}} = \frac{\mu}{I\hbar} \text{ rad s}^{-1} G^{-1}.$$

Since $(h/2\pi) \cdot \nu \cdot 2\pi = h\omega$ and $g_N \beta_N = \mu(\text{erg } G^{-1})/I$, the relation $h\nu = g_N \beta_N H$ may be written

$$\hbar\omega = \frac{\mu}{I} \cdot \text{H}$$

wherefrom

$$\omega = \frac{\mu}{I\hbar} H = \gamma H.$$

Hence $\nu = \gamma/2\pi \cdot H$.

Tables listing the properties of the nuclei that are of importance for NMR spectroscopy [481, 534, 564] usually include the magnetic moment μ (in nuclear magnetons) and the nuclear spin, from which one can readily obtain $g_N = \mu/I$ and the magnetic moment (in erg G^{-1}) equalling $\mu\beta_N$, and sometimes the gyromagnetic relation in rad · s^{-1} · G^{-1} or $\gamma/2\pi$ in s^{-1} · G^{-1} (i.e., Hz^{-1} · G^{-1}).

From the resonance condition the frequency at which nuclear resonance with a given $g_N = \mu/I$ or $\gamma/2\pi$, observed in a given magnetic field (often in the field $H = 10,000$ G) is determined. For example, in the case of the proton:

$$\mu = 2.793 \text{ nuclear magneton}; I = 1/2; g_N = 5.586;$$

$$\nu = \frac{g_N \beta_N}{h} H = \frac{5.05038 \cdot 10^{-23}}{6.6252 \cdot 10^{-27}} g_N H = 0.7623 \cdot 10^3 g_N H = 0.7623 \cdot 10^3 \cdot$$

$$5,586 \cdot 10000 = 42 \cdot 10^6 \text{ Hz} \approx 42 \text{ MNz}.$$

Or simpler: $\gamma/2\pi = 4257$ Hz $\cdot$ G^{-1} (Table 11) and then

$$\nu = \frac{\gamma}{2\pi} H = 4257 \cdot 10\ 000 = 42.57 \text{ MHz}.$$

4.1.1 Types of Nuclei Viewed from the Standpoint of NMR

Let us consider the following characteristics of the nuclei: (1) the spin and the presence of the magnetic moment; (2) the presence of the quadruple moment; (3) the value of the magnetic moment; (4) natural abundance of the isotope.

The presence of the quadrupole moment in the nucleus points to a deviation of the charge distribution from the spherical i.e., the nucleus then appears not in the shape of a sphere, but as an ellipsoid of rotation, elongated or flattened out. The quadrupole moment value is written down as eQ, where e is the charge unit, Q the measure of deviation from the spherical symmetry (Fig. 62), given in barns (10^{-24} cm^2), i.e., in units of measurement of the nucleus area, with the radius of the nucleus of the order of $r_N \approx 10^{-12}$ cm.

All the nuclei with a zero spin have no magnetic and quadrupole moments, and yield no NMR signals. They include atomic nuclei with an even atomic number Z and even mass number A, in particular such commonly encountered nlclei as C^{12}, O^{16} (O^{18}), Si^{28}, S^{32}, and Ca^{40}. The possibility of investigating the NMR spectra of nuclei with the spin differing from zero is conditioned by the value of the magnetic moment (more exactly by the value of μ/I) and by the natural abdunance.

The nuclei yielding the NMR signal can be conveniently subdivided into those with the spin $I = 1/2$, without quadrupole moment, and those with the spin $I \geqslant 1$, having quadrupole moment. The most intensive NMR spectra (for instance, shown in Fig. 63) come from the nuclei H^1, F^{19}, P^{31} with the spin $I = 1/2$ and without quad-

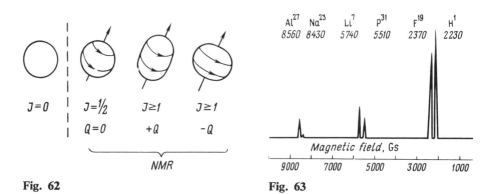

Fig. 62   Fig. 63

Fig. 62. Types of nuclei. I, nuclear spin; Q, quadrupole moment

Fig. 63. Resonance values for magnetic field with the frequency of 9.5 MHz for the nuclei of H^1, Li^7, F^{19}, Al^{27}, P^{31} [540]

Table 11. Properties of the nuclei most suitable for investigating by the NMR method

Isotope		Natural	Spin	μ, nucl.	$\gamma/2\pi$	Relative intensity		Quadru-
Z	A	abund-	I	magnetons	Hz. G^{-1}	with equal		pole
		ance				fields	frequen-	moment
		%					cies	10^{-24} cm^2
1	H^1	99.986	1/2	+2.79255	4257	1.000	1.000	—
3	Li^7	92.7	3/2	+3.2559	1655	0.29	1.94	−0.012
4	Be^9	100	3/2	−1.1774	598.7	0.014	0.7	+0.02
5	B^{11}	81.17	3/2	+2.6886	1367	0.165	1.60	+0.036
9	F^{19}	100	1/2	+2.6283	4007	0.83	0.94	—
11	Na^{23}	100	3/2	+2.2171	1126.7	0.093	1.32	+0.11
13	Al^{27}	100	5/2	+3.6408	1110	0.21	3.04	+0.156

The NMR frequency for the field H is ν (Hz) $= \gamma/2\pi \cdot H$ (G).

rupole moment, the nuclei Li^7, Be^9, B^{11}, Na^{23} with the spin 3/2 and Al^{27} with the spin 5/2, all of them experiencing quadrupole splitting (see Table 11).

In NMR the distance between the spin sublevels is very small (of the order of $h\nu = g_N\beta_N H$) and, therefore, the occupancies of these sublevels are very close to one another: for protons the occupancy of the lower state with $m_I = + 1/2$ is merely 1.0000066 times greater than that of the higher state with $m_I = - 1/2$. Because of this quite insignificant difference in the occupancy, the NMR-transitions with absorption occur. The intensity of such transitions is low and, for this reason, NMR is a method intended for investigating nuclei contained in great numbers, i.e. of looking into major components of the materials and whenever possible, in specimens of considerable bulk (or the order of 0.5 cm^3).

NMR is commonly observed in diamagnetic substances. The presence of paramagnetic admixtures in the substance, or investigating NMR of paramagnetic ions (Mn^{55}, Cr^{53}, etc.) creates special conditions.

4.1.2 Two Types of NMR-Investigations

Two basic branches exist in NMR spectroscopy: (1) high-resolution NMR employed in investigating liquids and (2) broad-line NMR observed in solids and, therefore, of preponderant interest for us.

These two branches differ sharply. At the root of the difference lies the width of the NMR lines: in solids it is thousands of times larger than in liquids. This does not mean that high-resolution NMR has any advantage, for a different kind of information on luquids and solids is transmitted through dissimilar mechanisms of interactions that find their expression in NMR spectra.

This distinction extends also to the range of the substances undergoing investigation (solids or liquids), equipment (the construction of the broad-line NMR spectrometers is simpler, but they require special arrangements and devices for orienting the crystal, sweeping the field over a wider range), to the theory of the phenomena and the methods of calculation.

The cause of the narrowing of lines in liquids as compared with solids, and for the different features specific to their NMR spectra is explained by the following mechanisms of interactions inherent in them.

4.1.3 Principal Mechanisms of Interactions in NMR

Although NMR is based on transitions between the nuclear levels, for the spectroscopy of solids and liquids these are of importance only so far as they reflect the structure and properties of the matter, and play the role of a probe furnishing information on the matter. Therefore, it is necessary to consider the interaction of the magnetic and quadrupole moments of the nucleus with other atoms of the crystal or the molecule.

In the complex NMR phenomenon four principal types of interaction are distinguished.

1. Direct interaction of a given nucleus with other neighboring magnetic nuclei of a crystal or a molecule, termed magnetic dipole interaction (the nuclei appear as magnetic dipoles), or a direct spin–spin interaction (the interaction of the nuclei is described as interaction of the nuclear spins).

Two aspects of this interaction may be distinguished. It causes broadening of the resonance lines: in solids each nucleus finds itself not only in the external field H, generated by the magnet, but also in a local field built up by all the adjacent magnetic nuclei of the specimen. With distances r_i from a given nucleus to the neighboring nuclei, the intensity of the field generated by them lies within the range of $\pm 2\mu/r_i^3$; for a proton with $r_i \sim 1\text{Å}$ this amounts to 57 G. Hence, the resonance is observable within the ranges of the magnetic fields of the order of 50 G, and this results in broadening of the lines.

In solids this magnetic dipole–dipole (spin–spin) interaction is the main factor of the environmental influence on a given nucleus. (In the calcite $CaCO_3$, a crystal consisting almost entirely of nonmagnetic nuclei, the NMR of C^{13}, with natural abundance of 1.1 %, yields extremely narrow lines.)

In liquids and gases the observed picture is quite different. Here molecular motion manifests itself in the rotation and translation of the molecules with a frequency of the order of $10^{10} - 10^{12}\text{s}^{-1}$, which is much faster than the frequency of the nuclear precession in the NMR (10 MHz = 10^7 Hz = 10^7s^{-1}). Therefore, the magnetic field of the surrounding nuclei rapidly fluctuates, time-averaging to zero, and this eliminates the cause of broadening of the lines observable in solids with a fixed position of the nuclei, and the NMR lines become narrow.

In solids the width of the lines commonly lies within the range of 1–100 kHz in terms of frequency, or within 1–100 G in magnetic field expression, whereas in liquids it is less than 1–100 Hz (inferior to 0.1 G).

The other aspect of the magnetic dipole–dipole interaction has to do with the fact that it is just this interaction which determines the structural application of the NMR.

With the nuclear spin $I = 1/2$ one should observe a single resonance line, but the interaction with other nuclei produces its splitting. Inasmuch as this interaction diminishes rapidly with distance, it is usually possible to single out pairs, three, or four nuclei, called two-, three-, or four-spin systems for which it is possible to determine the distance between the nuclei, the direction of the vector connecting them and the number of nonequivalent positions. To do so it suffices for the spacing in a given pair or a group of the nuclei to have the relationship 1:2 to the distance to the next near-neighboring nuclei.

2. Interaction of the quadrupole moment of the nucleus with the crystal field gradient (for nuclei with $I \geqslant 1$) is the second most important type of interaction of the nucleus with the environment in NMR of solids, which is also not observed in liquids.

These two types of interaction, of paramount importance for solids, do not occur in NMR of liquids. But at the expense of what is the interaction of the nucleus with its environment accomplished in liquids? It is only owing to narrowing of the NMR lines in liquids (with the magnetic nuclear dipole–dipole interaction averaged almost to zero) that the other two types of interaction can be observed, leading as they do to such negligent shifts and splittings of the resonance lines, which are much smaller than the line width in solids, but are distinctly seen in the high-resolution NMR of liquids.

Both types of the influence produced by the environment on the resonant nucleus are realized in liquids via the electron shell of the nucleus.

The electron magnetic shielding that causes the chemical shifting of the resonant lines is conditioned by the fact that the external magnetic field acting upon the specimen is shielded by the electron shell of the atom. This results in the effective field on the nucleus becoming modified: $H_{eff} = H_0 (1 - \sigma)$, where σ is the screening (shielding) constant which characterizes the electron shell surrounding the nucleus in a given compound. Inasmuch as the distribution of the electron density varies in different environments, so does also σ and the resonance frequency, and it is this which leads to a chemical shift of the lines. The value of σ increases from 10^{-5} for the proton to 10^{-2} for heavy atoms; the value for σ is commonly given in the 10^{-6} units (in millionth parts).

An indirect spin–spin interaction is also accomplished through the electronic shells, but this is not a change wrought by them in the external magnetic field, but rather a way of transmitting the effect of the magnetic field from the neighboring nuclei, since these neighboring nuclei create a magnetic field affecting their electron shell, which, in its turn, influences the electron shell around the resonant nucleus, and this changes the magnetic field on the nucleus in question. Thus, for instance, when obtaining a high-resolution spectrum for ethyl alcohol CH_3CH_2OH it shows, as a result of shielding (chemical shift), not a single line of proton resonance, but three lines from protons in the groups CH_3, CH_2, and OH (with their intensities ratio 3:2:1). When recording at a still higher resolution, each one of these three lines is seen to split because of the indirect spin–spin interaction, i.e., CH_3 is split into three lines, CH_2 into four, while that of OH remains single.

The distance between lines CH_3–CH_2–OH is of the order of 50 Hz, and the spacing between lines in the group of CH_3 and CH_2 is less than 5 Hz. (Compare them with the line width in the NMR of solids, which is of the order of 1–100 kHz i.e., 100–100,000 Hz.)

4.1.4 Spectra of Nuclei with $I = 1/2$ (H^1, F^{19}) in Solids

In these nuclei, structural investigations are possible resulting in determining the distance between the near-neighboring nuclei, orientation of the direction connecting them, the presence of nonequivalent orientations and nonequivalent positions of these nuclei.

The problem is solved in different ways in the case of a monocrystalline specimen (Figs. 64, 65) or a powder (Fig. 66). If one succeeds in obtaining resolved doublets (for a two-spin system, i.e., a structure in which it is possible to single out two near-neighboring nuclei, whose interaction with other nuclei is much less intensive), or triplets or quadruplets, then it is the orientation dependence between the lines that is measured (see Fig. 65). An orientation with which the maximum splitting is observed corresponds to the coincidence of the direction linking the nuclei with the direction of the magnetic field.

In the case of a two-spin system, the local magnetic field is determined by the following expression, from which the internuclear distance r_i is found:

$$H_{\text{loc}} = \pm\ 3/2\ \mu_I/r^3\ (3\cos^2\theta - 1).$$

The distance between the doublet components equals the double value of H_{loc}.

If, because of the presence not of one to two, but several near-neighboring nuclei, a monocrystal spectrum shows an unsplit line, or if a polycrystalline specimen is subject to measurement, the second moment of the absorption line ΔH^2, i.e., the standard deviation from the center (H_{average}) of the resonance line is determined:

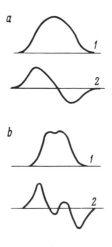

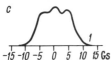

Fig. 64 a-c. Shapes of lines in the spectra of proton resonance and the resonance of F^{19} and of other nuclei with spin $I = 1/2$ in the presence of a single magnetic nucleus (**a**) pairs of nuclei [as in a water molecule H_2O (**b**)] and three nuclei (**c**). *1*, Absorption curves; *2*, absorption line derivatives

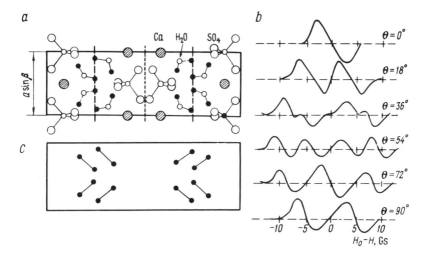

Fig. 65 a-c.   Orientation dependence of the monocrystal proton resonance spectra; a structure of gypsum $CaSO_4 \cdot 2H_2O$; b proton–proton vectors singled out from the entire structure (the nuclei of other gypsum atoms are non-magnetic or have too low natural content); c orientation-dependence of the spectrum [550]. From this orientation-dependence the direction of proton–proton vectors and spacing between hydrogen nuclei are determined

$$\Delta H^2 = \int_0^\infty (H - H_{av})^2 f(H)dH,$$

which is found graphically from the measured resonance-absorption curve.

The experimental value of the second moment and, in the case of a monocrystal, also of its orientation dependence are contrasted against the value calculated on the ground of a known or trial structure value ΔH^2 according to the formula

$$\Delta H^2 = 3/2 \frac{I(I+1)}{N} g^2_N \beta_N \sum (3 \cos^2\theta - 1)^2 r_i^{-6},$$

where I is the spin of the resonant nucleus; θ is the angle between the direction of the magnetic field and the vector connecting the resonant nucleus with the nucleus whose contribution is being evaluated; r_i is the distance between the nuclei; N is

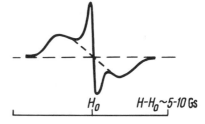

Fig. 66.   Proton resonance spectra of powders. Narrow lines of free water and broad lines of hydroxyl groups are singled out. From these the second moments are calculated by means of which the proton–proton distance is determined

the number of the nuclei included in the summation. Since the value of the second moment is inversely proportional to the sixth power of the internuclear distance, it suffices to include in the calculation the nuclei lying within the sphere over a distance of 6–7 Å.

Examples of such calculations for nuclei of the same type are works dealing with gypsum $CaSO_4$ $2H_2O$ [526, 534, 550], Rochelle salt, where the proton–proton directions for 16 molecules of crystallization water have been determined [535], for $Ca(OH)_2$ [523], $Mg(OH)_2$ [507]. The second moments for the lines of H^1 and F^{19} determined by nuclei of a different kind were calculated, for example, in the case of amblygonite $LiALPO_4$ (F, OH) [540].

4.1.5 Spectra of Nuclei with $I \geqslant 1$ in Solids

If nuclei possessing a quadrupole moment occur in a crystal field of cubic symmetry they then yield a NMR spectrum consisting of a single line (as do the nuclei with a spin of 1/2 without quadrupole moment). However, in crystalline fields of an axial and a lower symmetry a shifting of the nuclear levels takes place, which leads to the appearance of $2I$ lines: the central line and quadrupole satellites (Fig. 67).

In the case of nuclei with $I = 3/2$ (Li^7, Be^9, B^{11}, Na^{23}, etc). three lines, and in that of nuclei with $I = 5/2$ (Al^{27}), five lines are observed. The quadrupole interaction constant may be written as

$$e^2qQ/h,$$

where eQ is the quadrupole moment of the nucleus; eq is the gradient of the crystalline field (q is gradient of the field in terms of the charge units and e is the electrostatic charge); h is Planck's constant, introduced here to convert ergs into Hz (see Table 7).

The quadrupole interaction constant is often written as eQq, denoting with $q = eq/h$.

The quadrupole splitting of the nuclear magnetic resonance lines is of the same kind as the splitting in the nuclear quadrupole resonance (see Chap. 4.3) and as in the Mössbauer spectra (see Chap. 1.3).

The electric (crystalline) field gradient q equals $\partial V/\partial z^2$, in case of axial symmetry or $\partial^2 V/\partial x^2$, $\partial^2 V/\partial y^2$, $\partial^2 V/\partial z^2$ in orthorhombic symmetry, where V is the crystalline field potential. The value of q represents the same part of the potential, proportional to the coefficient A_2^0 in the lattice sums, which is calculated from the lattice sums and forms part of the Mössbauer, optical, EPR and NMR parameters.

Experimentally, the quadrupole interaction constant e^2qQ/h (its x, y, z axes components) is subject to measurements in NMR. Commonly, its z axis component e^2q_zQ/h (or eQV_{zz}/h) and the asymmetry parameter $\varDelta = (V_{xx} - V_{yy})/V_{zz}$ are adduced. The value of e^2Qq/h varies in different crystals from 20–30 kHz to 10–20 MHz. With the known exact value of the quadrupole moment it becomes possible to determine the value of the crystal field gradient, and to use it in evaluating the chemical bond and contrast it with the calculated values for q.

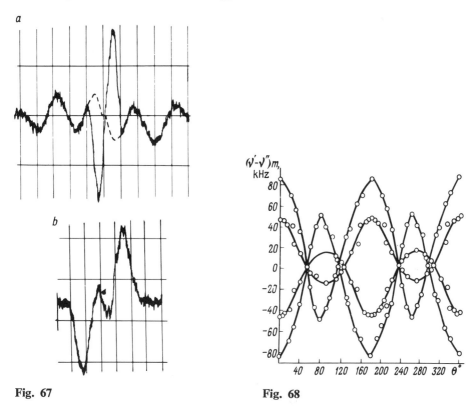

Fig. 67   Fig. 68

Fig. 67 a and b. NMR spectrum of Li[7] in amblygonite. Central line and two satellites (Li[7] $I = 3/2$, number of lines $2I$) should be observed. The presence of two pairs of satellites (**a**) and splitting of the central line (**b**) point to the distribution of Li between two nonequivalent positions in the structure of amblygonite; an equal intensity of both pairs of satellites shows this distribution to be disordered [540]

Fig. 68. Orientation-dependence of distances between the satellite lines during rotation of the crystal with respect to the direction of the magnetic field (on an example of two pairs of satellites Li[I] and Li[II] in the spectrum of amblygonite). From this dependence the value of the quadrupole splitting and the orientation of principal eQq/h tensor axes are determined [540]

    The magnitude and position of major axes of the electric field gradient tensor are found from the distance between the quadrupole satellites, and from the dependence of this distance on the crystal orientation with respect to the magnetic field (Figs. 67, 68). The methods used for this calculation are set forth in detail conformably to the classic example of spodumene [561, 580]; also the calculation of the quadrupole interaction constants for amblygonite is described [540].

    In the presence of two and more nonequivalent positions each line of the quadrupole spectrum dissociates into two and more components.

The symmetry, orientation, and value of the crystal field gradient are determined for each position. These observations are made use of in determining order–disorder relationships of atoms in a structure.

4.1.6 High-Resolution NMR-Spectroscopy in Solids

In the 70s, new prospects have appeared for the development of NMR in solids. Basic limitations of NMR methods, such as low resoltuion and poor sensitivity, can now be overcome by the development of pulse methods and Fourier spectroscopy [509] as well as by employing methods of dynamic polarization of the nuclei [511].

In contradistinction to ordinary stationary NMR methods with which a specimen placed in a magnetic field experiences the effect of the radio-frequency field throughout the whole period of observation, with the pulse method, short bursts, that is oscillation pulses of a definite frequency, are employed, the observation of the signal being made on termination of the pulse. To transform the accumulated series of such signals into a spectrum, calculation methods based on the ways of a rapid Fourier transformation, suitable for handling with small-size computers, have been devised.

The methods used for dynamic polarization of the nuclei (of O^{17}, Ti^{47}, Ti^{49} in the presence of 0.2% of Cr^{3+} in rutile [511], for example) represent a powerful mode of observing very weak nuclear transitions. As a result of this it becomes possible to observe the NMR from such nuclei as O^{17}, C^{13}, Si^{29}, from samples not enriched with them, and from the impurity contents of the elements, to secure their high-resolution spectra (Fig. 69), and to measure the anisotropy of chemical shifts and the crystal field gradients at the sites of these nuclei in the structure of crystals.

4.2 Nuclear Magnetic Resonance in Minerals

The NMR phenomenon was discovered in 1946 by F. Bloch and his co-workers and by E.M.Pursell and his co-workers. In 1955 the first publications on NMR in minerals (feldspars, clays, borates) made their appearance. In the years 1960 to 1970 the following basic trends in the development of NMR in mineralogy took shape.

4.2.1 Types and Behavior of Water in Minerals; Structural Position of Protons

An investigation of protons, which render the most intensive NMR signal, helps ascertain the structural position of the hydrogen atoms in crystals, the orientation of proton–proton vectors, and the distance between protons. As in the case of IR spectroscopy, NMR is helpful in educing various types of "water" in crystals and, additionally allows one to obtain information on the mobility of "water" in crystals, on the nature of its movement, on the direction of the precession axes of water molecules, on the features distinguishing its relation and links with the framework,

Fig. 69. O^{17} spectrum in rutile TiO_2 with Cr^{3+} admixture obtained with dynamic polarization (with natural O^{17} abundance of 0.04% without enrichment); nuclear O^{17} spin $I = 5/2$ [511]

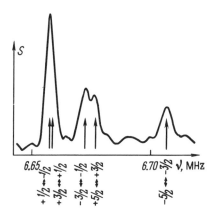

and on intermolecular and intramolecular interactions These investigations include a study on the dependence of the spectrum upon the orientation of the crystal with respect to the external magnetic field, and upon the splitting between the lines in the case of monocrystals, the measurement of the second moments in that of the polycrystalline specimens, the width of the lines and spin-lattice T_1 associated with it, and the spin–spin T_2 relaxation time (usually by the spin-echo method), as well as the temperature-dependence (on liquid helium temperatures up to those of dehydration or dehydroxylization), on the moisture content dependence, and on that of paramagnetic admixtures, etc.

To individual groups of minerals and to forms of water specific for them characteristic values of the proton resonance lines width $\Delta H(G)$ are common.

The most interesting problems are associated with NMR investigations of the types and behavior of water in clay minerals and in sheet silicates in general [505, 516, 582], in zeolites (Fig. 70) [485, 487, 488, 504, 512–514, 541, 551], of water in channels of the beryl structure [558], of chrysotile-asbests (Fig. 71) [548], of intersticial water in opal [545] and of the part played by protons in the ferroelectric properties of colemanite [489, 490, 581].

Zeolites. The special behavior of the "zeolite" water, and its practical applications as a molecular sieve prompted investigations of the proton resonance in natural and synthetic zeolites.

Two types of zeolite can be distinguished by proton resonance. The first includes natrolite, analcime, and thomsonite, which at room temperature yield a spectrum characteristic of crystalohydrates with a fixed disposition of the water molecules (hence, these are not "true" zeolites). Natrolite was subject to detailed studies, whereby four different proton–proton vectors and the position of protons in the structure were ascertained, and a structural explanation of the features attending the presence of water in it is given.

To the second type belong the remaining zeolites studied, characteristized by very narrow lines (0.1–0.2 G) that correspond to a great mobility of water, similar to that in adsorbed or capillary water. In cubic zeolites a single narrow peak is observed, in the others a doublet, except for harmotome, which has a quadruplet, due

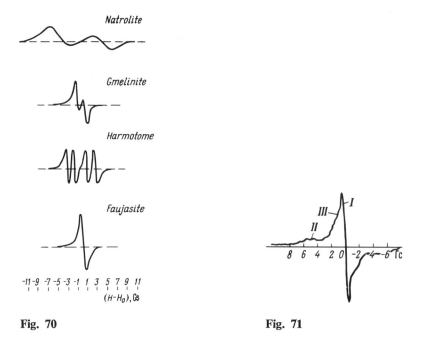

Fig. 70   Fig. 71

Fig. 70.   Zeolites proton resonance spectra

Fig. 71.   Chrysotile-asbestos NMR spectrum. I, Narrow lines of free water filling inner cavities of fibrous tubules; II, III, broad lines from two OH group types [548]

to the presence of two nonequivalent positions of water. The narrow doublet points to pair-wise grouped molecules of water. The nature of the water mobility has been studied in detail in desmine (isotropic reorientation), in harmotome (precession about the twofold axis, which precesses around an axis fixed in the crystal). A weak bond between the water molecule, the alumino-silico-oxygen framework, and the collective nature of the water mobility has been ascertained. At low temperatures the spectrum is analogous to that of crystallohydrates (in desmine $\Delta H^2 = 27$ G^2) or of ice (in A and X sieves $\Delta H^2 = 34$ G^2). A different course of zeolite dehydration is traced on the NMR spectra. The relaxation and correlation times, measured by the spin-echo method, and their temperature-dependence enabled determination of the existence of nonequivalent states in synthetic zeolites, an exchange between them, the activation energy, and the coefficient of self-diffusion.

Sheet Silicates. In the structure of many sheet silicates (especially in clay minerals) the simultaneous presence of interlayer water, brucite, and gibbsite layers, of hydroxyl groups and crystallization water is in evidence. Accordingly, their proton resonance spectra are of a composite nature.

For most of the studied minerals of this group two superposed spectra are noted: (1) a broad line up to 3 or 6 G and (2) a narrow (0.1–0.9 G) singlet line superposed on it associated with the adsorbed interlayer water (see Fig. 66). The narrow singlet line (but not doublet) indicates a rapid ($\sim 10^6$ times s^{-1}) exchange of protons of adjacent molecules, the interlayer water representing a highly ionized medium, entrained by the rapid movement. The width of these lines slightly decreases with diminishing moisture content and essentially increases with a growing abundance of paramagnetic ions in the structure.

The broad lines have different widths: up to 3 G in pyrophyllite, talc, phlogopite, vermiculite, and chlorite, and up to 6 G in kaolinite, halloysite. Illite and montmorillonite, which show narrow lines only, but after drying also yield broad lines measuring up to 3 G. The width of these lines increases with diminishing moisture content, In the minerals under observation, however, only in sepiolite does the width become closer to the width of lines typical of hydroxides and crystallohydrates (11–12 G), whereas in the rest of the cases it is much smaller.

Opal. In contrast to crystallohydrates and hydroxides, where water is in a fixed position, in opal the interstitial water, as concerns its properties, stands close to the ordinary liquid water. At room temperature a single, very narrow peak (less than 0.1 G) is observed. With falling temperature (200 K and 77 K), in addition to the narrow line, a broad line appears that corresponds to the frozen water. However, water in opal becomes fully frozen at 4.2 K and then $\Delta H^2 = 27$ G^2 (in ice 36 G^2). From this it is estimated that inidividual pores contain 10–15 molecules. This indicates the size of the pores to be less than 12 Å, i.e., inferior to the size of dodecahedral "molecules" H$_2$O. Thus the properties of water in opal are determined by the properties of individual H$_2$O molecules and not of the dodecahedral ones [545].

Crystallohydrates. A classical example here is the investigation of gypsum (see Fig. 65). In monocrystals a fine structure represented by four lines that correspond to two different orientations of the proton–proton vectors was observed. The distance between protons in the water molecule (1.58 Å) and the O–H distance (0.98 Å) were measured on the ground of the minimal splitting, which, depending upon the orientation, varies from 22 G down to 0. Similar spectra were observed for a number of other crystallohydrates: MgSO$_4$ 7H$_2$O (14 orientations of the p-p vectors) [544] and for others.

Hydroxides. The study of Mg, Ca, Al hydroxides is of special interest inasmuch as their structure enters the structure of many sheet silicates as one of their major modules. A detailed investigation of portlandite powder was made on the material of artificial compound Ca(OH)$_2$. At room temperature the second moment ΔH^2 is 10.6 ± 0.7 G and at the temperature of liquid nitrogen it equals 12.4 ± 0.3 G. On an assumed model of the hydrogen position in trioctahedral Bernal-Megaw hydroxides the parameter p, has been calculated. It describes the degree of distortion of the hexagonal hydrogen atom networks equalling 0.62 ± 0.04 Å, which corresponds to the distance O–H of 0.99 ± 0.02 Å.

The position of hydrogen in the O–H . . . O linkage has been determined for the diaspore powder ($\Delta H^2 = 9.11 \pm 0.7$ G^2) [482].

For the brucite monocrystal [507] the second moment of lines for different orientations has been determined on the basis of an analysis into oriented spectrum changes, and from this proton positions, very close to those found for Ca(OH)$_2$, have been calculated.

The position of protons in the structure has been also investigated in paramagnetic crystals dioptase [577], azurite [536] vivianite [537, 566], erythrine [546].

4.2.2 Structural Applications of NMR

In contrast to other spectroscopic methods, which usually furnish information on the electronic structure of crystals, this type of NMR work includes purely structural investigations.

Determination of the proton-proton vector orientation and of the distance between protons was discussed in the foregoing text.

Coordination of B^{11} in Borates and Borosilicates (Fig. 72). In minerals boron occurs in a trigonal and tetrahedral coordination, and this affects the value of the electric field gradient and local symmetry at the sites of these atoms. Since the B^{11} nucleus possesses the quadrupole moment, this manifests itself in a change of the quadrupole interaction constant eQq and that of the asymmetry parameter η, on which the width and the form of the NMR spectra lines for B^{11} depend. The value for eQq can change from the values corresponding to a free B^{11} atom (5.38 MHz) down to zero, when boron is in an undistorted tetrahedral position. For boron in trigonal coordination the value of eQq is 2.5–2.6 MHz, with tetrahedral coordination in tetrahedral rings it is 0.3–0.7 MHz and in isolated tetrahedral 0.05–0.1 MHz. If boron is present in the mineral simultaneously in the tetrahedral and trigonal coordination, it produces a composite spectrum, consisting of both the tetrahedral and trigonal spectra. Since the B^{11} nucleus spin is 3/2, the spectrum should consist of three lines, the central and two satellites, provided all positions of boron are equivalent. In colemanite [525] 15 lines are in evidence, whose position changes depending upon orientation of the specimen relative to the magnetic field. They accord with six positions of boron in the structure: two in the trigonal coordination, yielding six lines, and four in the tetrahedral coordination, yielding nine lines (the central line consists of four unresolved signals). A detailed analysis with determination of the orientation of the electric field gradient tensor axes was

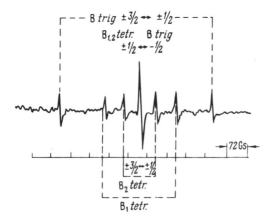

Fig. 72. NMR spectra of B^{11} in trigonal and tetrahedral coordination in kurnakovite [490] (one scale division equals 72 G)

made for colemanite [525], borax and tincalconite [499], inderite [559], danburite [532], and other borates [440, 500, 561, 563, 574].

Order–Disorder in the Structure of Minerals and the NMR Spectra. The AL^{27} and Na^{23} spectra observed in ordered and disordered feldspars differ sharply [496, 517–519]. In the case of albite a well-resolved spectrum is observable, which, because of a quadrupole splitting, consists of five Al^{27} (spin 5/2) lines and three Na^{23} (spin 3/2) lines. In microcline, because of twinning, the number of these lines is doubled. This indicates that the Al^{27} and Na^{23} nuclei have the same electric surroundings, i.e, each one of them is distributed in an orderly fashion. In sanidine, and also in heated albite and microcline the allowed lines are absent, which points to a disordered distribution of these nuclei. A similarity of the Al^{27} NMR spectrum in adular to the spectrum of microcline is indicative of the presence in the former of highly ordered, submicroscopically twinned domains.

In like manner an essentially ordered distribution of Al^{27} is shown in natural spinels, and its essentially disordered distribution in synthetic and heated natural spinels [408]. Distribution of Na^{23}, Al^{27} and Si^{29} in nephelines was also considered [494].

From the Li^7, H^1, F^{19} spectra in amblygonite $LiAlPO_4$ (F, OH) the existence of ordered and disordered specimens, as regards the distribution of Li therein, was ascertained [540]. Shifting of the hydroxyl group proton stabilizes the localization of Li in one of the structural positions, while fluorine makes for a disordered distribution of Li (Fig. 73, see also Figs. 67 and 68).

Other Structural Problems Solved by Means of NMR. These are such as making sure of preserving the short-range order in the disposition of fluorine in the X-ray amorphous metamict britholite and melanocite [547], determination of structural configurations of OH–F–V (where V is vacancy) and of OH–F–HO in apatite [544] and others.

4.2.3 Experimental Estimations of the Crystal Field Gradient

The eQq values obtained from the quadrupole splitting of the NMR spectra represent, in the case of such atoms as Li, Be, Na, Al, the sole method of securing information on the crystal fields for minerals of which they form a part. Here, the calculations of crystalline potential undergo an experimental verification.

The constants of quadrupole interaction, the asymmetry parameter, and the orientation of principal axes of the electric field gradient tensor have been measured for the following nuclei in minerals (B^{11}—see above).:

Li^6 and Li^7 in spodumene [501], Li^7 in amblygonite [540], and zeolites [515]; Be^9 in beryl [522] and chrysolberyl [570]; Na^{23} in albite [496], natrolite [562], faujasite [515], and nepheline [494]; Al^{27} in spinel [498], corundum [498], chrysoberyl [524], beryl [522], euclase [506], albite [496], adular [517], nepheline [494], cyanite [520], andalusite [521], sillimanite [569], zoisite [493], topaz [578], and in zeolites [515, 562].

Among other applications of NMR worthy of note is the possibility of assessing the P^{31} and F^{19} content in dispersed phosphates (phosphorites) [557]. A special

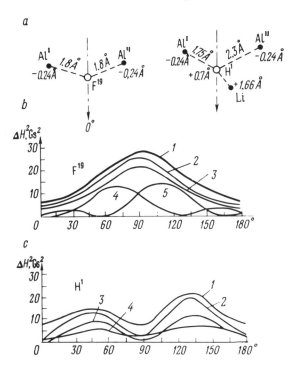

Fig. 73 a-c. Determining second moments of the NMR lines in amblygonite [540]. **a** F^{19} and H^1 disposition with respect to the nearest magnetic nuclei of Li^7 and Al^{27} in the amblygonite structure; **b** comparison of experimental (*1*) and calculated (*2*) orientation dependences of the second moment ΔH^2 of the F^{19}; curves *3,4,5* represent contribution of Al^I, Al^{II} and from both Al; **c** same for H^1

branch includes the application of the NMR in dealing with ferro- and antiferromagnetic crystals [576].

4.3 Nuclear Quadrupole Resonance

The NQR phenomenon, discovered in 1949 by H.G.Dehmelt and H.Krüger, has found application in investigating organic compounds (especially chlorine-containing, owing to the fact that the NQR signal is readily observed on the Cl^{35} nuclei) and inorganic crystals (see reviews on NQR) [502, 577]. The application of NQR in mineralogy is largely attributable to research work done by Penkov [522–557].

4.3.1 The Energy Levels Diagram and the Resonance Condition in NQR

Although NQR spectroscopy, like that of NMR, covers the same radio-frequency band of nuclear spectroscopy, and the qaudrupole splitting has already been discussed in the text dealing with the NMR, a marked difference exists between these two methods.

Nuclear levels in NQR (Fig. 74) are not magnetic levels of the nucleus, for here there is no magnetic field (the NQR set-up carries no magnet) producing orientation

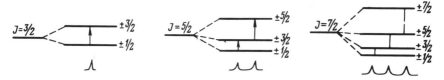

Fig. 74. Diagrams of quadrupole nuclear levels, arising under the effect of the crystal field gradient, and the NQR transitions for nuclei with spins $I = 3/2$ (As^{75}, Cu^{63}, Cu^{65}), 5/2 (Sb^{121}) and 7/2 (Sb^{123})

of the nuclear spin relative to this magnetic field and the resonance condition here does not provide for either the nuclear g_N factor or the nuclear magneton.

In NQR the interaction of the nucleus is interaction with the crystalline (electric) field, i.e., with the field which causes the appearance of optical absorption spectra and is described by the crystal field theory.

The quadrupole moment of the nucleus eQ interacts directly with this field. Consequently, the splitting of magnetic levels in NMR occurs as a result of the superposition of the external magnetic field, whereas in NQR the splitting of the quadrupole levels exists in the crystal itself, emerging in it under the influence of the crystalline field.

The interaction energy is calculated according to the formula

$$E_m = \frac{1}{4} eQq_{zz} \frac{3m^2 - I(I+1)}{4I^2 - I(I+1)},$$

where I is the nuclear spin; m is the nuclear magnetic quantum number; q_{zz} is gradient of the axial crystalline field (along axis z).

With $I = 3/2$ two quadrupole levels emerge in the axial crystalline field (see Fig. 74) with $m = \pm 3/2$ and $m = \pm 1/2$ and with energies

$$E_{\pm 3/2} = \frac{1}{4} eQq_{zz}; \quad E_{\pm 1/2} = -\frac{1}{4} eQq_{zz}.$$

The transition energy between levels $m = \pm 3/2$ and $m = \pm 1/2$ amounts to $eQq/4 - (-eQq/4) = eQq/2$ (selection rules: $\Delta m = \pm 1$). Hence the quadrupole resonance condition for a nucleus with $I = 3/2$ in an axial field is

$$h\nu = eQq_{zz}/2.$$

For a nucleus with the spin $I = 5/2$ three quadrupole levels with $m = \pm 5/2$, $\pm 3/2$ and $\pm 1/2$ energy in the axial field, producing two transitions. For the nucleus with the spin $I = 7/2$ there are four levels with $m = \pm 7/2$, $\pm 5/2$, $\pm 3/2$, $\pm 1/2$, and three transitions.

With the crystalline field deviating from the axial an asymmetry parameter $\eta = (q_{xx} - q_{yy})/q_{zz}$ is introduced in the expression for energy. For $I = 3/2$:

$$E_{\pm 3/2} = \frac{3}{4} eQq_{zz} \frac{\sqrt{1+\eta^2/3}}{4I(2I-1)} \text{ and } E_{\pm 1/2} = -\frac{3}{4} eQq_{zz} \frac{\sqrt{1+\eta^2/3}}{4I(2I-1)} ;$$

$$h\nu = eQq_{zz}/2 \sqrt{1+\eta^2/3} .$$

In the case of nuclei with spin $I = 5/2$ and higher (two and more transitions) both parameters eQq and η are determined from the correlation of the transition frequency, whereas in the presence of a single transition the spectra of the specimen in a superposed weak magnetic field are investigated for their determination.

The relationship between the behavior of quadrupole levels in the magnetic field and the NMR levels, complicated by the quadrupole interaction, is determined by a relative order of these interactions (see Figs. 61 and 74).

4.3.2 Quadrupole Nuclei and Requirements on the Study Substance

NQR spectra have been observed on the nuclei of $B^{10,11}$, N^{14}, S^{33}, $Cl^{35,37}$, $Cu^{63,65}$, $Ga^{69,71}$, As^{75}, $Br^{79,81}$, In^{115}, $Sb^{121,123}$, $I^{127,129}$, Hg^{201}, Bi^{209}.

For mineralogy the following requirements on the substance determining the choice of preferable nuclei are of essential importance: (1) local symmetry below the cubic one, i.e., a crystal field gradient should be observed; (2) a sufficiently high concentration of the nucleus, i.e., as in NMR this is a method for investigating basic components of the substance; the nuclei must have a high natural content, not less than 3%–10%; (3) low abundance of paramagnetic admixtures (Cu$^+$ with a filled $3d^{10}$ shell yields NQR signals, whereas the paramagnetic ions of Cu^{2+}, $3d^9$ are not investigated by this method). The specimens can be in the form of powers or monocrystals, their size about 0.2–2 cm^3.

The most convenient and important NQR studies of minerals proved to be the nuclei of three elements: As, Sb, Bi:

As^{75} (1000%)	$I = 3/2$	$eQ = +0.3$,
Sb^{121} (57.25%)	$I = 5/2$	$eQ = -1.3$,
Sb^{123} (42.75%)	$I = 7/2$	$eQ = -1.8$,
Bi^{209} (100%)	$I = 9/2$	$eQ = -0.4$,
Cu^{63} (68.9%)	$I = 3/2$	$eQ = -0.15$,
Cu^{65} (31.1%)	$I = 3/2$	$eQ = -0.15$.

Conditions for Observation of NQR Signals; Stationary and Pulse Methods. Since the NQR and NMR signals are in the same radio-frequency region (tens and hundreds MHz) there are many similar features in the equipment and procedures used in these two methods.

However, in NQR the resonance condition can be achieved only in one way, i.e., by changing the frequency. Then for a given isotope in different crystals the resonance frequency can assume values differing by tens and even hundreds of MHz and, therefore, the radio-frequency generator is re-adjusted within a band of ~ 2 to 1000 MHz.

The methods adopted in observation of NQR signals are divided into two groups [557]: stationary and pulse. With the first the radio-frequency field is generated continuously, with the second by short intensive pulses.

Most effective in investigating minerals are pulse methods, especially the method of the spin echo [557].

4.3.3 NQR Spectra Parameters

The principal parameter in NQR is the quadrupole interaction constant eQq. It is determined directly from the resonance frequency. Its meaning is the same as in Mössbauer spectra and in NMR. The crystal field gradient defined from it is composed of contributions made by two components: q_{val}, a contribution made by the deviation in the distribution of valence electrons from the spherical distribution, and q_{latt}, a contribution made by the distribution of the lattice charges.

In the As, Sb, Bi ($d^{10}s^2p^3$) atoms the s- and p-electrons are valent. Since the s-electrons have a spherical distribution, the value of q is basically determined in them by the number of imbalanced p-electrons. An analysis of the eQq constant enables one to judge on the sp-hybridization, on the ionicity-covalency, and on the degree of the structure distortion.

Additional information is obtained by measuring the spin–lattice relaxation time T_1 and the spin–spin relaxation time T_2, which can be most readily measured by the pulse methods, and also from measuring the width of the lines, which depends on the abundance of paramagnetic admixtures, defects and disorder in the distribution of the atoms in the structure.

4.3.4 Minerals Investigated and Data Obtained

Successes achieved by applying NQR in mineralogy are largely attributable to the fact that the As, Sb, and Bi nuclei form part of the major groups of sulfides and sulfosalts, for which it is difficult to obtain more complete information on the chemical bonding and features characteristic of their composition and structure by any other methods.

The studies covered the following minerals [491, 552–557]. As^{75}: realgar AsS, orpiment As_2S_3, lautite CuAsS, lorandite $TlAsS_2$, proustite Ag_3AsS_3, arsenolite, and clodelite As_2O_3; Sb^{121}, Sb^{123}: antimonite Sb_2S_3, chalcostibite $CuSbS_2$, miargyrite $AgSbS_2$, pyrargyrite Ag_3SbS_3, stephanite Ag_5SbS_4, bournonite $CuPbSbS_3$, franckeite $Sn_3Pb_5Sb_2S_{14}$, senarmontite, and valentinite Sb_2O_3; Bi^{209}: bismuthine Bi_2S_3, bismite αBi_2O_3, and sillenite γBi_2O_3; Cu^{63}, Cu^{65}: cuprite Cu_2O.

As to the characterization of chemical bonding, the NQR parameters appear to be the most important experimental substantiation for these groups of compounds (see [1]). The structural significance of the NQR data consists in the determination of nonequivalent positions (Fig. 75), of the order–disorder state and of the geometry of coordination groups. Important data are also obtainable on the forms of the impurities therein.

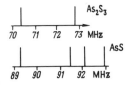

Fig. 75. Positions of NQR lines of two nonequivalent As[75] nuclei in orpiment, As_2S_3, and of four nonequivalent As[75] in realgar, AsS [552, 553]

5. Luminescence

5.1 Major Steps in the Development and the Present-Day State

Luminescence is common to an extremely wide range of objects of inorganic and organic nature, and synthetic materials and, for this reason, the mechanism of processes producing it is distinguished by great variety. Even the luminescence of such minerals as fluorite, sphalerite, and scheelite is caused by quite different types of processes.

A feature common to them is luminescence glow resulting from an emission transition of anion, molecule, or a crystal from an excited electronic state to a ground or other state with lesser energy. The emission transition is one of two possible ways of deactivating excited electronic states; the other being radiationless transitions resulting from interaction with the lattice or a transfer of energy to other ions.

The phenomenological aspects of luminescence find expression in its definition as a radiation excessive by comparison with thermal radiation (this distinguishes it from an equilibrium thermal radiation of heated bodies) and whose duration is much longer than the period of light vibrations, i.e., more than 10^{-10}s (this distinguishing it from other kinds of nonequilibrium glow, such as light scattering reflection, Čerenkov effect, and others, practically inertialess, with the time on the order of the light vibrations period of 10^{-15}s).

A unique property of luminescence which determines its applications is transformation of diverse kinds of electromagnetic and corpuscular emission, as well as of the electric, mechanical, and chemical energy into visible light.

In mineralogy, luminescence finds use for interpretation of, and as a method for investigating distinctive features specific to the composition and structure of minerals and for their genetic characteristics.

5.1.1 Applications of Luminescence in Mineralogy

A number of various factors, such as the progress of related branches (development of theories, discovery of new phenomena, construction of new devices and apparatus), an intensive technical and analytical application of luminescence, increasing of experimental possibilities, have all favored an extraordinary upswing in the significance of luminescence in the years 1950–1970, particularly in mineralogy, its deeper understanding as a phenomenon appearing in different concrete classes of compounds.

In this a supplement to the energy band theory of luminiscence, determining the crystallophosphors glow kinetics, were of prime importance the spectroscopic theories, and the crystal field and molecular orbitals theories. This affects not only the luminescence of predominantly ionic crystals, but also of typical crystal-lophosphors.

At the same time, kinetic theories of luminescence were being developed in great detail within the framework of the energy band theory.

The application of electron paramagnetic resonance to the study of photo-sensitive compounds helped to give concrete expression to the chemical nature of acceptor and donor levels in the band scheme of luminescence, to the part of the electron and hole trapping played in it.

Investigations into the stimulated emission of quantum generators—lasers and masers—provided a host of information on the fluorescence of rare earth ions, actinides, and some transition metal ions.

Extensive studies of EPR, optical spectra, and luminescence spectra of these ions in crystals yielded detailed schemes of the energy levels, and also highly circŭmstantial pictures presenting spectroscopic peculiarites of such model struc-tures as CaF_2, ZnS, $CaWO_4$, Al_2O_3, and others.

A factor such as an increased interest in lunar research stimulated an ap-pearance of a number of works on proton-luminescence, which had to serve as a model for possible lunar luminescence occurring under the influence of the solar wind. In addition to photo, cathode, and proton-luminescence also luminescence occurring under the effect of ions, X- and gamma-rays was subject to investiga-tions. Research works covered separate domains of electroluminiscence [652, 694], chemical luminescence [600], candoluminescence [821], luminescence of organic compounds (fluorescence), and bitumens (used in petroleum geology).

Fluorescent analysis has developed as a special method that finds application particularly in analyzing mineral raw materials [619, 790].

The construction of electron probe microanalyzers, and of a uv microscope made luminescence serves as one of the methods used conjointly with microscopy.

Technical applications furthered theoretical investigations of luminescence. At its root lay, in particular, conversions: (1) of invisible light into visible in lumines-cent lamps (phosphates, particularly of the apatite composition, with Sb and Mn, with Pb, Tl, Sn, willemite, wollastonite with Pb and Mn) and in compounds with long-term afterglow (ZnS-Cu); (2) electron excitation into visible light, in cathode-ray tubes (in radiolocation and television: ZnS and CdS with Ag); (3) X-ray emis-sion into visible light, in X-ray screens ($CaWO_4$, $BaSO_4$, ZnS, CdS, Ag); (4) nuclear radiation into visible light, in scintillators (alkali haloids with Tl).

These are joined by thermoluminescent dosimeters used in particular for reg-istration of nuclear explosions and installed on satellites ($CaSO_4$: Mn, LiF: Mn).

All these factors have created a situation propitious for an entirely new miner-alogical significance of luminescence. A primary pre-requisite for this is changing from qualitative observations to obtaining luminescence spectra. Photorecording at low temperatures makes it possible to obtain spectra with a good resolution of fine structure. The second condition is an accurate interpretation of lines and bands in the luminescence spectra and their assignment to definite electron transitions.

Under these conditions one can determine a whole set of impurity ions, mole-cules, and centers, causing the appearance of individual lines and bands, the valence of the ions, their coordination, the presence of the ions in nonequivalent positions, their local symmetry, the ways of compensating the charge, features related to the presence of other ions (stabilizers, quenchers, or ions promoting emission from some or other excited level), to the entering of the impurity ions in the structure of a given crystal and to conditions attending their formation (redox conditions, temperature and pressure, deformations), as well as to their radiation history.

When employed in conjunction with EPR, optical spectra, and an analysis of the electron-hole center formation, luminescence furnishes a good deal of important information for various comparisons of minerals, and for elucidating the supplementary characteristics of their genesis.

5.1.2 Major Steps in the History of Luminescence [599, 651, 687, 691, 717]

The phenomenon of luminescence has been known from ancient times in different civilizations. Mentions of the glow which may be attributed to bio- and chemoluminescence can be found in the Indian *Veddas* and in the Chinese "*Book of Odes* (second millenium B.C.), and in antique Greece and Rome. The first records of mineral luminiscence can be gleaned from works by Herodotus, Aristotle, Theophrastus, Strabo, and Pliny. It seems that Aristotle (384–323 B.C.) introduced the concept of glow of nonthermal origin. However, it is only guesswork as to what minerals were meant in these descriptions.

The first definite mention of mineral luminescence should be considered the glow of heated diamond (i.e., thermoluminescence) described by A.Magnus in 1280.

The preparation of the first luminescent material dates back to the sixteen hundreds. When searching for the philosopher's stone, Vinchenzo Cashporolo, alchemist from Bologna, obtained from the natural barite powder burned together with charcoal the "Bologna stone" which shone in the darkness. Through G.Galileo this sample came into the hands of Professor G.Ch.La Galla from Rome, who described it in the first publication on artificial phosphors.

In the 17th, 18th and 19th centuries, countless observations of natural and artificial phosphors were made.

Starting from the middle of the 19th century, luminescence became subject to systematic investigations: spectroscopic investigations were initiated by works of G.Stokes (1852) and kinetic ones (the laws of luminescence decay) by works of E. Becquerel (1859). The preparation of synthetic phosphorent wurtzite (T.Sideau, Paris, 1866) and then the synthesis of other sulfide phosphors provided luminescent materials that were being investigated for over a hundred years. Unlike D.Bruster and G.Gershel, who related the glow of solutions and fluorite to scattering of the light, G.Stokes established for the first time that the blue glow of fluorite comes as a secondary emission, which he called fluorescence.

The introduction of the phosphoroscope (1879) by E.Becquerel, and its improvement by E.Wiedeman (1888), who proposed the term "luminescence" and gave its definition, marked the beginning of investigating the laws of decay, which until the 50s of the 20th century were one of the major trends in experimental and theoretical research. It stems from the fact that phosphors of the ZnS, CdS type, subjected to most intensive studies, present very broad and structureless absorption and emission bands, whereas for decay it was possible to obtain quantitative data and simple relations.

Empirically, the laws of decay were established by E.Becquerel (1860) in the form of two different formulas: for compounds with a short-lived (< 1 s), and long-lived afterglow. Attempts at a theoretical substantiation of these formulas depended on the concepts of the mechanism of luminescence, and on the progress in other branches of physics.

At first, influenced by the progress made in thermodynamics and the mechanical theory of heat during the second half of the 19th century, these laws were formulated by comparing the mechanism of luminescence with the very different cooling of a heated body (E.Becquerel and then his son A.Becquerel, who, when looking into luminescence of potassium-uranyl-sulphate, discovered the phenomenon of radioactivity).

On the appearance of theories on ion activators of luminescence (M.A.Verneuil in 1886 working on ZnS with Mn, Cu), and influenced by the development of physico chemistry in the 90s of the 19th century, it was suggested that luminiscence be interpreted (the law of decay) as a dissociation into ions with subsequent recombination of the activator's molecules (E.Wiedeman and H.Schmidt, 1889–1912), based on the methods of physicochemical kinetics ("bimolecular" theory).

Following successful applications of Bohr's quantum theory in atomic spectroscopy, the P.Lenard's "theory of centers", according to which absorption and emission take place within the "centers", i.e., ion activators and ions of the ground substance, was given further justification as a transition of the "center" electron into an excited state and the reverse transition to the ground state attended by emission ("monomolecular" theory).

However, the lack of clear ideas as to the nature of the centers and use of Bohr's concepts in their simplest form without taking account of features specific to the spectroscopy of atoms in a crystal, and an attempt to apply this theory as a sole explanation for all kinds of crystals, and particularly of crystallophosphors by disregarding their kinetic specificity failed.

Present-day research into luminescence started from the middle of the 30s of this century. In works of S.I.Vavilov, V.L.Levshin, and others, two major mechanisms of luminescence are distinguished: (1) the glow of discrete centers ("monomolecular" mechanism; the absorption and emission processes proceed within the limits of individual ions and molecules in crystals; spontaneous and induced luminescence); (2) the glow of crystallophosphors ("bimolecular" mechanism); the absorption of light is accompanied by internal ionization and photoconduction; recombinational luminescence.

At the same time the band theory of luminescence was developed. By the 50s the treatment of principal positions in this theory was completed, and in the years 1950–1960 was being continually applied in interpreting the glow of crystallophosphors (especially of the ZnS type, and also of alkaline earth sulphide phosphors, those of alkaline haloids, silicates of the willemite type, phosphates of the apatite type, etc.).

These investigations, however, were subject to two basic limitations. First, they remained empirical in that the nature of the "electron capture" levels was not considered (but it is exactly the nature of the glow centers that presents the greatest interest for mineralogists); second, investigations covered mainly crystallophosphors and not crystals with discrete centers (whereas for mineralogists the luminescence of sphalerites is a much less important case than is the luminescence of discrete centers, such as Mn^{2+}, RE^{3+}, etc. in silicate, carbonate, phosphate, and other minerals).

Nearly all the monographs published so far on luminescence consider a limited number of compound types, mainly crystallophosphors.

Systematic research into the luminescence of discrete centers began in the 50s of this century, when optical spectroscopy of solids received new impetus by the use of the crystal field theory. The luminescence of rare earth ions in CaF_2 and other fluorides of the alkali earths had become the best- studied type of the luminescence of discrete centers. The discovery and successful applications of laser and maser served as a starting point for the most interesting and highly detailed investigations into the spectroscopic features (including also luminescence) of laser crystals [708], to which, for example, such systems as $CaF_2:RE^{3+}$, $CaF_2:RE^{2+}$, $CaWO_4$: RE^{3+} belong. During the decade that followed, the scope of research work done in this direction became comparable to the lengthy period of investigations covering the luminescence of crystallophosphors.

Spectroscopic investigations covered, for instanse, also such systems as $CaCO_3$: Mn^{2+}, $ZnS:RE^{3+}$, ZnS-Cu (the latter in works of Birman [610] with calculations made within the framework of the molecular orbitals theory). Various luminescence theories began to take shape (591, 657, 658, 664, 668, 687, 752, 753, 797, 798, 817].

To sum up the major traits and points in the history of the luminescence study:

1. Extensive investigations were carried on in the 17th–19th centuries, but an insurmountable obstacle prevailed: unlike other physical phenomena known for a long time, luminescence belongs to the phenomena of the quantum nature and of all the quantum phenomena luminescence was the first to be observed and studied experimentally long before Bohr, but could not be comprehended.

2. The current understanding of the nature of luminescence was due to a transition from the general quantum laws to a combination of minutely elaborated theories: spectroscopic (systems of energy levels), on the one hand, and semiconducting (kinetics) processes, on the other.

3. The history of luminescence in general may be considered only in its earliest stages, since already by the end of the 19th–beginning of the 20th century, it breaks up, forming histories of individual branches, such as luminescence of solids, organic compounds, fluorescent analysis, chemi,-electroluminescence, crystallophosphors, luminescence of rare earths, etc. In its turn, each branch of the research history breaks up into concrete sections, concrete systems and relationships, and for each of them a more of less extensive literary coverage exists. Therefore, from history one has to turn to selection of references, and even in these the researcher must often limit himself to the latest publications in which previous works are listed.

From ancient times minerals have been a convenient object for investigating the nature of luminescence, and to this day the attention of physicists is focused on them (natural spinel, fluorites, sphalerite-wurtzite, and others). In the last decades, research has been carried on almost exclusively on artificial analogs of the minerals CaF_2, ZnS, $CaWO_4$, $CaMoO_4$, $PbMoO_4$, $Ca_3(PO_4)_2ClF$, Zn_2SiO_4, and others. In mineralogical research on luminescence as such the following mutually overlapping stages are distinguished.

1. Detection of luminescent minerals and qualitative characterization of their luminescence. Utilization of luminescence in identification of ore and rock-forming minerals [604, 637, 730].

2. The band theory that dominated in the understanding of luminescence up to 50s to 60s of the 20th century could not shed sufficient light on problems concerning the nature of the mineral luminescence of interest to mineralogists, while spectroscopic theories of luminescence began developing only in the 50s to 60s. For some minerals, however, detailed empiric characteristics have been obtained, the luminescence spectra being assigned more or less convincingly to different ion activators (particularly to rare earths yielding characteristic line spectra). To these belong reviews by Przibram [791].

3. Current works on luminescence of minerals with their spectroscopic characteristics and interpretation [680–683, 785–788, 830–834] uncovered new potentialities for luminescence in mineralogy. However, they have also shown that much still has to be done to achieve a sufficiently complete investigation of luminescence in almost all minerals.

In mineralogy luminescence appears to be merely a method of determining the nature of the activator, of its possibly complete characteristics, as well as that of its host crystal. At the same time, low and extremely low contents of ion activators, whose presence it is often difficult to ascertain in any other way, the dependence of features specific to the glow on the presence and distribution of other admixtures in the structure, as well as of other defects make luminescence serve as a specific feature of the mineral in a given complex, and this may be made use of to establish various correlations, and as an indicator of the formation mode.

5.2 General Concepts, Elementary Processes, Parameters

In diverse classes of compounds even individual elementary processes making up the phenomenon of luminescence differ widely. It is rather difficult to set apart characteristics common to the whole of the phenomenon, for in their concrete definition one has to deal with multiform mechanisms of the processes.

Obviously, the systematics of the luminescence types is needed first of all. Classification according to the methods of excitation (photo, cathode, X-ray luminescence etc.) or to the duration of glow (fluorescence and phosphorescence) is based on external features that are of secondary importance.

More rational is the systematics based upon the substance of the phenomenon, on the mechanism of processes of which it is made. Such systematics, moreover, matches the accepted types of luminescent systems (rare earths in crystals, transition metal ions, semiconductor compounds of the ZnS type combined with various activators, etc.). Below, general information on and the systematics of the luminescence types are considered, while the luminescence of concrete systems is discussed in the next section (see Sect. 5.3).

5.2.1 Theoretical Bases Necessary for Understanding the Processes of Luminescence

With the application of the band scheme to interpretation of luminescence, and especially with the application of the crystal field theory, it has become clear that the theory of luminescence is based on more general theories of solid-state physics.

Two groups of theories on which the description of luminescence is based are distinguished: spectroscopic and semiconductor. To the spectroscopic theories may be referred the crystal field, molecular orbitals and energy band theories. The energy levels and energy bands considered within the framework of such theories serve as a basis for interpretation of the luminescence spectra of ions and crystals to the extent to which the same levels and bands describe optical absorption spectra of ions and crystals. Diversity of types of luminescence in crystals is of the same nature as are the differences in the absorption spectra (crystal field spectra, charge transfer spectra, spectra of rare earths, transition metals, etc.). Data on the absorption spectra of concrete crystals make up a good part in the description of the luminescence centers. In addition to the energy levels patterns subject to consideration are shifting of levels (in the model of configuration curves), emission and radiationless transitions, the duration of glow, etc. (see below).

Concepts based upon the band theory of semiconductors are necessary for interpreting those types of luminiscence when there occurs not just a simple excitation of impurity or principal ions of a crystal, but their internal ionization, i.e., a detachment of an electron, its entering the conduction band and transfer to other ions or defects of the crystal. These theories explain the kinetics of the crystallpohosphors' glow, the ionization phenomena, the electron-hole processes, and photoconduction.

In nearly all reviews on crystal luminiscence the presentation of the theoretical basis of luminescence is limited to the band theory and glow kinetics related to it, of practical importance for crystallophosphors. Then the nature of the glow centers, of defects with which various levels are associated remains unknown.

If such empirical schemes help explain the kinetics of the glow of artificial phosphors, for minerals this is quite insufficient. First, it is exactly the nature of impurities and defects that most interests the mineralogists, and second luminescence of the majority of minerals is determined by the patterns of the ion activator energy levels, described within the framework of spectroscopic theories, and in many cases in no need of reference to the band theory. In the 60s investigations of lasers stimulated spectroscopic research into luminescence.

Recently it has been shown that even in the case of the type ZnS phosphors, complete interpretation is possible by combining the band theory with the ligand field theory [610].

5.2.2 Absorption, Luminescence, Excitation Spectra: the Scheme of the Experiment and Energy Levels

The phenomenon of luminescence can be conveniently considered by comparing it with optic absorption (Fig. 76).

Absorption Spectra. When illuminating a crystal with monochromatic light with a varying wavelength, the ions with unfilled shells that make part of crystals (or solutions) pass from the ground to the excited state. For instance, the Cr^{3+} ion whose levels are shown in Figure 76 passes from the ground state $^4A_2$ to the excited states ^2E, $^4T_2$, $^4T_1$. This is attended by the appearance of an absorption band in the optical spectrum.

Luminescence Spectra. Ions in a crystal that have passed into an excited state return to the ground state (see Fig. 76) through radiationless transitions (from levels $^4T_1$, $^4T_2$) or by way of emission transitions (from level ^2E). Emission transition is seen as a glow of the crystal, and is registered in the form of a band in the luminescence spectrum. Unlike the absorption spectrum (Fig. 76), where three bands are observed which correspond to transitions to three energy levels of Cr^{3+} this luminescence spectrum has only one single narrow emission band. This is due to the fact that here broad $^4T_2$ and $^4T_1$ levels are not emissive. The energy absorbed by the ion, causing transitions to these and other levels, breaks up through radiationless transitions till it reaches emission level ^2E.

In Figure 76 the process of excitation is shown for the case of the ion in a crystal. It is possible in the same manner to demonstrate excitation attended by a transition from the ground levels in the scheme of molecular orbitals or in the energy band scheme. In the case of excitation with X-rays, electrons, protons, etc., more complex phenomena occur beyond the bounds of the ion (ionization, trapping of electrons and holes, their displacement, shifting of ions, etc.), but in the end all this comes back to the same emission levels.

The position of the band in the luminescence spectrum does not depend on the method of excitation, being determined by the interlevel spacing (but this or that band may become manifest in the spectrum or be absent, depending upon the mode of excitation). In Figure 76b a single emission band is shown. In the diagram of energy levels, however, there can be several emission levels, and transitions from each one of them can be not only to the ground state, but also onto the intermediate levels. The ground state can be split into sublevels and then transition from each emission level will be onto the ground state sublevels. This can result in the appearance of complex luminescence spectra consisting of many bands, this being characteristic of lantanides and actinides.

Excitation Spectra. The first part of the arrangement for observation of the luminescence excitation spectra (Fig. 76c) is the same as in the absorption spectra (Fig. 76). In point of fact a similar diagram is representative of the luminescence spectra shown in Figure 76b, where excitation can be induced by monochromatic light in the band of the ion absorption, or by means of UV radiation with a spectrum broad enough to overlap the ion absorption bands. Ultraviolet radiation is necessary to excite the ion up to energy levels lying above the emission level, which is usually in the visible or near IR region.

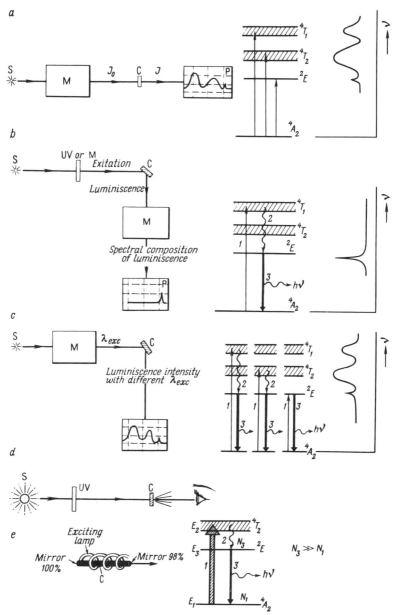

Fig. 76 a-e. Absorption, luminescence, excitation: schematic presentation of the experiment, energy levels, spectra. **a** Absorption spectrum; **b** luminescence spectrum; **c** excitation spectrum; **d** visual observation; **e** laser emission. *S*, light source; *M*, monochromator; *C*, crystal (sample); *R*, registration of spectra, UV-photofilter; *1*, excitation; *2*, radiationless transitions; *3*, emission

The difference in Figure 76a, b, and c consists in the registration of phenomena occurring due to excitation. In the instance of emission spectra subject to analysis it is the spectral composition of glow, i.e., determination of the ion level from which radiation occurs.

In excitation spectra one can observe a glow as a result of the transition from the sole emission level in this example, E. However, here it is not the luminescence spectrum that is registered, but its overall intensity which depends upon the absorption band in which excitation of the ion occurs. Therefore, the excitation spectrum is a replica of the absorption spectrum of the same ion (cf. Fig. 76a with c) but the position of the ion's levels is determined by the intensity of the glow, and not by the intensity of radiation that passed through the crystal.

The excitation spectrum shows that for an effective luminescence in the crystal to manifest itself not only is the presence of emission levels is necessary, but also of upper levels with a sufficiently intensive absorption. This also accounts for the variance in luminescence intensity with different modes of excitation, in the near or far UV or in the region of a concrete absorption band. The excitation spectra are made use of not only for choosing the region of luminescence excitation, but also for determining the pattern of the ion's energy levels, when its concentration is too small to observe the absorption spectra. An example illustrating the excitation and luminescence spectra is offered in Figure 77.

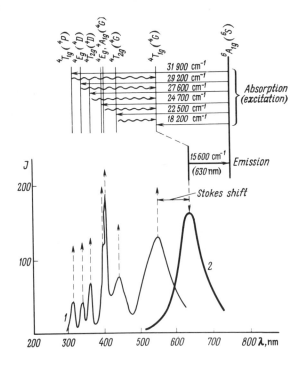

Fig. 77. Excitation (*1*) and emission (*2*) spectra of Mn^{2+} in calcite [763] and energy levels diagram for Mn^{2+} ion in calcite; assignment of the observed transitions

5.2.3 Energy Level Patterns and Configuration Curves Diagrams

By using an energy levels diagram, as in Figure 76, three elementary processes proceeding between the ion's states are described: (1) excitation (absorption), (2) emission, (3) radiationless transitions. If interaction between the ion and the lattice is weak, the diagram of levels, i.e., the same from which the absorption spectra are defined, will suffice to describe the luminescence. Thus for rare earth ions RE^{3+}, whose f-electrons are shielded against the interaction with the lattice, one and the same diagram of energy levels is used in assigning emission and absorption transitions that take place here naturally, with an equal energy, i.e., the wavelength of the emission band matches that of the absorption band.

Should the interaction of the ion and the lattice prove to be fairly strong, the energy levels diagram defining the absorption spectra will not suffice to describe the ion's luminescence. Thus for transition metal ions, or those of the Tl^+, Pb^{2+} type, F-centers, and other systems a shift toward the long-wave side of the emission band in relation to the corresponding absorption band is observed. This relation is known as the empirical Stokes law. In such cases one should, apart from the above-mentioned three elementary processes: excitation, emission, and radiationless transitions, also take acount of yet one more, i.e., the interaction of the ion with the lattice ions surrounding it.

Excitation and emission are separated by a time interval, i.e., by the lifetime of the excited state (of a given level). The time interval separating absorption and emission suffices for the system "ion-lattice" to adapt itself to the given excited electronic state that manifests itself in shifting of the surrounding (ligand) ions to a new equilibrium position with a new distance between the luminescent ion and other ions surrounding it and in transition to the lowermost of the vibrational levels that correspond to the given electronic level.

Since there is a change in the interionic distance, it becomes necessary to introduce yet one more coordinate, known as the configuration coordinate, the one that reflects the disposition of ions, or more exactly the average distances between ions in each electronic state. The dependence of the energy of these states on the distance between ions is described by configuration curves. The minima on these curves correspond to the equilibrium distances for the given state. Therefore, in place of a unidimensional diagram of energy levels (in the form of straight lines) we shall change over to a bidimensional pattern (energy-interionic distance) of the same levels in the form of configuration curves (Figs. 78, 79).

These curves may be calculated theoretically or plotted semiquantitatively from absorption spectra (in Fig. 78d the energies of transitions with absorption are obtained from the absorption spectra, from emission spectra, and from diagrams of the position of the energy levels versus crystal field intensity Dq (see Fig. 78b). The positions of the configuration curves minima corresponding to the mean distance $Mn^{2+} - O$ are different for each state: the distribution of the electron density Mn^{2+} varies in the states of $^6A_1$, $^4T_{1g}$, $^4T_{2g}$ and, therefore, ionic radii of Mn^{2+} and, consequently, distances $Mn^{2+} - O$ in these states are dissimilar. The disposition of the configuration curves minima (see Fig. 78d) is associated with the slope of the energy levels, and depends on the intensity of the crystal field Dq (see Fig. 78b). The ^4E_g,

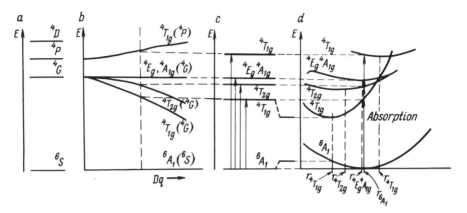

Fig. 78 a-d. Configuration curves of ion in a crystal **a** free Mn^{2+} ion; **b** Mn^{2+} ion energy levels in a crystal as a function of the crystal field strength Dq; **c** Mn^{2+} ion energy levels for a given Dq value; **d** change in the electron state depending upon distance r Mn^{2+}—O in different states

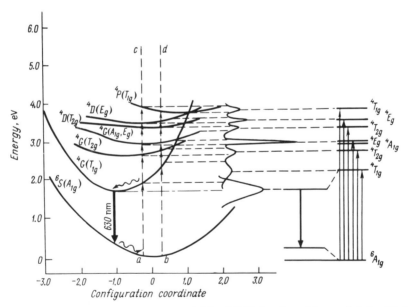

Fig. 79. Configuration curves model for Mn^{2+} ion in calcite explaining Stokes shift of the absorption band (after [765])

$^4A_{1g}$ levels run parallel to the ground state level $^6A_1$ (i.e., are independent of Dq), and the position of the ^4E_g, $^4A_{1g}$ curve minimum coincides with the position of the ground state minimum. The $^4T_{1g}$ (^4P) level is inclined (see Fig. 78b) upwards from the axis of abscissa, the $^4T_{2g}$ and $^4T_{1g}$ (^4G) levels are tilted downward, of these $^4T_{2g}$ sloping more gently than $^4T_{1g}$ (^4G). Accordingly, the minima of these curves (see Fig. 78d) lie on different sides from the ground state, and $^4T_{1g}$ (^4G) farther than $^4T_{2g}$ (^4G). The stronger the interaction of the ion with the lattice, the more the position of levels depends on Dq, the greater shifting of ions when passing into the excited state, the Stokes shift, and the width of the line. This is why in the emission spectra of the rare earth ions (RE^{3+}) there is no Stokes shift and the lines are narrow, while in the Mn^{2+} spectra the Stokes shift is large and the bands broad.

The sequence of events in luminescence is determined by the time during which individual elementary processes occur (Fig. 80). Excitation (absorption) associated with changes in the states of electrons whose mass is much smaller than that of the nucleus proceeds faster ($\sim 10^{-15}$ s) than the displacement of the nuclei can be accomplished. However, the time of staying in the excited state, i.e., the lifetime of the level (on the order of 10^{-8} and more s) is sufficient to produce displacement of the nuclei as well as for vibrational transitions ($\sim 10^{-12}$ s). This is followed by emission.

Thus the whole process leading up to luminescence may be represented separated in time into four steps A → B, B → C, C → D, D → A (see Fig. 80).

Different types of configuration curve relationships explain the cause of the appearance of luminescence with the Stokes shift (Fig. 81a) or without it (Fig. 81b) and also of radiationless transitions (Fig. 81c).

The temperature dependence of luminescence is determined (see Fig. 80) by intersection of the configuration curves and the value ΔE, the energy difference between the configuration curve minimum and its intersection with the subjacent state: the rise of temperature is conducive to radiationless transition B′→ C → B and to falling intensity of emission.

Figure 80 shows vibrational levels for the ground and the first excited electronic state (horizontal lines n_{gr} and n_{exc}). The distance between them is on the order of $100 \, cm^{-1}$. Vibrational transitions are attended by emission of photons, which manifests itself at low temperatures in the form of a vibrational structure of broad bands or in the vibrational satellites of narrow emission lines (in the case of Cr^{3+}, for instance) and at room temperature in broadening of the bands and emission lines. Transitions from the excited state with $n_{exc} = 0$ to the ground state on the vibrational level $n_{gr} = 0$ are called zero-phonon or nonphonon transitions.

5.2.4 Kinetics of Ion Luminescence in a Crystal; Fluorescence and Phosphorescence

Processes resulting in luminescence are described on the ground of a scheme of the ion energy levels in a crystal. However, in order to explain shifting of the emission band with respect to the absorption band (Stokes shift) or the absence of such shifting, and also to account for the mechanism of emission and radiationless transitions, it proved necessary to change over from the routine scheme of levels describing absorption spectra to a scheme of configuration curves.

Further, to underatand why some ions (with corresponding pattern of energy levels) are luminescent, and the others are not, why some levels of these ions are

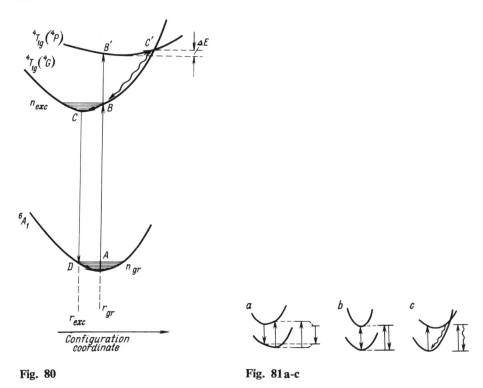

Fig. 80   **Fig. 81 a-c**

Fig. 80. Absorption and emission in a configuration curves model depending upon distance between ions in the ground and excited states, on an example of Mn^{2+} ion. A-B, Absorption as a result of transition from ground state $^6A_1$ to excited state $^4T_{1g}$ with transition energy E_B-E_A; during this transition ($\sim 10^{-15}$ s) position of 0^{2-} ions around Mn^{2+} has not enough time to change. B-C, Relaxation. Excited state $^4T_{1g}$ (^4G) has a lifetime on the order of 10^{-8} s, i.e., absorption A-B and subsequent emission C-D are separated by this time interval; during this time the 0^{2-} ions surrounding the excited Mn^{2+} ion with changed wave function shift from the equilibrium position that corresponds to ground state $^6A_1$ with distance between Mn^{2+}-0^{2-} ions (r_{gr}) to an equilibrium state corresponding to excited state $^4T_{1g}$ (^4G) with different distance Mn^{2+}-0^{2-} (r_{exc}), (i.e., r_{gr}-r_{ex}); there is enough time for vibrational transitions B-C (vibration period on the order of 10^{-12} s) into a state with minimum vibration energy (n_{exc}: vibration sublevels of the electron level $^4T_{1g}$). C-D, Emission resulting from transition $^4T_{1g} \rightarrow {}^6A_1$, with energy E_C-E_D. D-A, Vibrational transition to a lower vibrational state of ground level and return of ions to equilibrium position. C-B, Radiationless transition from state $^4T_{1g}$ (^4P) to $^4T_{1g}$ (^4G); (the energy difference $\varDelta E$ is small).

Fig. 81 a-c. Three principal types of configuration curves interrelations. **a** Luminescence with Stokes shift, emission band displaced toward longwave side by comparison with absorption band (for example Mn^{2+}); **b** luminescence without Stokes shift: emission and absorption lines coincide (for example, Cr^{3+}, RE^{3+}); **c** radiationless transition (luminescence quenching)

emissive, and the others radiationless, on which factors the intensity and duration of glow depend, one has to take into consideration an additional characteristic of the levels, i.e., their lifetime.

Thus, approaching an increasingly complete description of luminescence, we pass through the following stages: energy levels, configuration curves, and lifetime of the levels.

In the case of absorption spectra, reaching the excited state in transition from the ground state is the final stage of our studies, and the further fate of this excited state is left out of consideration. The scheme of energy levels in no way takes account of the duration of the existence of excited levels, and of their possible decay. The lifetime of the ground state from which, and only from which all transitions with absorption are made is infinite. Conversely, in the case of luminiscence the excited states are initial states from which transitions attended by emission take place. The characteristics of these excited states determine features specific to the luminescence of ions.

The major characteristic of excited states is their lifetime, i.e., the average stay of the ion in a given excited state.

The lifetime of a level τ_m is related to the probability A_{mn} of transition from level m to the level n and, through its mediation, to the oscillator strength for absorption band f_{nm}. Relations among τ, A, and f vary in the case of a free ion and in that of the ion in a crystal (Fig. 82). Here A and B are Einstein's coefficients of spontaneous emission (A) and absorption (B) [256].

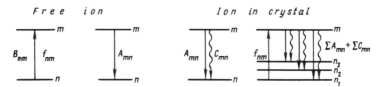

Fig. 82. Probability of transitions between levels $n \rightarrow m$ and $m \rightarrow n$ and the m level lifetime. *In free ion and in crystal*: B_{nm}, probability of transition with $n \rightarrow m$ absorption; f, oscillator strength for absorption band $n \rightarrow m$; A_{mn}, probability of emission transition $m \rightarrow n$; $\sum A_{mn}$ is sum total of probabilities for emission transitions $m \rightarrow n_1$, $m \rightarrow n_2$, $m \rightarrow n_3$; $A_{mn} = (K. \lambda^2) f_{mn}$. *In a free ion*: τ, life time of level m; $\tau = 1/A_{mn}$. *In an ion in crystal*: C_{mn}, probability of radiationless transition $m \rightarrow n$; C_{mn}, sum total of probabilities for radiationless transitions $m \rightarrow n_1$, $m \rightarrow n_2$, $m \rightarrow n_3$; τ_m, lifetime of level m; $\tau_m = 1/A_{mn} + 1/C_{mn}$ or $\tau_m = 1/\sum A_{mn} + \sum C_{mn}$; η, luminescence quantum yield; $\eta = \sum A/ (\sum A + \sum C)$

In a free ion, lifetime τ_m of excited state m is inversely proportional to the probability of A_{mn} transition from this level, i.e., $\tau_m = 1/A_{mn}$. The probability of A_{mn} transition is linked with the oscillator's strength f_{mn} through relation [256]

$$A_{mn} = \frac{8\pi^2 e^2 \nu^2}{3m_e c^3} f_{nm} \text{ or } f_{nm} = \frac{3m_e c^3}{8\pi^2 e^2 \nu^2} A_{mn}.$$

Substituting here atomic constants e and m_e, i.e., the charge and mass of the electron, velocity of light c ($3 \cdot 10^{10}$ cm/s^{-1}) and replacing frequency ν with wavelength $\lambda = c/\nu$, we obtain (λ in cm, A in s^{-1}).

$$A_{mn} = 0.22 \frac{\nu^2}{c^2} f_{nm} = 0.22 \frac{1}{\lambda^2} f_{nm} \text{ or } f_{nm} = 4.5 \frac{c^2}{\nu^2}.$$

Hence, for characteristic values of the oscillator strength we obtain the values for the probabilities of A_{mn} transition (in s^{-1}) and linked with them lifetimes $\tau_{mn} = 1/A_{mn}$ (in s). If transition from given level m occurs not over a single but over several subjacent levels, the lifetime then determines the complete probability of transition onto all these levels with $\tau_m = 1/\sum A_{mn}$ (see Fig. 82b). The lifetime is a characteristic of level m, and does not depend on the final state into which the ion passes, while the probability of transition, as well as the oscillator strength, features individual transitions.

Interaction of the ion in a crystal with ions (ligand) surrounding it (i.e., with "lattice") manifests itself in a greater or smaller relative amount of radiationless transitions. The probability of radiationless transitions is the probability of luminiscence decay (C in Fig. 82). The lifetime of the ion's levels in a crystal is then reduced due to radiationless transitions additional to the emission ones

$$\tau_m = \frac{1}{\sum A_{mn} + \sum C_{mn}}$$

This lifetime determines an experimentally measured value of the duration of glow (or afterglow continuing after excitation has been stopped). Excitation is accomplished by means of a photoflash lamp, the light flash lasting on the order of a few μs, the duration of afterglow being measured by using tautometers, fluorometers, and phosphoroscopes [620]. If with usual values of oscillator strength $f = 10^{-6} - 10^{-2}$ the glow time of free ions lasts from 10^{-2} to 10^{-6} s, the presence of radiationless transitions in a crystal reduces these values still further. For example the measured glow time for Sm^{2+} in CaF_2 is $2\mu s$ ($2 \cdot 10^{-6}$ s), for Nd^{3+} in $CaWO_4$ 130 μs ($1.3 \cdot 10^{-4}$ s), for Cr^{3+} in Al_2O_3 3000 μs ($3 \cdot 10^{-3}$ s).

The fraction of emission transitions ($\sum A_{mn}$) with respect to the sum total of emission and radiationless ($\sum A_{mn} + \sum C_{mn}$) transitions is defined as quantum yield η:

$$\eta = \frac{\sum A_{mn}}{\sum A_{mn} + \sum C_{mn}}.$$

In different wording, the quantum yield is defined as a ratio of the number of photons emitted from a given level to the number of photons excited to a given or other superjacent levels, i.e., to the number of photons absorbed by a given or overlying levels. The spectrum of the luminiscence excitation (see Fig. 77) obviously represents a dependence of luminiscence yield versus wavelength at which the excitation is effected.

By apparent quantum yield is meant the ratio between the number of photons emitted to the number of photons incident on the crystal. To pass from an apparent to the true yield, one has only to take account of the proportion of energy absorbed by a given ion with respect to the whole of the incident energy

$$E_{abs} = (1 - R)\frac{K - K_0}{K}(1 - e^{-Kt}),$$

where R is the crystal reflection coefficient; K and K_0 are absorption coefficients of the ion in a crystal and of pure crystal, respectively; t is the crystal's thickness.

The relative luminescence yield is often defined as a ratio of the luminescence brightness to the energy absorbed by the ion in a crystal. Finally, sometimes instead of the quantum yield, the energy yield is indicated, that is, the ratio of luminescence energy to absorbed energy. Let us consider the relationships among oscillator strength f, the probability of emission and radiationless transitions A_{mn} and C_{mn}, lifetime τ_m and quantum yield n on an example of Eu^{3+} and Cr^{3+} luminescence.

In the Eu^{3+} emission spectrum (Fig. 83) not a single band, but a series of narrow lines arising during transitions from the $^5D_0$ emission level to all the lower ones are in evidence. It is possible, however, to contrast these emission transitions against oscillator strength only for a single pair of levels, the initial $^5D_0$ and the ground levels $^7F_0$ (the $^7F_1$ and other levels are unoccupied and from them no transitions with absorption occur). From the measured value (based on the absorption spectrum) of $f(^7F_0 \rightarrow {}^5D_0) = 1.4 \cdot 10^{-10}$ is found that of $A(^5D_0 - {}^7F_0) = 0.22\ 1/\lambda^2 f = 0.027\ s^{-1}$ [801].

The probability of emission transitions from $^5D_0$ is determined not only by this $^5D_0 \rightarrow {}^7F_0$ transition, but by the sum total of probabilities for $\sum A$ transitions to all the lower levels. This sum total is found from the relation $I_0 : I_1 : I_2 \ldots = A_0 : A_1 : A_2 \ldots A(^5D_0) = 67.5\ s^{-1}$.

Should the radiationless transitions be lacking this, the $\sum A(^5D_0)$ value would have determined the lifetime of the $^5D_0$ level. However, measurement of the afterglow time gave $\tau(^5D_0) = 1.87 \cdot 10^{-4}$ s, or $1/\tau = 5350\ s^{-1}$ as against $\sum A = 67.5$ s^{-1}. Th large difference between these values is due to the probability of radiationless transitions $C\ (^5D_0)$:

$$\frac{1}{\tau(^5D_0)} = A(^5D_0) + C(^5D_0);\ 5350\ s^{-1} = 67.5\ s^{-1} + C(^5D_0);$$

$$C = 5282.5\ s^{-1}.$$

Because of a very large fraction of radiationless transitions quantum yield η amounts to only 1 %:

$$\eta = \frac{\sum A(^5D_0)}{\sum A(^5D_0) + \sum C(^5D_0)} = \frac{67.5}{5350} = 1.25\%.$$

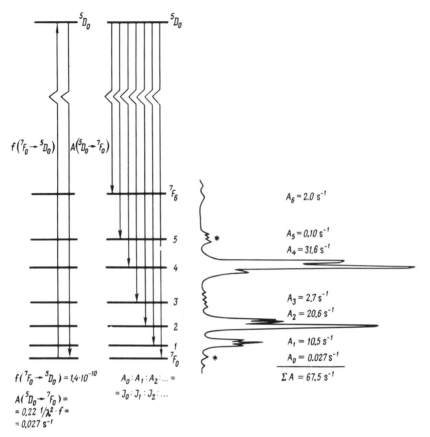

Fig. 83. Energy levels diagram and fluorescence spectrum of Eu^{3+} in $Eu_2(SO_4)_3\, 8(H_2O)$ after [801]; (for determining fluorescence quantum yield). f, oscillator strength; A, probability of emission transitions; I, integral emission intensities (* intensity scale enlarged ten-fold)

In the emission spectrum Cr^{3+} in Al_2O_3 (see Fig. 76 and Sect. 5.3) for line R (~ 693 nm) oscillator strength $f \approx 10^{-6}$, life time (varying as a function of Cr^{3+} concentration) $\tau = 15\ \mu s$ (0.015 s). Then $A = 0.22\ 1/\lambda^2 f = 0.47 \cdot 10^2\ s^{-1}$. $(A + C) = 66.6\ s^{-1}$; quantum yield

$$\eta = \frac{A}{A + C} = \frac{0.47 \cdot 10^2}{0.666 \cdot 10^2} = 70\% .$$

The luminescence decay law determines diminution with time of the number of emitted photons (i.e., decrement in the number of excited atoms), this being designated as $N_i(t)$ or dN_i/dt. Such a diminution is equal to the number of initially excited atoms N_m (photons in excited state) multiplied by the probability of the A_{mn} decay of this excited state in a time unit (s^{-1}) or divided by the lifetime of each of the excited photons in the state m with $\tau_m = 1/\sum A_{mn}$, i.e.,

$$N_i(t) = -\frac{dNi}{dt} - N_m A_{mn}, \text{ or } = \frac{N_m}{\tau_m}.$$

This may be rewritten as $dN_i/N_i = -A_{mn} dt$. By integrating this expression we obtain $\ln N_i = -A_{mn}(t) + \text{const}$. At the starting instant of time $t = 0$, i.e., before turning off the exciting flash of the photoflash lamp $N_i = N_m$. Then $\text{const} = \ln N_m$ and the expression $\ln N_i = -A_{mn} + \ln N$. Hence $\ln (N_i/N_m) = -A_{mn}t$. The decay law may then be written as

$$N_i = N_m e^{-A_{mn}t} \text{ or } N_i = N_m e^{-t/\tau},$$

in other words in this case decay proceeds by following the exponential law.

During time t equalling the lifetime of excited level τ_m, i.e., $t = \tau_m$, $N_i = N_m$ $e^{-t/\tau} = N_m/e$, the number of excited photons will diminish by $e = 2.718$ times.

By considering probability C_{mn} of radiationless transitions for the ion in a crystal, the decay law may be written as

$$N_i = N_m e^{(-A_{mn}+C_{mn})}.$$

In a like manner the decay law may be represented not by the number of diminishing photons N_i, but by a decreasing emission intensity within time $I(t)$. At the initial time period $I_m = N_m/\tau_m$. The decay proceeds according to expression

$$I(t) = I_m e^{-At} \text{ or } I(t) = I_m e^{-(A_{mn}+C_{mn}t)}.$$

With the probability of transition A_{mn} is associated radiation power P:

$$P = h\nu_{mn} A_{mn}$$

and the emission intensity I_{emis} (of free ion)

$$I_{emis} = N_m h\nu_{mn} A_{mn}.$$

Here the number of excited ions N_m is determined by the effectiveness of excitation, i.e., by the absorption coefficient (oscillator strength) of the excitation band coinciding with the emission level. The higher the absorption coefficient (oscillator strength), the more effective is excitation.

Also understandable is the existence of a direct relationship between the emission intensity and the probability A_{mn} of the emission transition: the shorter the lifetime, i.e., the time of the photon emission, the more intensive its glow. Conversely, the greater the probability C_{mn} of a radiationless transition, the lesser the intensity of glow, since in that case some of the excited ions passing to a subjacent or ground state by giving off emission diminish, while others make the transition without "fluorescence". In this way a dual significance of the excited state lifetime becomes apparent: if it is short-lived because of the properties intrinsic in this state, the intensity is great; if it decreases due to interactions favoring radiationless transitions, the intensity (and quantum yield) diminishes.

Oscillator strength f of a given transition, and the probability of transition A_{mn} associated with it determines, other conditions being equal, the intensity of emission: the greater f is, the higher the intensity. However, an interaction with the lattice in the event of transitions with great oscillator strength can lead to a reduced emission intensity due to radiationless transitions, or result in these levels becoming radiationless. A weak interaction with the lattice and a greater oscillator strength of allowed transitions in ions RE^{3+} are factors leading to an intensive luminescence of these ions.

In the duration of afterglow are distinguished: short-lived glow: fluorescence (10^{-4}–10^{-6}s and shorter), slow fluorescence (10^{-5}–10^{-2}s), quick fluorescence (10^{-2}–0.5 s), and protracted glow: phosphorescence (0.1–1 s and more).

However, the duration is merely an external, and, for the afterglow period of 10^{-2}–10^{-5} s, an ambiguous characterization of two fundamental mechanisms of luminescence: discrete centers (ions, molecules, radicals) and crystallophosphors. Therefore, this division is better made according to the mechanism of glow, i.e., transitions within the confines of the ion or molecule levels, in the case of fluorescence, and recombination emission with participation of the conduction band and transfer of charge, in the case of phosphorescence. With the mechanism of luminescence are linked the duration of glow and the decay law, exponential in fluorescence, and hyperbolic or still more complicated in phosphorescence.

5.2.5 Transfer of Energy in Luminescence: Sensitization and Quenching

If two or more ion activators are present in a crystal, or if the crystal itself possesses the capacity of luminescence, an interaction can take place between these ions finding its expression in their luminescence spectra.

Absorption spectra of two different ions concurrently present in a crystal are simply a superposition of each of their absorption spectra.

In emission spectra of two simultaneously present ion activators, the position of emission lines of each of them does not change in the presence of the other, i.e., superposition of emission spectra from two ions occurs, but the following changes are also possible: (1) the luminescence spectrum intensity of one ion can gain in strength at the expense of the diminishing intensity of the other; (2) an ion not luminescent at a given concentration in a given crystal becomes luminescent in the presence of another ion in a different crystal; (3) luminescence of an ion can be observed under conditions of an excitation (in the absorption band of another ion) in which it is not luminescent without the presence of another ion; (4) in some cases one can see an intensified luminescence of one ion with complete quenching of the other.

These changes in luminescence of one ion in the presence of the other are due to transfer of excitation energy from one to the other. Luminescence of ions excited as a result of the energy transfer from other ions excited in the absorption band is termed sensitized luminescence, and proceeds confromable to the scheme:

$$S^* \longrightarrow A^*$$
$$\uparrow \qquad \downarrow$$
$$S \qquad A$$

here S is the sensitizor (energy donor); A is the activator (energy acceptor); asterisks denote their excited states.

The excitation energy transfer differs from the transfer of the charge, i.e., from the transfer of an electron or a hole proceeding with luminescence of semiconducting crystals, and considered in the following chapter; with transfer of energy excitation does not as yet induce ionization.

Sensitization is employed to raise the effectiveness of laser radiation [708]. One of the factors limiting the effectiveness of the excitation light transformation into laser radiation in lasers on solids with RE^{3+} is the absence of suitable excitation bands in RE^{3+} ions owing to the presence in them of merely narrow and weak absorption lines associated with forbidden f–f transitions. This limitation is overcome through incorporation in the crystal of a second impurity ion with a suitable absorption band. A common activator such as Mn^{2+} is also in need of sensitization with only weak absorption bands. Conversely, the ions of Cr^{3+}, RE^{2+} and others have, alongside emission levels, other levels onto which transitions occur which manifest themselves in the appearance of broad intensive absorption bands suitable for inducing luminescence.

The simultaneous existence of two or several ion activators in natural minerals is accepted as nature-sent, resulting from geochemical conditions existing at the time of mineralization. The permanence or variations in the relation of the co-existing ion activator concentrations is expressed in distinctive features of observed luminescence of minerals.

The substance of the sensitized luminescence consists essentially in that the energy absorbed in the absorption band of one ion (sensitizer) can be emitted in the emission band of another ion (activator). The transfer of energy from sensitizer to activator is accomplished through the following main types of transfer (Figs. 84–86) (1) emission-reabsorption, of rare occurrence, (2) resonance radiationless and (3) nonresonance radiationless.

1. The emission-reabsorption type of the energy transfer ("cascade" luminescence) implies emission of light by a single ion (primary luminescence) and its absorption (reabsorption) and emission (secondary luminiscence) by the other ion (see Fig. 84). Both ions behave as independent systems, and do not interact directly. A condition necessary for this way of transfer to manifest itself is the closeness of the emission energy of one ion to the absorption energy of the other. The patterns of the ions' energy levels must have two pairs of levels at close distances. Furthermore, it is necessary that in one of these patterns such a pair would correspond to a sufficiently intensive emission transition, and in the other, consisting of the ground level and of another level away from the former at a distance pre-set by the first pair, would conform to a sufficiently intensive absorption band, i.e., both ions should be activators and have sufficiently intensive absorption bands, while the luminescence emission band of one of them must be overlapped by the absorption band of the other.

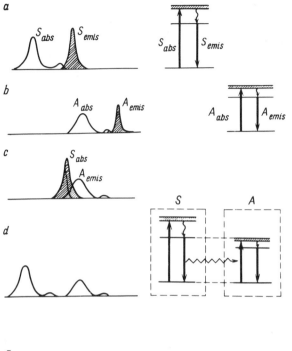

Fia. 84 a-d. Emission-reabsorption mechanism of energy transfer in sensitization of luminescence. **a** Sensitizer's absorption and emission spectra (S_{abs} and A_{emis}); diagram of levels; **b** the same of activator (A_{abs} and A_{emis}); **c** overlapping of the sensitizer's emission band and activator's absorption band as a condition for the energy transfer; **d**, excitation spectrum and diagram of levels for activator; luminescence in the pres ence of sensitizer

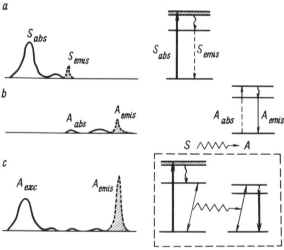

Fig. 85 a-c. Resonance radiationless mechanism of energy transfer in sensitiza tion of luminescence. **a** Sensitizer's absorption and emission spectra (S_{abs} and S_{emis}); **b** the same of activator; **c** excitation spectrum and diagram of levels for luminescence of activator A in the presence of sensitizer S; strong absorption S induced intensive luminescence A

The intensity (quantum yield) of the secondary luminescence will be the higher, the greater the luminescence quantum yield, the absoprtion intensity of each one of them and the more complete overlap of the emission band of one and of the absorption band of the other is. The duration of afterglow (lifetime) and decay curves of each of the ions are independent of the presence of the other.

In the luminescence spectrum of the ion-sensitizer it is only the intensity of the

Fig. 86. Nonresonance radiationless mechanism of energy transfer

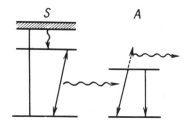

line absorbed by the activator that diminishes, the others remaining unchanged. Then, with changed concentration of the activator, it is not just the intensity of this sensitizing line that decreases, but its shape is also subject to change for with incomplete overlap with the absorption band part of this line is "engorged" (absorbed) by the activator. With increasing concentration of the sensitizer, a linear growth of the luminescnece of the activator intensity is observed. In as much as the emission of light becomes apparent at a much greater distance than that of direct interaction of ions, the effective transfer of energy in the case of the emission-reabsorption mechanism depends on the size and shape of the sample (diminishing in a powder by comparison with monocrystals).

An example of an emission-reabsorption transfer is the intensification of Nd^{3+} luminescence in the presence of Sm^{3+}, Eu^{3+}, Tb^{3+}, Dy^{3+}.

The main absorption band of Nd^{3+} about 580 nm (transition $^4I_{9/2} \to {}^4F_{3/2}$, $^4F_{5/2}$; see Fig. 114) is the principal band of the laser radiation excitation (1.06 μm transition $^4F_{3/2} \to {}^4I_{9/2}$). At the same time, the luminescence emission lines of Sm^{3+}, Eu^{3+}, Dy^{3+} lie in the neighborhood of 580 nm. Consequently, Nd^{3+} absorbs a part the luminescence of these ions and its own luminescence increases; this absorption by neodymium "engorges" some of these ions emission lines, thus changing the shape of the latter.

2. The resonance radiationless mechanism of the energy transfer ("cross-relaxation" in laser terminology) is effected between interacting ions behaving like a single system (see Fig. 85). A condition necessary for this mechanism of transfer to show itself is coincidence or a close distance between energy level pairs of the ion sensitizer and the ion activator. An essential difference consists in that the transition between the levels of the sensitizer is not necessarily emissive, whereas the activator's absorption corresponding to the pair of levels close to the first one may be of a very low intensity. Then the emission of the activator does not necessarily occur as a result of transition between the pair of levels to which the energy of the sensitizer is transmitted. The ion sensitizer (energy donor) cannot by itself be an effective activator of luminescence, while the ion activator (energy acceptor) may not have sufficiently intensive absorption bands suitable for inducing luminescence (excitation of laser radiation). However, the broad absorption bands of the sensitizer and emission transitions of the activator comprise together a system capable of effective luminescence.

Inasmuch as sensitization is of importance for just such activators devoid of suitable absorption bands, and owing to less stringent demands on the pair of ions, the resonance mechanism of the energy transfer is of paramount importance and is most prevalent.

Here the process of transfer is effected without primary emission resulting from dipole–dipole or dipole–quadrupole interaction between ions, the energy transfer probability depending upon the type of interaction. With dipole–dipole interaction the probability of transfer is proportional to r^{-6} (where r is the mean distance between interacting ions which in this case should not exceed $\sim$ 30Å), and with dipole–quadrupole interaction to r^{-8} (not more than $\sim$ 10–12 Å).

Here, the process of luminiscence is of an additive nature and a reduced duration and decreased quantum yield of the luminescence of the sensitizer (partial or complete quenching) are compensated by a longer duration and greater quantum yield of the activator's luminescence. The lifetime of the sensitizers from which the energy is transferred is determined, apart from the probability of emission and radiationless transitions, by the probability of the energy transfer to the ion activator.

The energy transfer time (and transfer probability, inversely) is determined from the difference between the lifetime (afterglow duration) of ion-sensitizer luminescence without any admixture of other ions, and that of the ion sensitizers in the presence of the activator τ_0 to which "leakage" of energy occurs according to formula

$$\frac{1}{\tau_{\text{tran}}} = \frac{1}{\tau} - \frac{1}{\tau_0} \, .$$

A change in the lifetime is unambiguously indicative of a resonance radiationless mechanism of the energy transfer.

Here, the decay curves are not purely exponential; they may be represented as a sum of exponential curves. Changes in the luminescence of each of the ions with their joint presence are traced with the aid of temporal sweeps of the luminescence spectra, i.e., of the spectra obtained over different time intervals comparable to the lifetime of emission levels for these ions (see below energy transfer $Cr^{3+} \rightarrow Nd^{3+}$).

The probability of the energy transfer increases proportionally to the diminution of the mean distance between ions (r^{-6} or r^{-8}), reduction of the lifetime of sensitizer in an excited state and to a greater overlap of the sensitizer's emission and activator's absorption.

With interaction of ions the intensity of the entire luminescence spectrum of the sensitizer decreases (but without any change in the shape of the emission lines). Because of short critical distances between the sensitizer and activator ions, the effectiveness of sensitization does not depend either on the size or shape of the sample.

3. The nonresonance radiationless mechanism leads to the transfer of energy between ions in the event of a substantial nonconcurrence of distances between the levels of an ion transferring the energy (upper) and the levels of an ion receiving the energy (lower). The difference between these distances goes either to the lattice in the form of a phonon spectrum or to a third ion which has a pair of levels supplementing this difference (see Fig. 86).

This mechanism of the energy transfer involves an exchange interaction between ions and comes into play over very short distances (up to 7–8 Å, over one–three anions). Sensitization can be accomplished with the help of admixture ions (impurity sensitization), or with the aid of ions entering the composition of the crystal itself (lattice sensitization).

Lattice sensitization can take place with excitation in absorption bands of various types:

1. In the intrinsic absorption band of the crystal. In most oxygen compounds, however, the absorption edge lies in the remote UV region, beyond the limits of energy generated by habitually used sources of UV excitation.

2. In the absorption band associated with the transfer of the charge; for instance in tungstates and molibdates the WO_4 or MoO_4 charge transfer band begins at about 300 nm; with excitation induced by 2537 Å UV radiation a broad emission band is observable, in pure $SrMoO_4$, for example, of around 450–650 nm and a maximum of 530 nm. In a Tb^{3+}-doped $SrMoO_4$ the band of its intrinsic fluorescence related to MoO_4 radical overlaps $^5D_4$ level of Tb^{3+} ion and sensitizes its glow through a resonance energy transfer, whereas the $^5D_3$ level of the same Tb^{3+} not overlapped by the MoO_4 emission band is not excited (Fig. 87).

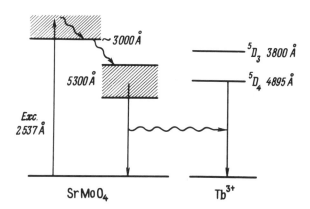

Fig. 87. Luminescence sensitization of Tb^{3+} in $SrMoO_4$ lattice Excitation with UVemission of 2537 Å in absorption band $SrMoO_4$: chagre transfer band MoO_4^{2-} (Stokes shift). Energy transfer to level ^5D of Tb^{3+} increases its emission to 4845 Å: $^5D_3$ level is not sensitized

3. In the RE^{3+} absorption bands representing basic crystal ions or contained in the crystal in large concentrations. For instance, in CeF_3, owing to interaction between ions, the Ce^{3+} ion levels appear in broad bands, transition $f-d$ declines down to 22,000 cm^{-1} (as compared to $\sim$ 31,000 cm^{-1} in a doped Ce^{3+} ion) and there occurs a direct transfer of energy from the CeF_3 lattice as well as an emissive energy transfer due to an intensive $d \rightarrow f$ Ce^{3+} fluorescence toward Nd^{3+} and Pr^{3+} ions contained in CeF_3 as an admixture. Fluorescence of Tm^{3+} and Ho^{3+} admixtures in garnet $Y_3Al_5O_{12}$ and in $(Y_{0.5}Er_{0.5})$ Al_5O_{12} increases 100–and 40–fold, respectively.

For such common activators as Mn^{2+} and RE^{3+}, devoid of intensive absorption bands, the admixture sensitization is of the greatest importance.

For these the following types of ions serve as sensitizers:

a) Tl^+, Pb^{2+}, Bi^{3+} mercury-like ions with electronic configuration of $5d^{10}6s^2$, In^+, $Sn^{2+}Sb^{3+}$ $(4d^{10}6s^2)Ga^+$, Ge^{2+}, As^{3+} $(ed^{10}, 4s^2)$, Ag^+, In^{3+}, Sn^{4+} $(4d^{10})$, Cu^+,

Ga^{3+}, and Ge^{4+} ($3d^{10}$). These ions have intensive absorption bands in the UV region and display emission bands in the near-neighboring UV and visible region. Excitation in the absorption bands of these ions leads to transfer of energy to Mn^{2+} ion (or to other activators).

A typical example is sensitization of luminescence of Mn^{2+} in calcites (natural and synthetic) activated by Pb^{2+} or Tl^+ [808]. When excited with UV light (250 and 360 nm) $CaCO_3:Mn^{2+}$ do not emit any light or emit it very scantily. Following excitation with electrons (cathode-luminiscence) one can observe an orange-red luminiscence. With UV light excitation $CaCO_3$: Mn^{2+} and Pb^{2+} display a similar bright orange-red luminescence.

The absence of $CaCO_3:Mn^{2+}$ luminescence without Pb^{2+} sensitizer with UV excitation is explained by the presence in this region of only very weak forbidden Mn^{2+} absorption bands (see Fig. 77), i.e., calcite with Mn^{2+} is transparent for these emissions.

On addition of Pb^{2+} there appears a new strong absorption band in which excitation is transmitted to Mn^{2+} that emits luminescence. The Mn^{2+} luminescence in calcite without sensitizer in excitation with electrons is due to the energy transfer by the calcite lattice excited with electrons (see p.   , *Methods of Luminescence Excitation*). The Mn^{2+} luminescence spectrum does not depend on the sensitizer or the mode of excitation. Natural halites from marine common salt deposits containing about 0.0003 % of Mn and around 0.01 %–0.001 % of Pb demonstrate red luminescence which may be regarded as a glow of Mn^{2+} sensitized with Pb^{2+} [772].

Sensitization of calcium phosphate with Mn^{2+} sensitized with Sb^{3+}, i.e., $Ca_3(PO_4)_2:Mn^{2+}$, Sb^{3+} finds wide use in luminescent lamps.

b) Ions of transition metals [with unfilled (vacant) d-shells]. $Cr^{3+} \rightarrow Nd^{3+}$. In Y-Al-garnet ($Y_3Al_2Al_3O_{12}$) with an admixture (about 1 %) simultaneously of Nd^{3+} and Cr^{3+} [725] (see also energy transfer of $Cr^{3+} \rightarrow Nd^{3+}$ in glasses [711]) with excitation in absorption bands of only Nd^{3+} normal fluorescence of Nd^{3+} alone is in evidence. With excitation in absorption bands of Cr^{3+}, first, fluorescence not only of Cr^{3+} is observed, but also fluorescence of Nd^{3+} arising due to the transfer of the charge from Cr^{3+} to Nd^{3+}; second, the duration of Cr^{3+} fluorescence diminishes, and that of Nd^{3+} increases, pointing to an additive nature of their luminescence and, consequently, to the resonance radiationless mechanism of the energy transfer from Cr^{3+} to Nd^{3+} (Fig. 88): $\tau_{Cr} = 8.1$ μs; $\tau_{Nd} = 0.230$ μs; $\tau_{Cr-Nd} = 3.5$ μs; $\tau_{Nd-Cr} = 0.240$ $\mu s + 3.5$ μs. Here τ_{Cr} is the lifetime of luminescence of Cr^{3+} in Y-Al-garnet without Nd^{3+}; τ_{Nd} that for Nd^{3+} without Cr^{3+}; τ_{Cr-Nd} that for Cr^{3+} in the presence of Nd^{3+}; and τ_{Nd-Cr} for Nd^{3+} in the presence of Cr^{3+}. Hence is determined the time of the energy transfer τ_{tran} to the $^4F_{3/2}$ level of Nd^{3+} ion (see Fig. 88):

$$\frac{1}{\tau_{Cr}} + \frac{1}{\tau_{tran}} = \frac{1}{\tau_{Ce-Nd}} \; ; \quad \tau_{tran} = 6.2 \; \mu s.$$

The energy transfer time is rather long by comparion with the duration of fluorescence of Nd^{3+} ($\tau_{Nd} = 0.230$ μs). Therefore, with a short-pulse excitation, the luminescence of Nd^{3+} in the presence of Cr^{3+} will not differ from normal

Fig. 88. Resonance radiationless transfer of energy from Cr^{3+} to Nd^{3+} [724]. *1*, Excitation in Cr^{3+} absorption bands $^4T_2$ and $^4T_1$; *2*, radiationless transitions in Cr^{3+} ion (relaxation); *3*, normal Cr^{3+} ion luminescence; *4*, energy tranfer $Cr^{3+} \rightarrow Nd^{3+}$; *5*, radiationless transitions in Nd^{3+} ion; sensitized luminescence of Nd^{3+} (excited in emission bands of Cr^{3+})

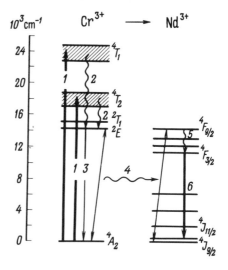

fluorescence of Nd^{3+}, but with a longer excitation the chromium sensitization will show. In this case the Nd^{3+} luminescence decay curve in the presence of Cr^{3+} consists of two components: a rapid (as without Cr^{3+}) and a lengthy ($\tau_{Nd-Cr} = \tau_{Cr} = 3.5 \mu s$) one.

The energy transfer from Cr^{3+} to Nd^{3+} does not proceeds directly from the absorption band (excitation band) of $^4T_2 Cr^{3+}$, but there first occurs a rapid (on the order of 0.001 μs) radiationless transition to $^2E Cr^{3+}$ and only from this the energy is transferred to $^4F_{3/2}Nd^{3+}$.

$Mn^{2+} \rightarrow RE^{3+}$. Although Mn^{2+} has only weak absorption bands, a transfer of energy from Mn^{2+} to ions of RE^{3+} is seen: Nd^{3+}, Sm^{3+}, Eu^{3+}, Tm^{3+}. The absence of Mn^{2+} luminescence in natural apatites, where the presence of Mn^{2+} is ascertained from the EPR spectra, is due to the energy transfer from levels $^4T_2$ and $^4T_1 Mn^{2+}$ onto $^4G_{5/2}$, $^6H_{9/2}$, $^6H_{7/2}$, $^6H_{5/2}$, Sm^{3+}.

c) $RE^{2+} \rightarrow RE^{3+}$ is sensitization of trivalent rare earth ions with divalent ones. The presence of intensive RE^{2+} absorption bands associated with allowed $f \rightarrow d$ transitions favors their utilization for excitation in the presence of RE^{3+} levels fit for the transfer of energy. An energy transfer from a divalent to trivalent ion of one and the same rare earth element, $Tm^{2+} \rightarrow Tm^{3+}$ (see Fig. 89) or an energy transfer between RE^{2+} and RE^{3+} ions of different elements are possible.

d) $UO_2^{2+} \rightarrow RE^{3+}$ is sensitization with uranyl of the trivalent rare earths luminescence [598].

e) $RE^{3+} \rightarrow RE^{3+}$. The energy transfer between different trivalent rare earth ions is one of the most common manifestations of sensitized luminescence. Because of the presence of a large number of levels in the visible and near-neighboring UV and IR regions, most of RE^{3+} ions can interact with one another, transmitting or receiving excitation energy. With the joint presence of several RE^{3+} ions they behave in this respect as a "system of communicating vessels".

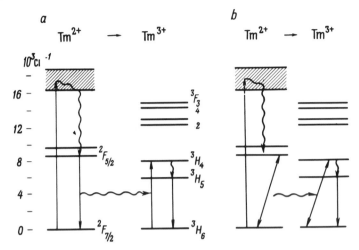

Fig. 89 a and b. Two possible ways of energy transfer from T_m^{2+} to T_m^{3+}. **a** Emissive-reabsorptive; **b** radiationless [658]

The energy transfer, however, occurs only from levels with a considerable lifetime, i.e., from metastable levels to the share of which falls up to 80% to 90% of the summary ion luminescence yield. This is understandable, since the energy transfer time comprises only a part of the level lifetime spent in emission and radiationless transitions from it and for energy transfer.

In considering practical possibilities of sensitization RE^{3+} ions with other RE^{3+} ions, it suffices in most instances to confine oneself to considering the energy transfer from long-living (metastable) donor levels. These will be (see Fig. 114): Pr^{3+} $^1D_2$; Nb^{3+} $^4F_{3/2}$; Sm^{3+} $^4G_{5/2}$; Eu^{3+} $^5D_0$; Gd^{3+} ^6P; Tb^{3+} $^5D_4$; Dy^{3+} $^4F_{9/2}$; Ho^{3+} $^5I_7$; Er^{3+} $^4I_{13/2}$; Tm^{3+} $^3H_4$; Yb^{3+} $^2F_{5/2}$.

In receiving the transfer energy by the ion-acceptor it is not necessary that the levels to which the energy is transferred should be long-living and emissive. The energy can be transferred to any pair of the acceptor levels the distance between which is close to the energy transmitted by the ion-sensitizer. Regardless of the fact to which pair of the acceptor levels the energy is transferred they become deactivated by means of rapid radiationless transitions to radiation levels from which emission occurs.

Typical examples are the following combinations of the RE^{3+} pairs of rare earth ions.

$Sm^{3+} \rightarrow Yb^{3+}$ (Fig. 90): radiationless resonance energy transfer (multipole transfer). Sensitization of Yb^{3+} luminescence with the Sm^{3+} ion was observed in sodium and rare earth tungstates resulting from the energy transfer of $^4F_{5/2}–^6H_{15/2}$ transition to a $^2F_{5/2}–^2F_{7/2}$ Yb^{3+} transition practically matching the former by its energy. Here the energy transfer is irreversible since it occurs from overlying Sm^{3+} levels down to underlying Yb^{3+} levels.

$Er^{3+} \rightarrow Ho^{3+}$, Tm^{3+}: a radiationless nonresonance transfer of energy. In $CaMoO_4$ activated simultaneously with Er^{3+} and Tm^{3+} or with Er^{3+} and Ho^{3+},

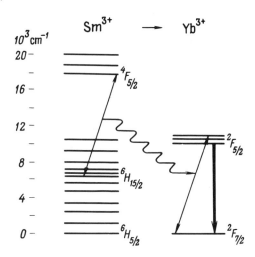

Fig. 90.   Resonance radiationless transfer of energy from Sm^{3+} ion to Yb^{3+}

the luminescence spectra represent superposition of the spectra of each of these ions, but here relative intensities of Er^{3+} and Tm^{3+} (or of Er^{3+} and No^{3+}), present together, change by comparison with the intensity of the spectrum of each one without admixture of the other ion. Thus the luminescence intensity of Tm^{3+} (0.5 at. %) increases 2.5-fold in the presence of Er^{3+} (0.75 at. %).

In the meantime the difference of Er^{3+} and Tm^{3+} energies (see below Fig. 114) amounts to 900 cm^{-1} (in the case of Er^{3+} and Ho^{3+} 1400 cm^{-1}). This implies that the energy transfer from Er^{3+} to Tm^{3+} is a nonresonance one, and that part of the energy is given off in the form of phonons to the molibdate lattice, whose fundamental vibration frequency is about 850 cm^{-1}, which nearly matches the difference of energy transmitted from Er^{3+} to Tm^{3+}, whereas the doubled amount (1700 cm^{-1}) stands close to the difference of energies in the case of $Er^{3+} \rightarrow Ho^{3+}$.

$Nd^{3+} \rightarrow Yb^{3+}$ is a variant of a nonresonance energy transfer (temporal spectral sweeps Figs. 91–93). The energy of $^4F_{3/2}-^4F_{9/1}$ Nd and $^2F_{5/2}-^2F_{7/2}$ Yb^{3+} transitions stand very close to each other, but in silicate glasses and tungstates Na and TR, in molibdates La and Na, and in fluoroberyllate glasses, no overlap of the Nd^{3+} emission and Yb^{3+} absorption spectra is demonstrable (see Fig. 91). However, a reduced luminescence time of Nd^{3+} in the presence of Yb^{3+} shows that the energy from Nd^{3+} is transferred to Yb^{3+} in a radiationless way [849]: $\tau_{Nd} = 151$ μs; $\tau_{Nd-Yb} = 54$ μs; $\tau_{Yb} = 350$ μs; $\tau_{Yb-Nd} = 750$ μs.

Here the energy transfer is accomplished in the same manner as in $Er^{3+} \rightarrow Ho^{3+}$, Tm^{3+}, in a nonresonance way with the release of the excess energy of the lattice in the form of phonons (see Fig. 92). Temporal sweeps of Nd^{3+} and Yb^{3+} luminiscence spectra show the kinetics of their glow to be in conformity with their varying luminescence time (see Fig. 93).

$Nd^{3+} \rightarrow RE^{3+}$(Fig. 94): the direction of the Nd^{3+} ion interaction with other rare earth ions. Measurement of the lifetime (duration of glow) of Nd^{3+} in $Na_{0.4}$

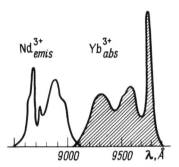

Fig. 91. Interrelations of Nd^{3+} emission spectrum and of Yb^{3+} absorption spectrum in fluoroberyllate glass at 77 K [849]. Nd^{3+} emission and Yb^{3+} absorption overlapping is practically absent (narrow Nd^{3+} and Yb^{3+} lines are shown here on an enlarged scale)

$Gd_{0.46}$, $Nd_{0.02}$, $RE_{0.02}WO_4$, i.e., present conjointly with one of RE^{3+} [849] is indicative of the nature of the Nd^{3+} interaction with these other rare earth ions (see Fig. 94).

From the comparison of the lifetimes it follows that the presence of Pr^{3+}, Sm^{3+}, Eu^{3+}, Tb^{3+}, Dy^{3+}, Ho^{3+}, Er^{3+}, Tm^{3+}, Yb^{3+} reduces the lifetime of Nd^{3+}, i.e., produces its partial quenching (decay). A comparison of the energy levels of these ions (see Chap. 5.3, Fig. 114) enables one to see that the said ions have their levels arranged in a way satisfying the condition of the resonance transfer of the energy from Nd^{3+} to these ions.

The Y^{3+}, La^{3+}, Ce^{3+}, Gd^{3+}, Lu^{3+} ions have no energy levels in the near-neighboring IR region where Nd^{3+} emission occurs (see Fig. 114) and, therefore, are inactive with respect to the Nd^{3+} luminiscence (a diminished Nd^{3+} lifetime in the presence of an additional number of Nd^{3+} ions stems from concentration quenching, see Fig. 99).

Hence, Nd^{3+} plays the part of a sensitizer in regard to most RE^{3+}, whose presence causes its partial quenching. No energy transfer to Nd^{3+} from other RE^{3+} has been observed.

$Er^{3+} \rightleftarrows RE^{3+}$: sensitization of some ions, quenching of others, and a dual interaction with the third type of ion. Measurement of the lifetime and of a relative intensity of the Er^{3+} luminescence in the presence of one of the remaining RE^{3+} ions, as well as measurements of the luminescence time of these RE^{3+} in pair with Er^{3+} indicate a different direction of the nonradiative Er^{3+}–RE^{3+} energy transfer: (a) Yb^{3+} increases the intensity of the Er^{3+} luminescence; (b) Er^{3+} adds strength to the luminescence of Ho^{3+} and Tm^{3+} which quench it; (c) Pr^{3+}, Nd^{3+}, Sm^{3+}, Eu^{3+}, Dy^{3+} transmit energy to the Er^{3+} ion (which, consequently, quenches their light emission), but Er^{3+}, in its turn, transmits energy to the underlying radiationless levels of those ions which accordingly quench the Er^{3+} glow (Fig. 95).

$Yb^{3+} \rightarrow Tb^{3+}$ (Fig. 96): a collective and cumulative transfer of energy, the anti-Stokes luminescence. In the CaF_2 crystals containing simultaneously Tb^{3+} and Yb^{3+} (in high concentrations—3% of Tb and 7% of Yb), following their excitation

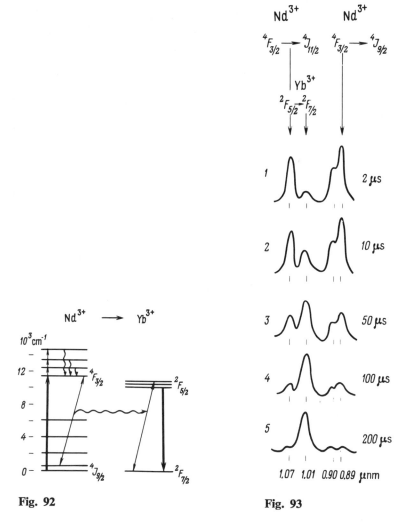

Fig. 92   **Fig. 93**

Fig. 92. Energy levels diagram illustrating radiationless nonresonance $Nd^{3+} \rightarrow Yb^{3+}$ energy transfer. Narrow emission and absorption lines fail to overlap due to noncoincidence in the difference of closely lying level pairs $^4F_{3/2}-^4I_{9/2}$ of Nd^3 ion and $^2F_{5/2}-^2F_{7/2}$ of Yb^{3+} ion

Fig. 93. Time sweeps of luminescence spectra in an interacting system $Nd^{3+} + Yb^{3+}$ in $Na_{0.5} Gd_{0.46} Nd_{0.62} Yb_{0.02} WO_4$ [849]. *1*, Spectrum in 2 μs after a pulsed flash; in evidence are intensive Nd^{3+} lines and a very weak Yb^{3+} line; *2-5*, gradual decay of Nd^{3+} luminescence and intensification of Yb^{3+} luminescence

in the IR region in the band of 0.98 nm $Yb^{3+} {}^2F_{7/2}-{}^2F_{5/2}$, luminescence can be seen in the UV-blue region of 380–490 nm from the $^5D_3 Tb^{3+}$ level and in the region 490–680 nm from the $^5D_4 Tb^{3+}$ level.

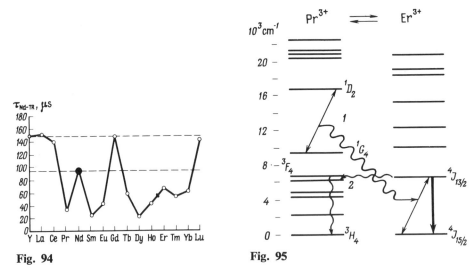

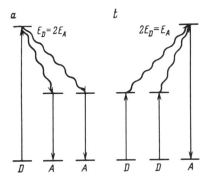

Fig. 94.

Fig. 95

Fig. 94. Lifetime (duration of luminescen ce) of Nd^{3+} in $Na_{0.5}$ $Gd_{0.46}$ $Nd_{0.02}$ $RE_{0.02}$ WO_4 in the presence of one RE^{3+} co-activator ion [849]

Fig. 95. Bilateral interaction $Pr^{3+} \rightleftarrows Er^{3+}$. *1*, Resonance radiationless energy transfer from $^1D_2 - {}^1G_4$ of Pr^{3+} ion to $^4I_{13/2} - {}^4I_{15/2}$ Er^{3+} ion that sensitizes its luminescence; *2*, partial decay of Er^{3+} luminescence owing to a reverse energy transfer from Er^{3+} to radiationless level $^3F_4$ Pr^{3+}

Fig. 96 a and b. Cooperative excitation of two acceptor ions: **a** owing to donor's energy exchange; **b** by summation of two donor's ion energies [781]

From the level $^5D_4$ of Yb^{3+}, ion luminescence is excited as a result of a resonance collective energy transfer of two ions Tb^{3+} ($^2F_{7/2}-^2F_{5/2} = 10{,}200$ cm^{-1}) to the Yb^{3+} ion (see Fig. 96). Luminiscence from level $^5D_3$ of the Yb^{3+} ion (27,000 cm^{-1}) is explained by the cumulative mechanism of excitation, caused by a summary energy of the three Tb^{3+} ions: first there occurs collective excitation of the $^5D_4$ level by two Yb^{3+} ions, then additional excitation to the $^5D_3$ Tb^{3+} level by yet another Yb^{3+} ion.

Fig. 97 a and b. Redistribution of different transition intensity. **a** Gd^{3+} in CaF_2 luminescence spectrum: transition $^6I \rightarrow {}^8S$ predominates, whereas transition $^6P \rightarrow {}^8S$ has an intensity six times smaller; **b** Gd^{3+} in CaF_2 with presence of Pr^{3+}: quenching of Gd^{3+} transition from ^6I and intensification of transition from ^6P occurs as a result of energy transfer $(^6I - {}^6P)$ $Gd^{3+} - (^3F_2{}^3H_6 - {}^3H_4)$ Pr^{3+} [746]

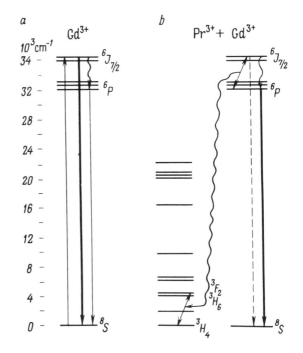

The luminescence intensity from the $^5D_4$ level increases proportionally to the square, and from the $^5D_3$ level to the cube of the intensity of light incident on the crystal (0.98 μm) (two and three Yb^{3+} ions participating in excitation, respectively).

$Pr^{3+} \rightarrow Gd^{3+}$ (Fig. 97): redistribution of various Gd^{3+} transitions intensity resulting from interaction with Pr^{3+}. In $CaF_2:Gd^{3+}:Pr^{3+}$ a resonance radiationless interaction of the $^6I-{}^6P$ Gd^{3+} transition with the $^3H_4 - {}^3H_6{}^3F_2$ Pr^{3+} transition leads to the quenching of Gd^{3+} emission in transition of $^6I-{}^8S$ and, at the same time, to a greater emission intensity of the same ion in the $^6P-{}^8S$ transition [746].

A multiple energy transfer in the chains of ions leads to a change in the luminescence intensity of these coexisting ion activators, but in the last analysis they have only small effect because of concomitant losses in nonradiative transitions. Thus in K-B-silicate glass one can observe an energy transfer in the chain of ions [598].

$$UO_2^{2+} \rightarrow Nd^{3+} \rightarrow Yb^{3+}$$
$$\searrow Er^{3+} \quad \nearrow \searrow Er^{3+} \quad \nearrow \searrow Er^{3+}$$
$$Tm^{3+} \qquad Tm^{3+}$$

Luminescence decay is yet another manifestation of the excitation energy transfer.

Concentration Quenching (Self-Quenching). Changes of the luminescence intensity with growing concentration of the activator manifests itself in different ways for different activators. For most of the latter the intensity of glow passes through a maximum and then decays; for some of them the maximum is observed with the activator's content of about 1%, while for others with the concentration of one or two orders higher. Compounds of which such ions form part in the basic composition of the crystal are not luminescent ($GdCl_3$, $NdCl_3$, for instance). Other activators do not experience any concentration quenching (Mn^{2+}, for example) and are luminescent in their compounds (in $MnCO_3$, Mn_2SiO_4, for instance)

This depends on the mechanism of interaction between these ion activators of the same kind. The ways of interaction between ions of the same kind are basically the same as between ions of different kinds (these were considered earlier when analyzing sensitization).

1. The emission–reabsorption energy transfer between ions of the same kind can result in self-absorption. As a characteristic example may be cited self-absorption of the Cr^{3+} ion. The Cr^{3+} fluorescence consequent upon an emission transition of $^2E \rightarrow {}^4A_2$ by other Cr^{3+} ions becomes absorbed as a result of backshift with absorption of $^4A_2 \rightarrow {}^2E$ (see Fig. 76).

2. Much more common is concentration quenching resulting from nonradiative resonance multipole energy transfer between ions of the same type (Fig. 98). This mechanism of concentration quenching is inherent in ions of a number of trivalent rare earths (Fig. 99). A peak luminescence intensity and decay occur in this type with low concentrations of the activator.

3. An exchange mechanism of concentration quenching becomes apparent in the case of ions that fail to have coincident differences of the emission and radiationless levels of transitions (in the case of Eu^{3+}, for example). The maximum luminescence and quenching intensity are seen to occur with concentration one order higher than with a multipole resonance transition.

4. Interaction with the charge transition band produces quenching not only of other ions, but also of the very ion which has in the absorption spectrum a charge transfer band that shifts with rising concentration of the ion to the visible and near-lying IR region (Co^{2+}, Ni^{2+}, Fe^{2+}).

Self-quenching of all types does not affect the luminescence of other ions, provided their emission bands do not overlap the ion's absorption bands in which concentration quenching occurs. For instance, in $GdCl_2$ no Gd^{3+} luminescence is seen to occur due to self-quenching, but the Nd^{3+} and other TR^{3+} impurities making part of $GdCl_3$ are luminescent.

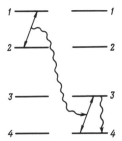

Fig. 98. Concentration quenching resulting from multipole energy transfer between ions of the same kind. The energy difference between emission levels (*1–2*) is close to the energy difference of radiationless transition (*3–4*)

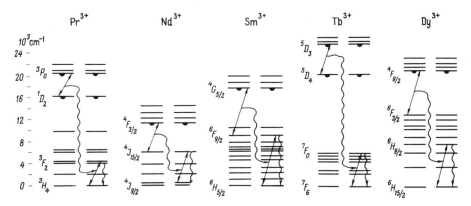

Fig. 99. Diagrams illustrating concentration quenching in RE^{3+}

Decay as a Reverse Aspect of Sensitization. This intensified glow discussed above of the ion-energy acceptor is attended by relaxation, i.e., partial quenching of the light emitted from the ion-energy donor. Unlike the next type of luminescence decay, this is of an selective nature, i.e., it is active between a pair of heterogeneous ions (donor–acceptor) between which an energy transfer takes place resulting from coincidence or closeness of the differences of energy levels. The possibility of a bilateral interaction, and the presence in the TR^{3+} ions of many radiationless underlying levels results in the decay (partial or complete) occuring with interaction of different ions more often that does an intensification of the glow. In particular, it is Pr^{3+} that has the strongest quenching effect of TR^{3+}.

Quenching with Ions with Intensive Charge Transfer Bands. The Fe^{3+} and also Fe^{2+}, Co^{2+}, and Ni^{3+} ions have charge transfer bands in the absorption spectrum, whose maxima lie in the UV region, but their intensity is so great that with appreciable concentrations the tail of these bands extends into the visible and even to (Fe^{3+}) the near IR region. As distinct from the preceding type of luminescence decay, the charge transfer band quenches the glow of all the activators whose emission bands are overlapped by it (i.e., this is nonselective quenching). Here too, however, one must distinguish between quenching with the Fe^{3+} ion and quenching with Fe^{2+}, Co^{2+}, Ni^{2+} ions. In the latter the intensity of the transfer band is much lower, and even with their significant concentrations one can in some cases observe luminescence of other ions in the "windows" between the d–d basorption bands; compounds with low Fe^{2+}, Co^{2+} Ni^{2+} concentrations are luminescent. The Fe^{3+} content of the order 0.n% drastically reduces the glow intensity of all other activators, and the proportions on the order of the first percent quench the luminescence completely. However, in the IR region not overlapped by the Fe^{3+} transfer band, luminescence of ions can also be seen in these cases. For example, in the Y-Fe-garnet, luminescence of the Ho^{3+} ions occurs in the region of 1–2 μm.

Quenching in "Pure" Crystals. In most inorganic crystals without any substantial amount of impurities, no luminescence is observed (NaCl, for instance), whereas some other pure crystals display the presence of luminescence ($CaWO_4$, for example). This stems from the competitive action in the speed of two processes: interaction with the lattice (relaxation) of the excited state (transition $B \rightarrow C$ in Fig 80 in the model of configuration curves) and energy transfer from one ion to the other. In $CaWO_4$ the transfer rate (through exchange) between WO_2^{4-} ions is low by comparison with the lattice relaxation rate. Therefore, in the excited WO_4^{2-} relaxation first occurs which leads to the Stokes shift that displaces the excited ion from the resonance position with surrounding ions.

In NaCl, excitation is associated with excitons that migrate rapidly all over the crystal till they reach the parts displaying some or other defects capable of giving rise to radiationless or emission transitions.

5.2.6 Representation of Luminescence in the Band Scheme and the Luminescence of Crystallophosphors

Earlier, when describing the mechanism accounting for luminescence, use was made consecutively of the theories of energy levels of ion activators, of configuration curves by which these levels ought to be represented if one takes into consideration changes in the distances between excited ions, and of the lifetime characterizing these states. However, all these events are enclosed within the limits of the ion activators levels split by the crystal field whose action is accounted for in this way.

To complete the discussion of basic theories and concepts with whose aid the luminescence phenomena are described, let us dwell upon the position of the ion activator levels with all their characteristic features in the band scheme of the crystal.

With small dimensions of the forbidden band, crystals possess properties of semiconductors and photoconductors, i.e., they have properties embodying the electron transfer of the impurity ions or of the ground substance to the conduction band. Conformable to the phenomenon of the semiconductors luminescence, crystals with a small interband spacing are endowed with features which, in their mechanism of luminescence, make them contrast with other luminescent crystals. Luminescent semiconducting crystals are called crystallophosphors[8]. Most typical among these and most important in the practical, theoretical and mineralogical sense are ZnS crystals containing various impurities (activators and co-activators) as well as other compounds of the IIA-VIB type.

The glow of the majority of luminescent crystals (non-semiconductors), and of the great majority of minerals is due to transitions within the limits of the ion activator levels, without participation of bands, and for this reason it may usually be considered without reference to the band scheme (Fig. 100). To such luminescent systems belong, for instance, $Al_2O_3 : Cr^{3+}$, $CaF_2 : TR^{3+}$, $CaCO_3 : Mn^{2+}$ and so on, i.e., minerals of all classes, except for sulphides, selenides, tellurides, sulfosalts, and diamond.

A different principle lies at the root of the excitation and light emission processes of semiconductor crystals of the ZnS type, of those considered with the help of

---

[8] The term crystallophosphors often designates crystals at large, capable of luminescence.

the band scheme (see Fig. 100). The number of such luminescent compounds, and especially of minerals, is limited, but their scope is enlarged by the fact that to their number most important commercial luminophors are referred. Moreover, the mechanism of processes active in them constitutes an inalienable part of a more general phenomenon that includes thermoluminescence, formation and transformation of the electron-hole centers, photosensitive electronic paramagnetic resonance, photoconduction, etc. Many general works on luminescence had pre-eminently as their background the material dealing exactly with crystallophosphors, and to avoid the idea that all the processes of luminescence follow the pattern of crystallophosphors one must understand features specific to their luminescence.

Phosphorescence as such represents glow with participation of band transitions; it always occurs in excitation with high-energy radiations, and not only in crystals of the ZnS type, but also in crystals of all other types ($NaCl$, $CaWO_4$, etc.), although in them it tells not so much on the luminescence spectrum of ion activators, but rather on the duration of their after-glow.

It will be recalled that the conduction band in these is considered approximately in terms describing the state of the Zn^{2+} ($3d^{10} 4s^0$) cation, i.e., it is represented by empty orbitals comprised largely of 4s-orvitals of the Zn^{2+} cation. The valence band is considered approximately in terms describing the state of the anion: S^{2-} ($3p$), i.e., it is represented by filled ($3p^6$) orbitals made up in the main of the 3p-orbitals of the S^{2-} anion.

Excitation of Crystallophosphors, Formation of Free Electrons and Free Holes. Excitation with UV light is not confined to turning the ions into the excited state, as, for instance, in the case of Cr^{3+} in Al_2O_3 (see Fig. 100). Here, in crystallophosphors ionization of impurity ions and ions of the ground substance takes place, i.e., detachment of an electron from them, and its transposition from filled valence band into an empty conduction band (Fig. 101). With the interband spacing of 3.83 eV in sphalerite ZnS a UV irradiation with a wavelength of less than 325 nm suffices to accomplish this, whereas with an interband distance of 7.4 eV in Al_2O_3 it requires 167 nm. Schematically this process may be put down as

$$S^{2-} (3p^6) + Zn^{2+}(4s^0) \rightarrow S^- (3p^5 \leftarrow e^+) + Zn^+ (4s^1 \leftarrow e^-).$$

Then the sulphur shell remains with a missing electron (shortage of a negative charge and an excess of a positive charge) which is designated as a positive hole e^+, i.e., $S^- = S^{2-} + e^+$. This hole does not belong to any single sulfur atom that transforms from S^{2-} ($3p^6$) into S^- ($3p^5$), as occurs with the formation of the hole center S^-, but it is distributed all over sulfur atoms of the ZnS crystal, when each of the S^{2-} atoms (yielding an electron and capturing a hole) turns for an instant into S^-. In like manner an electron transposed into the conduction band does not belong to a single Zn^{2+} ($4s^0$) + $e^- \rightarrow Zn^+$ ($4s^1$) atom, but is distributed among all atoms of zinc in the ZnS crystal, each one of which (by capturing an electron) for a moment changes from Zn^{2+} into Zn^+.

In this case one can speak of a free electron in the conduction band (from 4s-orbitals of Zn), and of a free hole in the valence band (from 3p-orbitals of S).

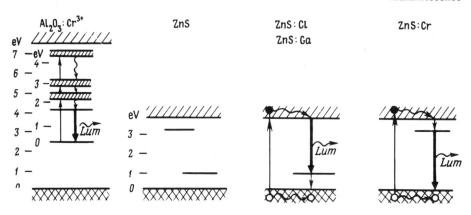

Fig. 100. A comparison of the ion activator energy levels position in a band scheme for different crystal types.

Al_2O_3: Cr^{3+}, excitation and luminescence occur as a result of transitions between the ion activator levels without participation of conduction and valence bands (transition from ground state to a level energy 4.85e V corresponds to $\sim$ 275 nm, but does not reach the conduction band).

ZnS, interband distance of 3.83 eV (320 nm) is divided by activators levels. Excitation is accomplished as a result of the electrons transfer from the valence band to the conduction band or (and) through ionization of the activator. Luminiscence proceeds according to two basic schemes:

$ZnS:Cl$ and $ZnS:Ga$, transition from conduction band to activator's level (Schön-Klasens scheme). Examples: $ZnS:Cl$ and $ZnS:Ga$ (see Fig. 104).

$ZnS:Cr$, transition: activator's level, valence band (Lambet-Klick scheme. Example: $ZnS:Cr$ (see Fig. 103)

A free electron and a free hole may be considered as an excited state of a single large Zn*S* molecule in which a single electron of sulfur bonding the 3p-orbitals changes into a state described by the antibonding 4s-orbital of zinc (see Fig. 101).

Position of Impurity and Vacancy Levels in the Forbidden Band; Donors and Acceptors; Formation of Localized Electrons and Holes. In the luminescence spectra of crystallophosphors are mostly seen one to two broad bands carrying as such no

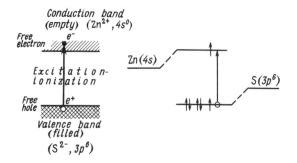

Fig. 101. Excitation of crystallophosphors in the band scheme and in the scheme of molecular orbitals

special indications sufficient for characterization of the sum total of processes that include not just the ion activator levels, but also the bands and levels of various capture centers (traps) of electrons and holes. Therefore, until recently articles and reviews on luminescence of crystallophosphors have of necessity been confined to empirical determinations of the levels position in the forbidden band, without indicating their physical meaning, and without ascribing them to concrete ions and centers. This was the cause of dissatisfaction with the band theory of the luminescence of crystallophosphors. Lately a combination of the band scheme with the theory of the ligand field, and also an insight into luminescence mechanism by using photosensitive electron paramagnetic resonance led to the understanding of this phenomenon, having provided concrete designation to the formerly "nameless' levels in the forbidden band.

Into the forbidden band of the band scheme of crystallophosphors come (see with the example of ZnS, Fig. 102):

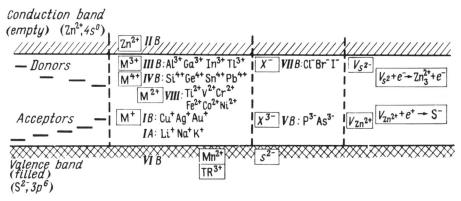

Fig. 102. Position of donor and acceptor ion activator levels in the band scheme of the IIB-VIB type compounds (on an example of ZnS)

a) Levels of defects (precursors, see Chap. 7), i.e., (1) impurity ions (activators, co-activators, quenchers), usually isomorphous, more seldom interstitial, (2) cation vacancies (V_{Zn2+}) and anion vacancies (V_{S2-}), 3) basic ions Zn^{2+} and S^{2-} near vacancies and impurity ions.

b) Electron and hole center levels forming at the expense of these defects during trapping of an electron or a hole by them in the course of excitation of luminescence: (1) of impurity centers, (2) of centers appearing at the expense of vacancies; (3) of levels due to the free electrons and holes localization (in conduction and valence bands) in the neighborhood of electron and hole centers appearing at the expense of impurities and vacancies.

As concerns their position in the forbidden band the levels of all these defects are divided into two groups: donors (nearby the conduction band), and acceptors (in the vicinity of the valence band). This division has a triple meaning.

1. In regard to semiconductor features (from where the designations of donors and acceptors have been adopted):

Donors represented by impurity ions with an excess positive charge or by excess cations (i.e., anion vacancies) give off an excess electron to the conduction band, for example: in substituting $Ga^{3+} \rightarrow Zn^{2+}$ in ZnS, two Ga electrons take part in the formation of a normal chemical bond and the third, excess one enters the conduction band; an excess positive charge arises also in substituting $Cl^- \rightarrow S^{2-}$ and, therefore, Cl^-, Br^-, and I^- in ZnS are donors, just as the sulfur vacancies.

Acceptors represented by impurity ions with a deficit of positive charge (carrying an excess negative charge), capture the lacking electron from the valence band, leaving there a positive hole (i.e., a shortage of an electron in the S^{2-} shell transferred from one sulfur ion to another), whose transposition causes a hole conduction. For instance: in replacing $As^{3-} \rightarrow S^{2-}$ an excess negative charge appears that results in trapping of an electron from the valence band.

2. As regards luminescence the role of the donor and acceptor levels is in a certain sense opposite to that played by them in the semiconduction processes, i.e., they are traps of electrons and holes[9].

Donors are electron traps; in a normal (uncharged) state they are empty (unoccupied), for example: in replacement of $Ga^{3+} \rightarrow Zn^{2+}$ in ZnS a Ga^{3+} ion has no p-electrons in the electron configuration; donor levels are levels of the electron's capture; with the capture of an electron (charged state of the donor) the valence of the donor impurity decreases; $Ga^{3+} + e^- \rightarrow Ga^{2+}$ (a p-electron appears in the electron configuration), i.e., for this time there arises a new ion Ga^{2+}, with a different position of its level in the band scheme. As distinct from free electrons (in the conduction band) here we have to deal with a localized (captured) electron.

Acceptors are hole traps; in a normal (uncharged) state they are filled: for instance, in substitution of $Cu^+ \rightarrow Zn^{2+}$ in ZnS the Cu^+ ion has a filled d-shell ($3d^{10}$); acceptor levels are the levels of the hole's capture; when they trap a hole (i.e., release an electron) their valence increases: $Cu^+ + e^+$ (release of e^-, i.e., release of a single d-electron) $\rightarrow Cu^{2+}$ ($3d^9$ instead of $3d^{10}$, a hole in the d-shell). As distinct from free holes (in the valence band) here we deal with a localized (trapped) hole. The M^{2+} ions in ZnS can be both donors and acceptors, depending upon the effective M^{2+} charge relative to Zn^{2+} and on excitation conditions, i.e., Fe^{2+} can in ZnS change into Fe^{3+} and into Fe^+.

3. As regards the electron-hole centers, a capture of an electron or a hole in excitation of luminescence by donor and acceptor defects may be considered as formation of electron and hole centers (see Chap. 7), referred to as photosensitive centers (i.e., they arise following excitation by light and gradually disappear after excitation has been turned off [641, 642, 656, 696, 709, 712, 806, 807, 813, 814, 826].

Donors are precursors (see Chap. 7) of photosensitive electron centers, i.e., these are defects at whose expense electron centers form on excitation of luminescence, for example: Ge^{3+}, In^{3+}, $Tl^{3+} + e^- \rightarrow Ga^{2+}$, In^{2+}, Tl^{2+} and Si^{4+}, Ge^{4+}, Sn^{4+}, Pb^{4+}

---

[9] In works on luminescence this terminology of donors–acceptors is more current (see Fig. 102), but sometimes donor impurities are referred to as acceptor (and vice versa) in the sense that during luminescence they capture an electron.

$+ e^- \rightarrow Si^{3+}, Ge^{3+}, Sn^{3+}$, and Pb^{3+}; vacancy $S^{2-} + e^- \rightarrow$ electron center ($Zn_i^{2+} + e^-$) (see Fig. 132).

Acceptors are precursors of photosensitive hole centers, i.e., defects at whose expense hole centers form on excitation of luminescence, for instance: $Cu^+ + e^+ \rightarrow Cu^{2+}$, vacancy of $Zn^{2+} + e^+ \rightarrow$ hole center S^- (see Fig. 132). This is attended by the formation of new electron and hole center levels in the forbidden band.

Mechanism Governing Luminescence of Crystallophosphors: Recombination of Free Electrons and Localized Holes or of Free Holes and Localized Electrons. Knowing the origin of levels in the forbidden band one can trace down the sequence of events leading up to the luminescence of crystallophosphors.

Two principal patterns of their luminescence correspond to the donor or acceptor nature of the impurity (activator) or the vacancy ("self-activated" crystals). The donor or acceptor character of the defect determines the further sequence and direction of the process.

The pattern corresponding to donor defects involves a capture by these defects of a free electron (see Fig. 103), i.e., formation of an electron center, for example: Cr^+ ($Cr^{2+} + e^-$) in ZnS that gives an EPR signal. Upon disappearance of this center, i.e., after it has given off an electron as a result of its recombination with the hole in the valence band, luminescence becomes apparent. This type accords with the Lambet-Klick scheme, where luminiscence appears in the course of an electron center recombination with a hole in the valence band, i.e., of a recombination of a localized electron with a free hole or of a transition "impurity or vacancy level (more exactly of an electron center on an impurity or a vacany) → valence band.".

Acceptor defects correspond to the Schön-Klasens scheme that provides for the capture of a free hole (with release of an electron) by these defects (see Fig. 104)

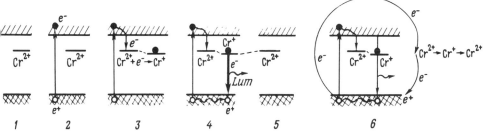

Fig. 103. ZnS:Cr luminescence mechanism pattern (Lambet-Klick type). *1*, initial position of Cr^{2+} ion activator level in the ZnS band scheme; *2*, UV light excitation-ionization of the ZnS ground substance; formation of free holes (e^+) and free electrons (e^-); *3*, trapping of a free electron from the conduction band by Cr^{2+} ($3d^4$) ion and formation of a photosensitive electron center Cr^+ ($3d^5$) giving an EPR signal; Cr^{2+} level is electron trapping level (electron trap); *4*, luminescence emission as a result of the electron transfer from Cr^+ to valence band, i.e., to $S^{2-} + e^+ \rightarrow S^-$ hole migrating toward Cr^+; luminescence occurs following recombination of localized Cr^+ ($Cr^{2+} + e^-$) electron "intercepted" by the activator that has for this time turned into an electron center, with a free hole S^- of the valence band; *5*, return to initial position—end of the cycle; *6*, the entire cycle of excitation, electron capture and recombination luminescence

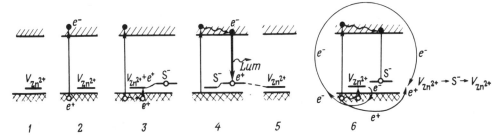

Fig. 104. Luminescence mechanism pattern for ZnS:Cl, ZnS:Ga (Schön-Klasens type). *1*, initial position of Zn^{2+} vacancy level in the ZnS band scheme (with substitution of $Cl^- \rightarrow S^{2-}$ or $Ga^{3+} \rightarrow Zn^{2-}$; in the absence of charge compensation there arises a zinc vacancy: "self-activated" ZnS); *2*, excitation with UV light; formation of free holes (e^+) and free electrons (e^-); *3*, trapping of free hole from valence band by zinc vacancy V_{Zn}^{2+}; since Zn^{2+} vacancy cannot give off the electron (capture hole) this is accomplished at the expense of S^{2-} ions surrounding the vacancy, i.e., the vacancy Zn^{2+} which has captured the hole is S^-, a photosensitive hole center giving an EPR signal; V_{Zn}^{2+} is a hole-capturing level (hole trap); *4*, luminescence emission resulting from recombination of free electron from conduction band with localized hole S^- ($S^{2-} + e^+$) that formed from the hole which has become for this time a hole center; *5*, return to initial position; *6*, the entire cycle of excitation, hole capture and recombination luminescence

with formation of a hole center, for example S^- (at the expense of vacancy Zn^{2+} in ZnS) that gives an EPR signal. Luminescence occurs in recombination of this hole center with an electron from the conduction band, i.e., of an electron with a localized hole, this being due to a transition "conduction band → impurity or vacancy level (more exactly of a hole center on an umpurity or vacancy)".

Take notice of the difference between the free hole $S^- \rightleftarrows S^{2-}$ and the localized hole S^- near the vacancy Zn^{2+}. In ZnS:Cr^{2+} there occurs first the capture of an electron and formation of the center $Cr^+ = Cr^{2+} + e^-$ and only thereupon and because of it the free hole (that fails to give an EPR signal and is distributed prior to this among all the S^{2-} ions, each one of which for a moment turns into S^-), approaches the Cr^+ site. However, the moment this free hole localizes in the form of S^- near Cr^+, their recombination takes place accompanied by emission of light.

Thus the specificity of the luminescence of crystallophosphors is determined by the following distinctive features:

1. Transitions determining their excitation and glow are not confined to the levels of a single ion (although this is so in the case of ZnS : RE^{3+}, or ZnS : Mn^{2+}, see Figs. 109 and 110) and are not confined to excitation alone, but are accompanied by ionization and formation of free electrons and free holes.

2. They do not recombine with each other, but free electrons do so with localized holes, and free holes with localized electrons. The luminescence here is of a recombination nature and occurs not as a result of "band–band" transitions or those between the impurity levels, but is consequent upon the "band–defect levels", and "defect levels–band" transitions (a direct "band–band" transition is seen to occur in particularly pure defectless crystals in the form of "edge luminescence", i.e., by

its wavelength close to the region of the intrinsic absorption of the crystal). Hence, in the presence of impurities the transition band–band is separated by impurity levels at which the electron makes "stop–overs", charging them (forming an electron center), and discharging them with emission of light.

3. Charged states of defects (twice changing the valence, for instance, Cr^{2+} $\rightarrow Cr^+ \rightarrow Cr^{2+}$) capturing an electron or a hole represent photosensitive electron or hole centers. Luminescence centers in crystallophosphors are short-lived photosensitive electron or hole centers recombining with free holes or free electrons. The luminescence of crystallophosphors is the glow of electron-hole centers that appears on excitation and disappears in the course of luminescence. The EPR fixes the appearance of the center, and luminescence its decay.

Other Patterns of the Luminescence of Crystallophosphors. Further diversification and complication of the crystallophosphor luminescence patterns occur due to the following basic factors.

1. A single level in the forbidden band may characterize an ion or vacancy only in cases when their excited levels lie above the edge of the conduction band or do not split because of the type of their wave function symmetry (for instance, s-electrons or filled s^2, p^6, d^{10} -shells) or that of their crystalline field. In the general case the state of ion activators is described by levels split by a crystalline (ligand) field. The levels of electron and hole centers that originate from the ion activators during their ionization are the levels of new ions, which are also split by the crystalline field in conformity with their electron configuration and symmetry of the center, the configuration interaction being capable of causing a displacement of the levels (Stokes shift).

2. The excitation of luminescence can be induced not only by a UV radiation of an energy exceeding the interband spacing, but also by a light of much inferior energy. This is attended not by ionization of the ground substance, but by that of impurity ions and vacancy whose levels are situated in the forbidden band, and whose distance from the conduction band requires less energy for their ionization. Moreover, a light bias at a different wavelength (often by IR radiation) affects the luminescence, causing redistribution of electrons and holes between traps and bands and between different traps.

3. The presence of not a single but several impurity ions also affects the luminescence of each one of them. Unlike sensitization and quenching through energy transfer without ionization, ionization and transfer of charge occurs here (negative charge of the electron, and positive one of the hole). Redistribution of the charge (through bands) among various donors and acceptors, and among traps of different depth produces an intensification of luminescence of some and decay of other activators, along with a change in the duration of phosphorescence.

4. Spatial rapprochement of donors and acceptors in the lattice, i.e., formation of associated donor–acceptor pairs, ensures the possibility of transitions between ground and excited donor and acceptor levels with changes in their charge, but without any direct participation of the conduction band.

5. The activator levels may not occur in the forbidden band but may overlap the valence band, being situated considerably below the edge of the valence band. At any rate the luminescence of Mn^{2+} and RE^{3+} in ZnS is not attended by a change in

their valence following excitation (which is shown with the aid of EPR and UV light bias). This implies that the energy of their ionization is greater than that of the interband distance. However, here too the luminescence of these ions which is close to their luminescence in crystals with a great interband distance is affected by recombination processes.

Some of these possibilities are shown on the luminescence diagrams for ZnS : Cu (Fig. 105), ZnS : Cu, Cr (Fig. 106), ZnS : Cu, Ga (Fig. 107).

Phosphorescence and the Law of the Decay of Crystallophosphors. The stage of the ground substance or ion activator ionization following excitation of crystallophosphors is usually separated from subsequent stages through a temporary capture of free electrons and holes by shallow electron and hole traps (Fig. 108). Shallow traps (with the distance to the conduction band edge on the order of less than 0.1 eV) can capture electrons and holes, and serve, as it were, in the capacity of a reservoir supplying luminescence for a more or less lengthy time.

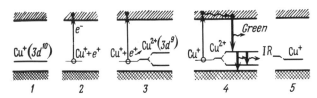

Fig. 105. Luminescence mechanism pattern for ZnS:Cu. *1*, initial position of Cu^+ level with filled $3d^{10}$ shell; *2*, excitation-ionization of $Cu^+ \rightarrow Cu^{2+}$; Cu^{2+} ($3d^{10}$) may be regarded as $Cu^+ + e^+$, i.e. as a hole center; *3*, splitting of Cu^{2+} ($3d^9$) levels by tetrahedral crystal field into two levels; *4*, luminescence due to transitions: free electron $\rightarrow Cu^{2+}$ level (green luminescence of ZnS: Cu), and between Cu^{2+} levels (IR luminescence), and between Cu^{2+} levels and valence band; *5*, recombination of free electron with hole center Cu^{2+} occurring in luminescence leads to restoration of the initial state Cu^+

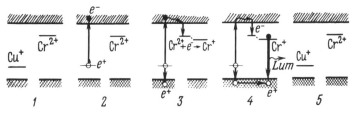

Fig. 106. Luminescence mechanism pattern for ZnS: Cu, Cr *1*, initial position of unassociated Cu^+ and Cr^{2+} ion levels; *2*, excitation-ionization of Cu^+; formation of a free electron in the conduction band and of a hole localized on $Cu^+ + e^+ \rightarrow Cu^{2+}$; *3*, capture of free electron by the $Cr^{2+} + e^- \rightarrow Cr^+$ ion; transfer of localized hole from $Cu^{2+}(= Cu^+ + e^+)$ to valence band: $[Cu^{2+} = Cu^+ + e^+] + S^2 \rightarrow Cu^+ + [S^- = (S^{2-} + e^+)]$ and migration of this hole which has become free to Cr^+; *4*, luminescence because of recombination of localized electron $Cr^+ = Cr^{2+} + e^-$ and of free hole; *5*, initial position of Cr^{2+} and Cu^+

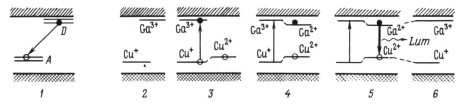

Fig. 107. Luminescence mechanism pattern of donor-acceptor pairs: ZnS:Cu, Ga. *1,* luminescence pattern of donor-acceptor pairs after Prener-Williams; *2,* starting position levels for associated ions Cu^+ ($3d^{10}$) and Ga^{3+} ($3d^{10}$) in ZnS (both ions are not paramagnetic); *3,* excitation-ionization of $Cu^+ \rightarrow Cu^{2+}$; *4,* capture of electron by Ga^{3+} with formation of Ga^{2+} ($3d^9$)—electron center; Cu^{2+} is a hole center ($3d^9$); Ga^{2+} and Cu^{2+} are paramagnetic and from both EPR signals are observed; *5,* recombination of the electron center Ga^{2+} and of the hole center Cu^{2+} with emission of luminescence; *6,* return to the initial state

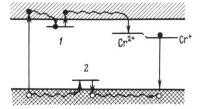

Fig. 108. Phosphorescence (ZnS:Cr, see Fig. 103) due to capture and giving off of electrons (*1*) and holes (*2*) by shallow traps

The glow of activators resulting from a heat release at a constant temperature of electrons and holes captured by shallow traps is phosphorescence. The glow of activators due to heat liberation with rising temperature of electrons or holes captured by shallow and deeper traps is thermoluminescence (see Chap. 6). The glow of activators secondary to optical liberation due to the effect of light of a certain energy on electrons or holes captured by shallow and deeper traps occurring under the effect of light is known as stimulated luminescence ("light flash").

The number of electrons or holes captured by traps determines the sum of stored light of the crystallophosphor, and its luminiscence is sometimes referred to as the release of the sum of stored light. The strong dependence of phosphorescence on the temperature is quite understandable, since phosphorescence is associated with heat release from the traps (longer after-glow at lower temperature and shorter at higher temperature).

The process of phosphorescence is conditioned by transitions not only with participation of the activator levels, but also of transitions between bands and the activator, with the transfer of the electron and hole involving temporary capture by traps, by the presence of traps with different depth, by possible repeat trappings, all reasons why the decay crystallophophors is of an extremely complicated nature and cannot be described by simple expressions.

In some cases the law of decay closely approaches the hyperbolic one: $I = At^{-\alpha}$ where I is the intensity of glow, t is time, A and α are constants. The law of decay is ordinarily approximately described by Becquerel's formula: $I = I_0 (1 + At)^{-\alpha}$ where I_0 is initial intensity.

The kinetics of the glow of crystallophosphors is considered in detail in special monographs [591, 664]. The capture of electrons and holes by shallow traps can take place also in crystals of the NaCl, $CaWO_4$, and other types. Upon their excitation with X-rays, electrons, and other high-energy emissions, a more or less lengthy phosphorescence can usually be observed.

Photoexcitation and Photoquenching of EPR and Crystallophosphors Luminescence. A feature specific to the glow of crystallophosphors that hinders their investigation is the fact that luminiscence spectra as such are not specific to each type of crystal, but are rather dependent upon the charge of the activator (of the donor or acceptor nature) than on its chemical individuality. One to two broad luminescence bands shift only slightly with a changed composition of activators of an equal charge, there being no drastic changes in the number and positions of bands in the spectrum as is the case, for example, in the RE^{3+} spectra changes. This is quite understandable, for the transition inducing luminescence is commonly made between a band and one of the activator's levels, and not inside the system of levels specific to each ion. Other experimental data, such as decay and thermoluminescence curves provide information merely on the emptying of traps, and on the position of the trap levels within the band scheme, and not on the levels of optical transitions enabling identification of the ion activators.

Application of the electron paramagnetic resonance (EPR) creates a new situation as regards standard demands on the description of luminescent systems in crystallophosphors [641, 642, 655–568, 696, 709, 712, 896, 806, 807, 813, 814, 826]. Instead of nameless designations, such as "luminescence centers", "capture centers", or "traps", determined merely by their position in the band scheme, it becomes possible to define them as concrete ions and vacancies and their charge states in the form of electron and hole centers. EPR permits obtaining all the information on the activator and its position in the crystal, such as charge, the state of bonding, the way of its incorporation in the structure, local symmetry with the pattern of levels ensuing therefrom, and the presence and ways of the charge compensation and concentration. Moreover, an observation of the EPR signal with light illumination furnishes information on the dynamics of the formation of luminescence centers and on the nature of the trapping centers, on the mechanism of the luminescence of crystallophosphors itself, associated with the capture and release of the charge. Finally, an investigation into photoexcitation and photoquenching enables the position of an impurity or vacancy ground level and electron-hole centers developing therefrom (i.e., their charge states) in the band scheme to be determined. This latter is of a more general significance, for if molecular orbitals are of decisive importance in identifying electron-hole centers from the EPR spectra (see Chap. 7), their position in the band scheme then determines their stability and mutual transformations.

In crystallophosphors the electron paramagnetic resonance may be observed in the initial state and on excitation of luminescence, usually with UV light.

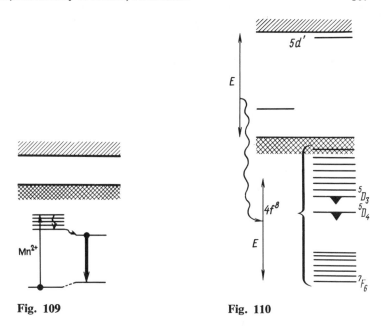

Fig. 109   **Fig. 110**

Fig. 109. Position of Mn^{2+} levels in ZnS; Mn^{2+} levels lie below the valence band edge [852, 853]

Fig. 110. Position of RE^{3+} levels in ZnS. Tb^{3+} in ZnS [588]: Tb^{3+} levels of $4f^8$ configuration lie below the valence band edge; the Tb^{3+} donor level is associated with excited configuration and $5d^1$-electron participation; excitation of the Tb^{3+} ion emission from $^5D_3$ and $^5D_4$ levels occurs as a result of a resonance energy E transfer

In ZnS, for instance, Mn^{2+}, RE^{3+} (Figs. 109, 110) is seen in the initial state EPR of. However, many activators and coactivators, such as Cu^+, Ag^+, Ge^{4+}, Pb^{4+}, Ga^{3+}, In^{3+}, Tl^{3+} (with filled d^{10}-shells) are not paramagnetic in their initial state. On being photoexcited and having trapped holes (Cu^{2+}, Ag^{2+} : d^9) or an electron (Ge^{3+}, Pb^{3+}, Ga^{2+}, In^{2+}, Tl^{2+} with an s-electron), they become paramagnetic and give an EPR signal.

The appearance of an EPR signal after excitation with a light illumination (ionization of the defect), and the dependence of its intensity on the wavelength of the excitation light are determined (according to the intensity maximum) by the distance of the initial impurity ion level from the conduction band edge (Fig. 111) or from the vacancies level. This spectrum of the EPR signal excitation coincides with the spectrum of the luminescence excitation.

The trapping of an electron or a hole by a defect (activator) and the appearance of the ensuing EPR signal can occur also with a UV light bias, and an energy greater than the interband distance. In this case, by acting upon the subsequently formed electron-hole centers with IR light of a wavelength equalling the distance of this center level from the nearest band, one can see quenching of the EPR signal and

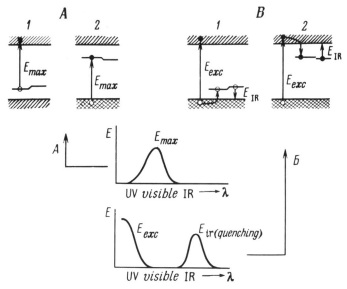

Fig. 111 A and B. Photoexcitation and photoquenching of EPR signal in ZnS crystallophosphors. **A** Photoexcitation of EPR (band scheme of photoexcitation spectrum); photoionization of impurity with formation of a hole center (*1*) (for example, $Cu^+ \to Cu^{2+}$, see Fig. 105) and of an electron center (*2*) (for example, Cr^{2+}); **B** photoquenching of EPR (band scheme and photodecay spectrum).
1, hole capture (for example, $V_{Zn}2+$, see Fig. 104) with UV excitation and quenching of EPR with IR light; *2*, electron trapping (for example, $Cr^{2+} \to Cr^+$, see Fig. 108) with UV excitation and quenching of EPR with IR light

concurrently decay of luminiscence. The spectrum of the EPR photodecay determines the position of the electron or hole center level, i.e., that of the defect in a charged state.

5.2.7 Methods of Luminescence Excitation

Luminiscence may be regarded as a way of converting various kinds of energy, such as optical, nuclear, electric, mechanical, and chemical, into light emission (detection of invisible radiations). This forms a background for a great variety of the ways for the application of luminescence. It is possible to set apart the types of luminescence depending upon the type of energy employed for their excitation. This is of decisive importance for singling out domains for the technical use of cathode luminophors, roentgen luminiphors, electroluminophors, and scintillators (neutorn and gamma-quanta counters of nuclear radiation).

As concerns the substance of the phenomenon of glow as such and the utilization of luminescence for determining all the peculiarities of the ion activator in a crystal (which is the principal objective of the luminescent investigation of minerals), this classification is of secondary importance, though it opens up some additional possibilities.

The following types of luminescence are distinguished[10] by the methods of their excitation.

Excitation by irradiation

A. *Electromagnetic irradiation (photo-, X-, gamma-rays):*
   Excitation without ionization
      Photoluminescence
   Ionization, formation of centers, excitation
      X-ray luminescence
      Gamma-luminescence
B. *Corpuscular irradiation (radioluminescence):*
   Formation of defects, ionization, formation of centers
   Excitation by charged particles
      Electron (cathode) luminescence
      Proton luminescence
      Ion luminescence
   With uncharged particles
      Neutron-luminescence
   Atom- and molecule-induced excitation
C. *Excitation by electrons trapped during irradiation and released from the capture centers on heating*
      Thermoluminescence

Excitation by a superimposed electric field
Electroluminescence (discharge glow)

Excitation with liberation of a part of the chemical reaction energy in the form of light energy
Chemiluminescence
Candoluminescence

Excitation during conversion of mechanical disintegration and crystallization energies
Triboluminescence
Crystalloluminescence

Excitation owing to biological processes
Bioluminescence

Consider some features peculiar to individual types of luminescence.

Photoluminescence, i.e., the glow under influence of excitation produced by light photons corresponding to the UV or visible regions of the spectrum. (In some

---

[10] One considers the emissions in one and the same region of the spectrum, visible and adjacent to it near-UV and IR, divided according to the method of their excitation. In the type of radiation emitted by the system are distinguished: (a) nuclear gamma-resonance fluorescence (Mössbauer emission spectra) arising as a result of transitions between intranuclear energy levels (see Chap. 1); (b) X-ray fluorescence arising during transitions between electron levels of inner, filled shells of free atoms and atoms in a crystal (see Chap. 2); (c) optical luminescence resulting from transitions between the outer electron levels.

instances an intensive luminescence can be induced also by IR light.) It is on the ground of the photoluminescence data that basic notions of luminescence have been considered in the preceding sections. This type of luminescence may be opposed to all other modes of excitation, for here occurs only excitation of ion activators without their ionization, and with no side complicating processes of the formation of electron-hole centers. Excitation can be completed electively for any level (excitation spectrum) which allows it to investigate the pattern of the activator's energy levels comprehensively in a great variety of ways. This mode of excitation is most simple as regards the necessary equipment and interpretation of the results.

In employing UV excitation with an energy greater than the interband distance, ionization of the ground substance takes place, accompanied by formation of electrons and holes, and the process of excitation acquires features similar to those occurring in excitation by ionizing irradiation.

General Features Distinguishing Excitation of Luminescence by Means of Ionizing Irradiation. High-energy radiation: electromagnetic (X-ray, gamma-ray) and corpuscular (with electrons, protons, neutrons, α-particles, recoil atoms, etc.) which while passing through crystals ionize atoms, and for this reason are known as ionizing radiation.

The types of luminescence arising under the effect of various kinds of nuclear radiation are put under a single heading of radioluminescence. Excitation of luminescence by ionizing irradiation forms part of a broader problem, the interaction of such emissions with the matter. The problem has diverse scientific and practical aspects, such as the effect of nuclear radiation on the matter and its properties (radiation chemistry), formation of electron-hole centers, fields of technical application. Here we shall dwell upon only one aspect of this phenomenon, on the utilization of ionizing radiation as a means of exciting luminescence. Its major features are as follows.

With all the diversity of processes occurring during irradiation, the first thing to be noted is that the way of achieving excitation of luminescence is always the same, although in utilizing various kinds of nuclear radiation the starting points differ, i.e., they can be nuclear reaction leading to the formation of charged particles and gamma-quanta or ionization of atoms in a crystal and formation of secondary electrons, etc. Secondly, the final result of the action exerted by irradiation and leading to excitation only after radiation energy has been spent in accomplishing processes of a wide variety of aspects and orders of energy (nuclear, bond breaking, formation of defects in the form of vacancies and interstitions, formation of centers) is always the same, that is the formation of electron-hole pairs. This stage may be regarded as absorption of the excitation energy by the lattice.

The next step involves transfer of absorbed excitation energy to some or other luminescence activator. In this respect radioluminescence may always be considered as luminescence sensitized by the lattice (but here the processes of the excitation energy absorption by the lattice and the process of de-excitation of the latter by the activator proceed separately.) The transfer of the excitation energy can be completed through an electron-hole recombination, or by means of the excition transfer, and it can also pass through an intermediate stage, storage of the light sum by the electron and hole-trapping levels.

The energy transfer can be short-term and result in momentary flashes (through excitons in scintillators: alkali iodides with Tl, or through recombination: in ZnS-based scintillators), or it can be of a fairly long duration in phosphorescent crystals with a stored light sum.

The glow induced by all types of ionizing irradiation arises as a result of transitions between levels of the same activators as in the case of photoexcitation. The position of the luminescence lines or bands of an activator in a given crystal does not change following excitation by nuclear radiation or by any other mode of excitation. Changes in the spectra can, however, appear owing to the fact that additional levels, nonradiative with UV excitation, become emissive and emission transitions occur onto extra levels along with redistribution of intensity in various lines. Furthermore, in many cases radioluminescence becomes manifest at the time when UV excitation either fails altogether to yield luminescence or produces a very weak glow. Conversely, in some instances the missing capability of an effective transfer of the excitation energy absorbed by the lattice results in the crystals which are strongly luminescent under UV illumination failing to be luminescent when exposed to ionizing radiation.

With radioluminescence an appreciably increased intensity of glow is possible, this being of paramount importance for minerals displaying weak glow. The luminescence intensities observed upon excitation by ionizing radiation can exceed by far the intensity of photoluminescence (but adding strength to the photoluminescence can be achieved also by employing powerful sources of UV excitation, expecially laser ones).

In addition to the luminescence observed in the activators both under UV and ionizing irradiation, extra luminescence bands stemming from electron-hole centers can appear, which can owe their origin to the already existing (earlier in crystals) defects (X- and gamma-ray excitation) or to those formed under the effect of corpuscular irradiation.

Technicalities of observing luminescence under the effect of ionizing irradiation are also essential. Sources of X-rays, as well as conditions of excitation, are more accessible. The sources of electrons, protons, and ions are part of special set-ups and require observation in vacuum.

The sources of neutrons and gamma-rays (nuclear reactors, radioactive isotopes) require for observation of the glow during irradiation the use of special light guides and bringing the spectrometer closer to the source of these types of radiation, to register luminescence. For this reason, in practice excitation with neutrons and gamma-rays is not used in observing luminescence at the time of irradiation. These types of irradiation are, however, employed in excitation of crystals with the purpose of accumulating the light sum in electron-hole centers which subsequently becomes manifest in the form of glow in thermoluminescence. The luminescence phenomenon is used for registration of neutrons and gamma-rays when they gain access into the scintillator, a radiolumnescent crystal.

Roentgenoluminescence is excited with the help of ordinary X-ray tubes and its intensity increases with rising voltage (40–50 kV) and amperage. In roentgenoluminescence many additional lines emerge in the spectra of rare earths as contrasted against optical excitation. With X-ray excitation of ruby there appear alongside the

Cr^{3+} lines also luminiscence bands associated with centers, arising secondary to irradiation, with a long-lived phosphorescence demonstrable after discontinuance of excitation. In the alkaline-haloid crystals excited by X-rays, a rapid decay component (on the order of 10^{-6} s) is in evidence, originating from the exciton energy transfer from the crystal to the luminescence center, and also a slow one, consequent upon the process of an electron-hole recombination. Roentgenoluminescence finds application in sorting and grading of diamonds.

Cathodoluminescence is excitation by electrons generated and accelerated in the cathode tube. It permits luminescence to be excited in the greatest number of minerals, and in this respect it is a more universal mode of excitation than optical and X-ray excitation. The penetration depth of electrons in a crystal is not great (about 0.002–0.003 mm for electrons with energy of 10^{-20} kV) and, therefore, the excitation density is high, and the glow can reach a degree of a brightness unattainable by all other modes of excitation. The brightness of cathodoluminescence is proportional to the accelerating voltage and amperage, but the increase of voltage is limited by the destruction of the specimen under electron bombardement. In the glow the share of the swift-decaying component (according to the exponent) is dominant with a very small proportion (less than 1%–2%) of long-term phosphorescence. Cathodoluminophors represent a most important field of technical luminescence application (oscillographs, TV tubes, radiolocation indicators, image converters).

Cathode- and photoluminescence are the most effective methods of observing the glow of minerals, the roentgenoluminescence playing the role of secondary importance [670, 818]. Gamma-ray and neutron excitation is used for accumulation with their aid of a light sum which is subsequently de-excited in producing thermoluminescence. The remaining modes of excitation are of significance only in dealing with particular cases. The cathode luminescence acquires additional importance in connection with the use of the X-ray microanalyzer in which an electron probe induces the glow of luminescent minerals [749].

Protonoluminescence is excitation of glow by positive ions of hydrogen with an energy on the order of 5 keV. In recent years it has attracted attention in connection with simulating possible luminescence processes on the moon under the effect of the "solar wind" (a stream of corpuscular particles, largely protons and nuclei of helium [684, 773]. In the protonoluminescence spectra of the most widespread rock-forming minerals (feldspars, quartz, pyroxense) intensive bands that are absent in photo- and cathodoluminescence spectra are in evidence.

Thermoluminescence is of dual significance. First it is a method of detecting electron trapping centers emerging upon irradiation (see Chap. 6). However, the liberation of electrons from traps on heating can be used as a mode of exciting the luminescence of ions. This mode of excitation has been utilized very effectively in investigating luminescence of rare earths in fluorite [767, 827]. The method may prove to be of value in observing the spectral composition of the low intensity glow in photoexcitation, and also in the presence of several activators, when the pattern of the energy transfer between them may change.

Electroluminescence [652, 660, 694] originates as a result of applying an electric field (of more than 300 V) with alternating voltage (countinuous glow) and direct-current voltage (a flash, quickly decaying, or continuous glow in monocrystals). A direct conversion of electrical energy into luminous energy that ensues therefrom is taken advantage of in luminescent lamps. Electroluminescence is usually common to activated semiconducting crystals of the ZnS type, and stems from processes in an electron transition to the conduction band and their subsequent recombination, i.e., here the recombination mechanism of the luminescence excitation comes into play. The electroluminescence spectra are analogous to the spectra obtained by the UV and other modes of excitation. Diamonds have been subject to a detailed investigation [674]; the glow of minerals placed in a condensator has also been under observation, of asbestos (anthophyllite, rhodusite, rhezhikite, chrysotile), vermiculite, barite, scheelite, calcite, and fluorite [718].

A precondition for observing electroluminescence is sufficient electric conductivity of the crystals. In crystals of low conductivity this can be produced by an activating ion. However, in the face of a forbidden band of considerable breadth an increased conductivity is not seen to occur even in the presence of impurities. In such cases the observation of electroluminescence is possible upon heating the crystal, provided the activator is not liable to decay under the effect of heat. Thus one could observe electroluminescence of Tb^{3+} in CaF_2 [602] upon heating up to $400°–500°C$ with a spectrum analogous to that of photoluminescence (like most RE^{3+}, the Tb^{3+} ion is not liable to undergo decay under the influence of heat).

Chemiluminescence [600] is a glow appearing during a chemical reaction. In chemical reaction the free energy is almost always released in the form of heat, and only in some individual reactions (oxidation of various substances with oxygen or hydrogen peroxide) in the form of light. This process is the reverse of photochemical reactions, i.e., of reactions proceeding under the effect of light. The result of the reaction is the appearance of a glow. Examples of this are oxidation of phosphorus, of unsaturated silicon oxide and, in nature, the luminiscence of some bacteria and plankton. It finds application in chemiluminescent analysis [600].

Candoluminescence [821] is a glow appearing under the influence of flame, but is not of thermal radiation; it is of a radical-recombination origin, i.e., adsorption of free atoms and radicals on the surface of crystals occurs here followed by their subsequent recombination with the surface of crystals. The adsorbed particles are not centers of the glow, but serve as a source of electrons and holes that transfer the energy released during reaction of the superficial recombination to the activator ions, the same which are active in photoluminescence. The mechanism of candoluminescence is considered from the standpoint of the chemisorption electron theory [821].

Triboluminescence [606] is a glow observed in disintegration of crystals, for instance under the effect of axial compression, appearing as momentary light pulses. The development of photoelectric methods for registering the spectrum of this flash, and the methods that enable it to avoid large expenditures of matter and time, make their detailed investigation possible. The spectra include, on the one hand, bands and lines characteristic of a given substance or activator, and nitrogen lines [383, 410, 415, 444 nm] emerging due to ionization of the nitrogen atoms in the air surrounding the crystal, on the other.

Triboluminescence has been observed in many minerals, such as arsenolite, clodetite, antimonite, aragonite, barite, celestite, fluorite, gypsum, mica, cryolite, quartz, rutile, halite, witherite, sphalerite, and alunite. Among triboluminescent crystals are encountered crystals endowed also with the property of emitting light during their crystallization from a solution or from a melt, called crystalloluminescence.

5.3 Types of Luminescent Systems in Minerals

Interpretation of luminescence of concrete crystals in mineralogy implies above all determination of the nature of the glow center and of features specific to its presence in a crystal so as to elucidate their dependence on conditions attending their formation. To this end it is necessary that the bands observed in the luminescence spectrum be ascribed to transitions between a system of levels describing the peculiarities of the electronic structure of the glow center in a given crystalline field, and in given conditions of the energy transfer between ions and between the center and the lattice.

The glow centers in minerals are always ions (impurity or basic ones), or electron-hole centers whose glow proceeds mostly without participation of the conduction band, and only in semiconductors, sulphides of the ZnS type, with its participation. For this reason, the systematics of the luminescent system types in mineralogy is most convenient to base on the typing of the electronic structure of the glow centers: transition metal ions, rare earths, actinides, heavy metals, electron-hole centers, and crystallophosphors of the ZnS type.

5.3.1 Transition Metal Ions; the Crystal Field Theory and Luminescence Spectra [633, 725, 739, 751, 789]

The significance of the transition metal luminiscence in minerals is determined by the fact that these metals include the most prevalent luminophor in nature, Mn^{2+}, the most wide-spread, strong quencher, Fe^{3+}, and Cr^{3+}, the ion-emitter largely used in lasers.

The energy level diagrams of transition metal ions in crystal fields form a basis for the overall consideration of the luminescence spectra in the same measure as the diagram of levels in the optical absorption spectra. Therefore, the energy level diagrams of the ions form a starting point for the analysis of their luminescence as well [633].

Among the levels are distinguished emission and radiationless states (depending upon the lifetime and their relation to the configuration curves model, see Chap. 5.2 and Fig. 78) and also levels onto which transitions yield broad absorption bands, these levels being used for excitation of luminescence in them.

The dependence of luminescence on the valence, coordination, and the state of chemical bonding is determined in the same way as are absorption spectra, by using crystal field or molecular orbitals parameters.

Chromium in Ruby, Emerald, Cyanite (Fig. 112), Spinel, Luminescence and Lasers [625, 661, 678, 708, 834]. These systems were chosen as an illustration in presenting basic notions of luminescence (see Fig. 76). Schematic diagrams of the Cr^{3+} levels in crystal fields of varying symmetry and absorption spectra of these minerals make it possible to understand the way in which fluorescence occurs in them, and why these systems form an ideal material for lasers. Excitation is effected by green (in the broad band of $^4A_2 \rightarrow {}^4T_2$), violet (in the broad band of $^4A_2 \rightarrow {}^4T_1$), and UV light, while the emission spectrum consists of a sharp doublet $^2E \rightarrow {}^4A_2$ (split through a spin-orbital interaction). It is exactly this doublet R_1 and R_2, yielding a faint red glow in the luminescence (or not observable at all because of self-absorption), that manifests itself in the form of a high-power ruby radiation.

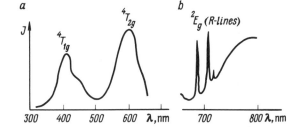

Fig. 112. a Excitation and **b** luminescence spectrum of Cr^{3+} in kyanite [834]

To secure generation of laser emission three conditions have to be met. (1) Excitation ("pumping") should be strong enough to make the occupancy of the emission level ^2E greater than the occupancy of the ground state $^4A_2$ ("occupancy inversion"), this being achieved by using lamps of 200 kw capacity. (2) The ruby crystal is made in the shape of a rod whose ends are placed between mirrors, one of which is semi-opaque. The spontaneous emission from the ^2E level becomes forced, i.e., it occurs under the effect of emitted light photons. The presence of the semi-opaque mirror enables a part of the generated light to be used in the form of laser emission, and the part reflected and remaining in the rod to be utilized for repeatedly building up an induced radiation by Cr^{3+} atoms, i.e., to produce an avalanche of photons. (3) The generation of a coherent laser emission occurs only when the threshold capacity value has been reached.

Because the ruby is used as a laser, the emission of the Al_2O_3: Cr^{3+} system has been studied in great detail [678, 708, 751].

With a Cr^{3+} concentration higher than usual in laser rubies, ($> 0.05\%$) additional N_1N_2 lines stemming from emission of Cr^{3+} ion pairs appear in the luminiscence spectrum.

Luminescence of Mn^{2+} [695, 751, 834]. The diagrams of the Mn^{2+} energy levels are shown in Figures 77, 78, 79, and 109, while the mechanism of luminescence was dealt with earlier with the example of calcite.

In the Mn^{2+} energy level diagram there are no broad absorption bands and, therefore, excitation is ordinarily effected by UV irradiation. An intensive glow is observed in cases of roentgenoluminescence and cathode luminescence, in particular, with the aid of an electron probe. The Mn^{2+} ions represent also most frequently luminescent centers in thermoluminescence.

Changes in the Mn^{2+} luminescence spectra in different minerals are first of all due to a change in coordination. With transition from octahedral to tetrahedral coordination, the pattern of levels for ions with the d^5-configuration to which Mn^{2+} belongs remains the same with only the strength of the crystal field Dq (and of parameters, B, C) being subject to variations. As against octahedron, in tetrahedron ($Dq_{tetr} = 4/9 Dq_{oct}$) the energy of all transitions increases with diminishing Dq. Accordingly, also the color of luminescence and the position of the only broad emission band that determines it become subject to change: in the octahedron the orange-red glow ($^4T_{1g} \rightarrow {}^6A_{1g}$ for example, calcite) is typical, whereas in the tetrahedron it is green ($^4T_1 \rightarrow {}^6A_1$ for example, willemite).

However, along with changed coordination one has to take account of variations in the strength of the crystal field, related to covalence, and in the length of interatomic distances as well as in the value of the Stokes shift. Moreover, the position of the luminescence band maximum and the glow color are affected by the Mn^{2+} concentration: in silicate glasses the color of the glow changes from green (with 0.05–0.1 MnO weight %) through a double (with green and red maxima) band (with 1%–3% of MnO) to a red one (with more than 3% of MnO). Green luminescence is related to single Mn^{2+} centers, and the red to luminescence of paired Mn^{2+} clusters. With elevated concentrations of Mn^{2+} (more than 0.05%) the luminescence spectrum is doubled, and the color of the glow is defined both by the position of each maximum and by correlation of their intensities, as well as by the degree of the bands' overlap.

The most intensive Mn^{2+} glow in thermoluminescence spectra occurs with its concentration on the order of 0.00n% Mn^{2+} [765]. Between Mn^{2+} and RE^{3+}, Mn^{2+} and Pb^{2+} a transfer of energy can take place [763, 772, 808]. Mn^{2+} acts as a sensitizer with respect to some rare earths, and as a result of this no Mn^{2+} glow may be observable in their presence.

From among other elements of the iron group titanium, with which the luminescence of quartz is presumably associated [755, 762, 834, 840], is possibly of great importance.

Quite unexpectedly the luminiscence of the Fe^{3+} ion with its low concentrations proved to be quite common [834] among minerals (Fig. 113).

Fe, Co, Ni ions are strong quenchers of luminiscence; the glow of Co and Ni described by the energy level diagram in the crystal field can be seen with their very low concentrations [795].

5.3.2 Rare Earths; Absorption and Luminescence Spectra [588, 626, 639, 658, 662, 706, 712, 737, 748, 759, 767, 792, 822, 842]

The spectroscopy of rare earths in crystals embraces to date a complex set of investigations that include closely related optical and luminiscent spectra, laser emis-

Fig. 113. Excitation and luminiscence spectra of Fe^{3+} in $LiAlO_5$ and adularia [834]

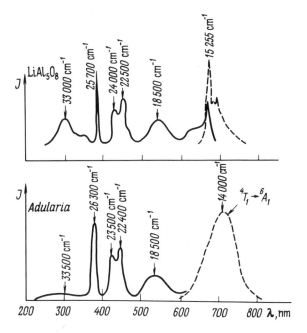

sion, formation of additive and radiation centers, EPR, and thermoluminescence. What unites them is the nature of phenomena proceeding between systems described by the patterns of the RE ion levels and their position in the band scheme, which encompass closely connected mechanisms of excitation, energy transfer, and the charge capture, thermal, optic, and radiation ionization, compensation of the charge, and formation of electron-hole centers.

In mineralogy this trend has a great potential thanks to its new possibilities, since all the trivalent and nearly all divalent ions of rare earths are identified in the luminiscence spectra of minerals and, according to the EPR spectra, some tetravalent rare earths, i.e., the spectra help not only to identify an element, but also to determine its valency unambiguously and, what is most important, also features specific for its position in a structure, such as compensation of the charge and different symmetry in the structural position of the ion, relation between the latter and electron-hole centers, as well as thermoluminescence.

The patterns of the RE^{3+} and RE^{2+} energy levels (Figs. 114–116) and common distinctive features of their absorption spectra help also to gain insight into the luminescence spectra of rare earths.

The Electron Configurations of the RE^{2+} and RE^{3+} Ions are listed in Table 12 where they are given together with electron configurations of neutral atoms. In their complete and general form they may be put down as

$$[1s^2 2s^2 2p^6 3s^2 3p^6 3d^{10} 4s^2 4p^6 4d^{10}]\boxed{4f^k}\,\boxed{5s^2 5p^6}\,\boxed{5d^1}\,\boxed{6s^2}.$$

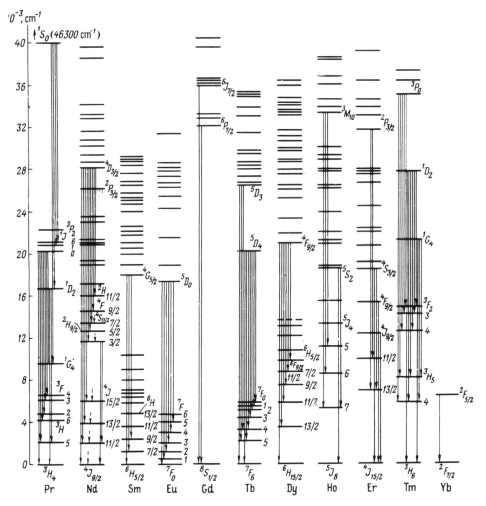

Fig. 114. Diagram of energy terms and emission transitions of RE³⁺ in CaF₂ due to X-ray excitation [848]

From among outer electrons were singled out:

Valence electrons: $6s^2$ whose detachment corresponds to RE²⁺ or $6s^25d^1$ (the sole $5d$-electron exists in atoms: Ce, the first in the RE series, Lu- the last and Gd, middle in this series), or to $6s^24f^1$, whose detachment corresponds to RE³⁺.

Shielding electrons: $5s^25p^6$ that form two completely filled shells $5s^2$ and $5p^6$, remaining unchanged when ions RE²⁺ and RE³⁺ form and shielding deeper-lying $4f$-electrons.

Unpaired electrons: $4f^k$-electrons that form part of the $4f$-shell in the stage of completion, these being related to the optical absorption, EPR, luminescence spectra, and to magnetic properties of rare earths.

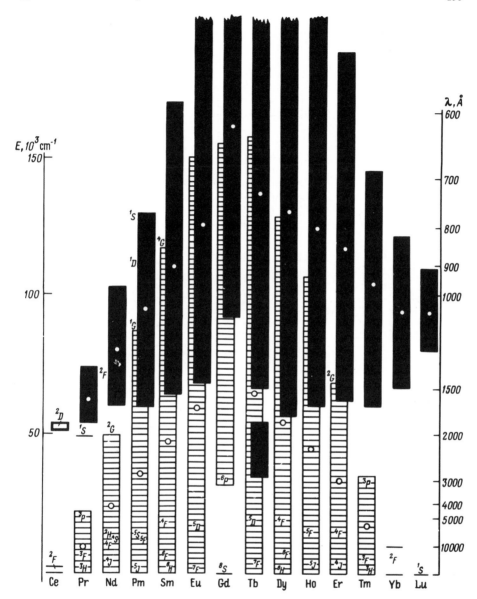

Fig. 115. Relative terms position of $4f^n$ (*white*) and $4f^{n-1} 5d$ (*black*) configurations for trivalent RE^{3+} ions [639]

For trivalent RE^{3+} ions in a briefly designated electron configuration it is only the number of 4f-electrons (bearing in mind also the constant presence of $5s^2 5p^6$ electrons), starting from $4f^1$ for the first RE^{3+} ion in the series of rare earth elements

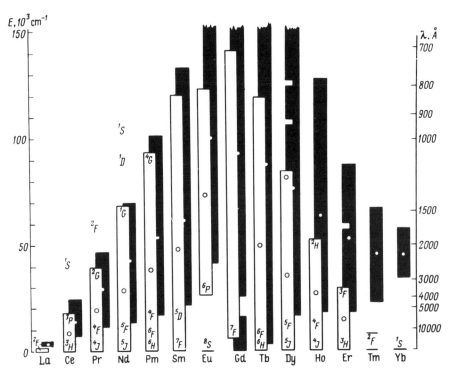

Fig. 116. Relative terms position of $4f^n$ and $4f^{n-1}$ $5d$ configurations for divalent RE^{2+} ions [639]

Table 12. Electron configuration of rare earth atoms and ions $4f^k$ [$5s^25p^6$]$5d^1$[$6s^2$]

Element	Atomic number	RE^0	RE^{2+}	RE^{3+}
Ce	58	$4f^1$ [$5s^25p^6$] $5d^1$ [$6s^2$]	$4f^1$ [$5s^25p^6$] $5d^1$	$4f^1$ [$5s^25p^6$]
Pr	59	$4f^3$ [$5s^25p^6$]-[$6s^2$]	$4f^3$ [$5s^25p^6$]	$4f^2$ [$5s^25p^6$]
Nd	60	$4f^4$ [$5d^25p^6$]-]$6s^2$]	$4f^4$ [$5s^25p^6$]	$4f^3$ [$5s^25p^6$]
Pm	61	$4f^5$ [$5s^25p^6$]-[$6s^2$]	$4f^5$ [$5s^25p^6$]	$4f^4$ [$5s^25p^6$]
Sm	62	$4f^6$ [$5s^25p^6$]-[$6s^2$]	$4f^6$ [$5s^25p^6$]	$4f^5$ [$5s^25p^6$]
Eu	63	$4f^7$ [$5s^25p^6$]-[$6s^2$]	$4f^7$ [$5s^25p^6$]	$4f^6$ [$5s^25p^6$]
Gd	64	$4f^7$ [$5s^25p^6$] $5d^1$ [$6s^2$]	$4f^7$ [$5s^25p^6$] $5d^1$	$4f^7$ [$5s^25p^6$]
Tb	65	$4f^8$ [$5s^25p^6$] $5d^1$ [$6s^2$]	$4f^8$ [$5s^25p^6$] $5d^1$	$4f^8$ [$5s^25p^6$]
Dy	66	$4f^{10}$[$5s^25p^6$]-[$6s^2$]	$4f^{10}$[$5s^25p^6$]	$4f^9$ [$5s^25p^6$]
Ho	67	$4f^{11}$[$5s^25p^6$]-[$6s^2$]	$4f^{11}$[$5s^25p^6$]	$4f^{10}$[$5s^25p^6$]
Er	68	$4f^{12}$[$5s^25p^6$]-[$6s^2$]	$4f^{12}$[$5s^25p^6$]	$4f^{11}$[$5s^25p^6$]
Tm	69	$4f^{13}$[$5s^25p^6$]-[$6s^2$]	$4f^{13}$[$5s^25p^6$]	$4f^{12}$[$5s^25p^6$]
Yb	70	$4f^{14}$[$5s^25p^6$]-[$6s^2$]	$4f^{14}$[$5s^25p^6$]	$4f^{13}$[$4s^24p^6$]
Lu	71	$4f^{14}$[$5s^25p^6$]$5d^1$ [$6s^2$]	$4f^{14}$[$5s^25p^65d^1$]	$4f^{14}$[$5s^25p^6$]

and ending with $4f^{14}$ for the last, 14th ion Lu^{3+} of this series, that is indicated. For divalent RE^{2+} ions the number of $4f$-electrons is, accordingly, one more, except for Gd-$4f^75d$ and Lu-$4f^{14}6s$.

For some RE ions tetravalent states (Ce^{4+}, Pr^{4+}, Tb^{4+}, etc.), differing from the trivalent ones in detachment of one more electron, and univalent ions which, as distinct from the neutral atom, have merely a single $6s$-electron, are noted.

RE^{3+} ions are isoelectronic with divalent RE^{2+} ions of the preceding element in the periodic system. Eu^{3+} and Sm^{2+} for instance, have the same $4f^6$-electron configuration.

Energy Levels of Trivalent RE^{3+} Ions in Crystals. Plotting a complete diagram of energy levels for the rare earth ions is done as follows.

1. From the electron configuration of the RE ions (see Table 12) a complete diagram of energy levels is determined unambiguously.

2. The deduction of terms possible for a given configuration is simple only for one to two f-electrons; for multi-electron f-configuration it has been considered in detail [256] and its final results are available in the table of possible terms for f-electrons [639, 653].

The cause accounting for a weak influence of the crystal field in rare earth compounds is the presence in the ions of $5s^25p^6$ outer shells beneath which $4f$-electrons lie. The mechanism of the $5s^25p^6$-electrons' action is explained not so much by their part in shielding, but rather by the fact that the wave functions of $4f$-electrons have a range that does not go as far as in $3d$-electrons, whereas $5s^25p^6$-electrons, conversely, enlarge the size of the rare earth ions, which are much larger than the ions of the iron group. Therefore, splitting of the rare earth ions levels by the crystal field reaches a magnitude on the order of 10^2 cm^{-1}, whereas in the group of iron this splitting amounts to 10^3–10^4 cm^{-1}. In rare earth ions the spin-orbital interaction is on the order of 10^3 cm^{-1} (from 600 to 2,500 cm^{-1}), and in the ions of the iron group it amounts to $\sim 10^2$ cm^{-1}.

In the rare earths' group, on the contrary, the free ion terms (for example, ^3H for the $4f^2$-configuration) first split into multiplet levels of the same free ion, due to the spin-orbital interaction ($L \pm S = I$, i.e., here quantum number I appears that was absent in the description of the iron group ions). For example, ^3H splits into three multiplet levels $^3H_6$, $^3H_5$, $^3H_4$. These levels are the same as in free ions. It is only after this that splitting of the multiplet levels by a weak crystal field occurs, determined not by an orbital quantum number (i.e., for the term ^3F of $3d^2$-configuration by the quantum number $L = F = 3$), but by the quantum number I (i.e., for the multiplet level $^3H_6$ $I = 6$, for $^3H_5$ $I = 5$ and for $^3H_4$ $I = 4$).

3. From terms possible multiplet levels are obtained. For instance, for term ^3H: (a) the designation ^3H corresponds to ^{2S+1}L, i.e., $L = H = 5$; $2S + 1 = 3$, i.e., there are three multiplet levels, orbital quantum number $L = 5$, spin quantum number $S = 1$, (b) total quantum number assumes the value from $I = L + S$ to $I = L - S$, i.e., from $I = 5 + 1 = 6$ to $I = 5 - 1 = 4$, in other words there are three multiplet levels: $^3H_6$, $^3H_5$, $^3H_4$.

Thus it is possible to obtain a complete set of multiplet levels that conform to a given electron configuration. The quantitative arrangement of these levels is determined experimentally: for free ions, from the emission spectra data [653]; for all the RE^{3+} ions introduced as impurity into the $LaCl_3$ crystals [639], from the absorption and fluorescence spectra data.

If in the iron group the levels of ions in crystals differ sharply from free ion levels, in the rare earth group the multiplet levels of the RE^{3+} ions in crystals match qualitatively and stand quantitatively very close to the free ion levels. Moreover, the diagrams of multiplet levels shown in Figure 114 for the RE^{3+} ions in CaF_2 [848] closely resemble the diagrams of the RE^{3+} levels in such crystals as $LaCl_3$ [626, 639, 678, 708], ZnS [588, 589], $CaWO_4$ [616, 706], and in glasses [711], all of which differ greatly in the coordination and the state of the chemical bonding.

4. Splitting of each one of the multiplet levels by the crystal field is determined by the quantum number I and the symmetry of the crystal field. The number of the crystal field levels into which each multiplet level splits in crystal fields of different symmetry is readily definable from Table 9. This table is common for all ions, since it is obvious that the splitting of multiplet levels by the crystal field, for example, of $^6H_{15/2}Dy^{3+}$ and $^4I_{15/2}Er^{3+}$, will be the same, for in both cases $I = 15/2$. In Table 9, for the sake of convenience the splitting for ions with an odd number of electrons (I assumes semi-integral values) and with an even number of elcetrons (I-integral values) is presented separately. The total degeneracy of the levels equals $(2I + 1)$, but with odd number of electrons there remain doublets (called "Kramers' doublets"), even in cases of the lowest symmetry.

For instance, the multiplet level $^3H_6$ with $I = 6$ and an even number of electrons ($2S + 1 = 3$; $S = 1$, two electrons) has a total degeneracy of $(2I + 1) = 13$, and is split by a cubic field into singlets (4×1) and triplets (3×3), by a trigonal field—into five singlets (5×1) and four doublets (4×2), while a rhombic field completely lifts degeneracy, producing splitting into 13 singlets (13×1).

To assign transitions in the absorption and fluorescence spectra of the RE^{3+} ions in practice it is possible to make use of Figure 114 (determination of multiplets) and Table 9 (determination of splitting by the crystal field).

Excited $4f^{k-1}5d$-, $4f^{k-1}6s$-, $4f^{k-1}6p$-Configuration of Trivalent RE^{3+} Ions. The considered terms and multiplet levels, both ground and excited, arise from the $4f^k$ electron configuration. However, with further excitation, one of the $4f$-electrons passes into the $5d$-state and further on into $6s$- and $6p$-states. From $4f^k$ there develop mixed $4f^{k-1}5d$-configurations (and further on $4f^{k-1}6s$- and $4f^{k-1}6p$-configurations. However, the terms that correspond to these mixed configurations have for trivalent RE^{3+} ions high energies that are accordant with a relatively far UV region (as distinct from the divalent RE^{2+}).

Energy Levels of Divalent RE^{2+} Ions in Crystals [639, 759]. Two specific features determine the similarity and difference of energy levels in di- and trivalent ions of rare earths.

1. Isoelectronic configurations of divalent RE^{2+} ions and trivalent RE^{3+} ions of the next elements in the periodic system determines a qualitatively similar pattern of terms and multiplet levels, i.e., for RE^{2+} the order of the $4f^k$-configuration levels and of excited $4f^{k-1}5d$-configuration levels is the same as for trivalent ions isoelectronic to them, shown in Figure 114.

2. Difference in the charge of the nuclei in isoelectronic RE^{2+} and RE^{3+} ions results, on the one hand, in an essential but relatively small and approximately

equal (~ 1.2) fall of energies of all levels derived from $4f^k$-configuration (i.e., the RE^{3+} levels diagram in Figure 114 "contracts" by $\sim 20\%$ with respect to isoelectronic RE^{2+}) and, on the other hand, it brings about a sharp drop in the energy of levels derived from mixed $4f^{k-1}\ 5d$-configurations. It is just this fall of the $4f^{k-1}$ $5d$ levels that causes a sharp difference in the absorption and luminescence spectra of the RE^{2+} and RE^{3+} ions.

General Characteristic of Optical Spectra for RE^{3+} and RE^{2+} Ions in Crystals. In the absorption spectra of the rare earth ions in crystals transitions are observable between levels arising within the limits of the $4f^k$-configuration (f–f-transitions) and the levels of $4f^k \rightarrow 4f^{k-1}\ 5d$ and other mixed configurations. These transitions are rather different. Transitions within the $4f^k$-configuration lead to the line spectrum consisting of weak narrow absorption lines revealing a fine structure.

The peculiarities of the line spectrum are explained by the nature of transitions determining it:

a) These are transitions between the states of one and the same configuration and, consequently, are parity-forbidden. This forbiddenness, strictly observed in the spectra of free ions, is less stringent for crystals and solutions owing to non-centrosymmetric interactions producing mixing of states of differing parity, and it is this that leads to the appearance of weak absorption lines of the rare earth ions in crystals and solutions, the oscillator strength being here on the order of 10^{-6} (from 10^{-8} to 10^{-5}).

b) These are also transitions within the $4f^k$-configuration well-shielded from the action of the crystal field, which preconditions: small splittings by the crystal field that reveals as a fine structure, small width of lines and slight displacements of these narrow absorption bands when passing from one crystal to another (on the order of 10–100 cm^{-1}) and from crystals to solutions. Changing symmetry of the crystal field in different compounds is a factor determining the appearance of a different number of fine structure lines and their relative intensity is determined by selection rules for transitions between the crystal field levels. The number of lines, however, is often not in keeping with that expected by the crystal field theory, this being usually due to the interaction with vibrational transitions. One should also consider possible nonequivalent positions of the ion due to a different manner of compensation of the charge.

Because of a small displacement of the same narrow absorption bands in different crystals, their assignment to transitions between multiplet levels (see Fig. 114) is ordinarily not difficult. However, it is another and a much more complicated thing to assign the fine structure lines so that they match transitions between the crystal field levels and to determine nonequivalent positions. In most cases this necessitates obtaining spectra at low temperatures. A comparatively large number of lines in the line spectra (except for the first and last ions in the rare earth series) is explained by the great number of multiplet levels terms, the transitions onto these levels falling into the visible and UV and IR regions.

Transitions onto the levels of mixed $4f^{k-1}\ 5d$-configuration, and also onto the $4f^{k-1}\ 6s$ and $4f^{k-1}\ 6p$ levels, result in the appearance of broad intensive absorption bands. These specfic features are due, first, to the fact that the said transitions occur between states with dissimilar electron configurations and, therefore, are parity-allowed. In these transitions the oscillator strength is of the order of $\sim 10^{-2}$–10^{-5}, i.e., by 3–4 orders higher than for the forbidden f–f transitions; effective absorption cross-section is here about 5.10^{-1} cm^2. Second, in these transitions d-electrons are involved, unshielded against interactions with the lattice, which makes for: (a) greater splittings by the crystal field (12,000–16,000 cm^{-1}), i.e., larger than splitting due to the spin-orbital interaction, this corresponding to the case of a strong crystal field; (b) greater half-width of the lines (on the order of 1500 cm^{-1}); (c) greater shifting of these bands in different crystals (on the order of 100–1000 cm^{-1}).

At low temperatures very narrow lines of electron-vibrational transitions appear superposed upon broad bands.

A relative ratio between the $4f^k$ and $4f^{k-1}$ $5d$-configuration levels energies (see Figs. 115, 116) specifies a sharply distinctive position of broad bands in the spectra of trivalent and divalent ions of rare earths.

In the RE^{3+} spectra broad bands fall into a relatively far UV region (from $\sim$ 33,000 cm^{-1} for Ce^{3+} to $\sim$ 71,000 cm^{-1} for Yb^{3+} and $\gtrsim$ 80,000 cm^{-1} for Gd^{3+} and Lu^{3+}) and, for this reason, the RE^{3+} ions yield only line spectra in the visible and adjacent regions. In the UV region are distinguished, for instance in the case of Ce^{3+} in CaF$_2$ (a) broad (half-width of 2000 cm^{-1}) intensive $4f \rightarrow 5d$ bands in the region 31,000–50,000 cm^{-1}; (b) weak broad CaF$_2$: Ce^{3+} bands between 60,000–77,000 cm^{-1}; (c) displacement of the CaF$_2$: Ce^{3+} absorption edge (as against the absorption edge of 80,000 cm$^-$ CaF$_2$ with no impurities) associated with the F$^-(2p^6) \rightarrow$ Ce^{3+} ($6s$) charge transfer band.

In the RE^{2+} spectra broad bands fall into the visible and near-UV regions. The narrow weak lines of the f–f transitions superimposed on these intensive bands almost fail to show up and, therefore, the RE^{2+} ions spectra are represented by broad bands which determine the color of crystals.

Energy Levels Diagrams and Luminescence Spectra of RE. From multiplet RE^{3+} levels (see Fig. 114) come principal transitions (observable in low-resolution spectra) leading to absorption and luminescence. In obtaining absorption and luminescence spectra with high resolution at low temperatures fine structure lines become evident in lieu of these narrow bands, and to interpret the former, one has to take account of the splitting of each one of the multiplet terms into levels arising under the effect of the crystal field of a corresponding symmetry.

Let us consider the general characteristics of the RE luminescence spectra.

1. The RE^{3+} and RE^{2+} spectra differ sharply. Although in both cases they are associated with two types of transitions: f–f and f–d, i.e., transitions between terms of the f-configuration, or between the terms of the f-configuration and terms of mixed f^{k-1} d-configuration in the RE^{3+} spectra the f–d transitions (in absorption and luminescence) occur in the UV region and do not overlap the f–f transitions, whereas in the case of RE^{2+} the f–d and f–f transitions lie close to each other and overlap. The RE^{3+} spectra stemming basically from f–f transitions are line spectra, and they consist of a large number of narrow and weak lines. The RE^{2+} spectra (f–d transitions) are broad intensive bands in absorption and luminescence.

2. The presence of a large number of lines in the RE^{3+} luminescence spectra is due to the fact that the lower multiplets (for instance ^7F in Eu^{3+}, see Fig. 114), are split into terms $^7F_6$, $^7F_5$, $^7F_4$, $^7F_2$, $^7F_1$ $^7F_0$.

Whereas the absorption occurs from a single ground level $^7F_0$ of this multiplet, luminescence transitions are made onto all the levels of these ^7F multiplets. Moreover, the transitions can be made not from a single excited level, but from several of them (for instance, in the case of Tm^{+3}, see Fig. 114, from $^3P_0$, $^1D_2$, $^1G_4$ onto all the $^3F_2$, $^3F_3$, $^3F_4$ and all the $^3H_5$, $^3H_4$, $^3H_3$ levels). As a result several groups of lines superimposed one upon another arise. In registering spectra with high resolution each one of the lines can become split into several lines because of their being split by the crystal field. With nonequivalent positions of one and the same ion in the crystal (for instance cubic, tetragonal, trigonal, and rhombic centers of Dy^{3+} in CaF_2) the spectrum represents a superposition of several spectra from each of the centers. In different classes of compounds or in diverse compounds, various types of emission levels can become manifest.

At the same time, the same transitions in the RE^{3+} spectra are a short way away from one another in different crystals with different symmetry and with only the fine structure of these transitions being subject to change.

3. The fine structure of the RE^{3+} absorption and luminescence spectra depends on the characteristics of the crystal (the presence of co-activators, local or nonlocal compensation of the charge and the manner of the ion's incorporation in the complex with an F-center, concentration of the activator), and on conditions attending observation (lowering of temperature produces narrowing of the lines, transitions in absorption spectra occur only from the lowermost level, etc.; diverse modes of excitation of luminescence give rise to emission transitions from different levels, tend to excite diverse types of centers) (Fig. 117).

4. Three individual cases are distinguished in the RE^{3+} luminescence spectra (657): (1) broad bands due to d–f transitions (Eu^2, Sm^{2+} in CaF_2); (2) line spectra of the f–f transitions (Dy^{2+}, Ho^{2+}, Er^{2+}, Tm^{2+} in CaF_2); (3) broad bands at room temperature and line spectra in cooling (Sm^{2+} in SrF_2 and BaF_2).

The displacement of broad d–f luminescence bands is as great as 100–$1,000\,cm^{-1}$ (for f–f the displacement of lines amounts to merely $\sim 10\,cm^{-1}$), the splitting by the crystal field reaches $12,000$–$16,000\,cm^{-1}$, i.e., it exceeds the spin-orbital splitting, and the spectrum is to be considered in like manner as the spectra of transition metals in strong crystal field.

In low-temperature spectra of some RE^{2+} (Sm^{2+}, Eu^{2+}) narrow lines with electron-vibration replices are seen superposed on the broad bands.

Ions	λ, nm	Intensity	
		in Y_2O_3	In La_2O_3
Sm^{3+}	565.8	5	1
Eu^{3+}	612.1	50	5
Gd^{3+}	315.7	5	2
Tb^{3+}	544.3	60	180
Dy^{3+}	573.3	100	3

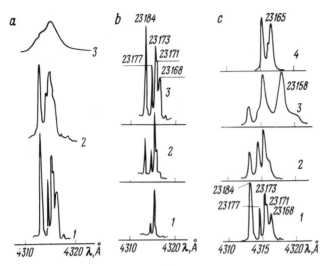

Fig. 117 a-c. The effects of temperature, concentration and impurities on the absorption spectrum of Nd^{3+} in CaWO$_4$ (transition $^4I_{9/2} \rightarrow {}^2P_{1/2}$ [585]. **a** Temperature dependence; 1:4.2 K; 2: 77 K; 3: 300 K; **b** concentration dependence: 1, 0.009% Nd; 2, 0.05% Nd; 3, 0.1% Nd; **c** impurity composition dependence: 1, 0.24% Nd + Na; 2, 0.3% Nd + Ni; 3, 1.9% Nd + V; 4, 0.57% Nd + Nb

5. Because of a different oscillator strength for the same concentrations of different RE^{3+} the relative intensity of their luminescence spectra is dissimilar [590]: Here the intensity of the line 573.3 of Dy^{3+} in Y$_2$O$_5$ is taken as 100.

Hence, it is seen that in different compounds, such as Y$_2$O$_3$ and La$_2$O$_3$ (and also with different methods of excitation) repartition of relative intensities occurs. From a large number of lines in the RE^{3+} luminescence spectra the number of high intensity lines is not too great, with about 80–90% radiation falling usually to the share of one to two lines.

From the diagram of the RE^{3+} levels are determined for each ion: (1) the regions of most intensive transitions with 80–90% fluorescence; (2) laser transitions; (3) most intensive absorption bands, i.e., "excitation" bands; (4) levels wherefrom energy transfer takes place.

Ultimate detectable RE^{3+} concentrations by means of the luminescence spectra constitute (in weight %): Sm^{3+} 6.10^{-6}, Eu 3.10^{-6}, Gd^{3+} 3.10^{-6}, Tb^{3+} 3.10^{-7}, Dy^{3+} 3.10^{-7} (with X-ray excitation, see [590]).

Feofilov [658] and Dicke [639] have summed up the results of investigations into the rare earth luminescence spectra made till 1965 to 1967. In the book by Dicke [639] a summary of the lines, energy levels and parameters of the RE^{3+} in LaCl$_3$ (C_{3h} symmetry), and a summary of literature published on the subject till 1967 (see also [783]) are set forth.

In the following years the spectra of rare earths in fluorite CaF$_2$ were subject to detailed examination [738, 767, 777, 792, 822, 848], scheelite CaWO$_4$ and powellite CaMoO$_4$ [616, 705, 784, 805, 812], LaCl$_3$ [626, 639], in silicate glasses [710, 711],

ZnS [588], and in many other compounds from the phosphate, sulphate, and oxide classes [639, 783].

Here we shall consider the most thoroughly studied rare earth luminescence spectra in minerals, examples being shown in Figures 117–125.

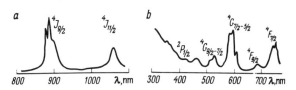

Fig. 118 a and b. IR luminescence of Nd^{3+} in apatite **a** and excitation spectra **b** [834]

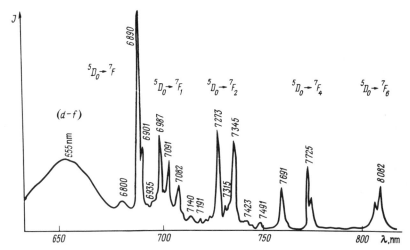

Fig. 119. Luminescence spectrum of Sm^{2+} in anhydrite (after [834])

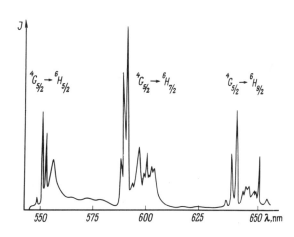

Fig. 120. Luminescence spectrum of Sm^{3+} in apatite [834]

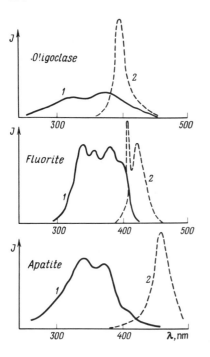

Fig. 121. Excitation (*1*) and luminescence (*2*) spectra of Eu^{2+} in oligoclase, fluorite, and apatite [834]

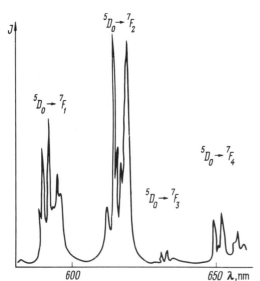

Fig. 122. Luminescence spectrum of Eu^3 in anhydrite (after [834])

Fluorite. Natural fluorites are always multiactivator systems and, therefore the spectra of synthetics samples should serve as a basis for interpretation of their luminescence spectra.

Fig. 123. Luminescence spectra and diagram of levels of Gd^{3+} in apatite, fluorite, and anhydrite (after [834])

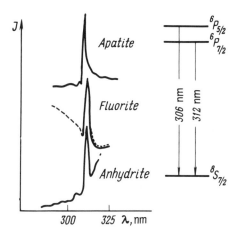

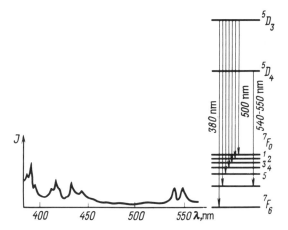

Fig. 124. Luminescence spectrum of Tb^{3+} in anhydrite (after [834])

Investigations into the spectra of rare earths in synthetic fluorites, which now represent a most extensively studied system, are divided in to two steps: observation and interpretation of the luminescence spectra with low and high resolution.

To the present, luminescence and absorption spectra of all the RE^{3+} and RE^{2+} in fluorite have been reproduced and interpreted [except for La, Ce, Gd, Tb which form in fluorite a complex center (RE^{3+} + F-Center) see Fig. 168]. With high resolution several types of incorporation in the structure, depending upon the mode of the charge compensation (see Chap. 7.3 and Fig. 168) can be distinguished for each RE^{3+} ion.

In natural fluorites all the RE^{3+} and RE^{2+} ions can be established from luminescence spectra [602, 697, 775, 779, 843], exceptions being Tb and Gd, which enter a complex center (see Chap. 7.3). The Gd^{3+} and Eu^{2+} ions are observed also in natural fluorites from the EPR spectra.

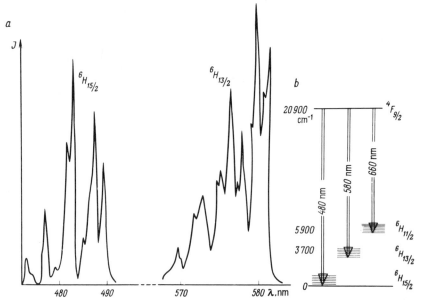

Fig. 125. **a** Luminescence spectrum and **b** energy level diagram for Dy^{3+} in zircon [834]

The visible glow of natural fluorites is more often due to the blue-violet lumine-scence of Eu^{2+} and to the yellow-green of Yb^{2+}, but in investigating them there is no need to confine oneself to the visible region.

Mn^{2+} determined from EPR spectra in natural fluorites does not yield any lu-minescence bands, apparently because of the fact that it acts as sensitizer for rare earths; it is also not observable in absorption spectra, owing to its usual low con-centrations, and low intensity of forbidden optical transitions.

The role of rare earths in coloration of natural fluorites seems to be limited to Sm^{2+}, Dy^{2+}, and Tm^{2+} (green color) and Eu^{2+}, and Yb^{2+}, since the absorption bands of the remaining RE^{2+} fall into the IR region, while the weak and narrow RE^{3+} lines with their usual concentrations in natural samples cannot exercise any essential influence on the color of the latter. For the study of the relations of RE^{3+} in natural fluorites with different compensation (symmetry), RE^{2+}, and electron-hole centers, complex observations combining low-temperature high-resolution luminescence and absorption spectra, the EPR spectra, especially low-temperature ones, and thermoluminescence are necessary.

Fluorites with Sm^{2+}, Dy^{2+}, Tm^{2+}, Nd^{3+}, Er^{3+}, Tm^{3+} and Ho^{3+} find use as materials for lasers [639, 708].

Apatite. In natural apatites the presence of Ce^{3+}, Sm^{3+}, Dy^{3+}, Eu^{2+}, Sm^{2+} and also of Mn^{2+} (see Figs. 118, 120, 121, 123) have been ascertained from lumines-cence spectra and Nd^{3+} and Pr^{3+} from absorption spectra [828, 834, 843]. The blue and violet glow of apatites comes from Ce^{3+} and Eu^{2+}, and various hues of pink, violet-pink and yellow-pink from Sm^{3+}, Dy^{3+}, the yellow luminescence being due to Mn^{2+}. However, the characterization of the luminescence of apatites according to the glow color is qualitative and does not permit identification of a complete set of luminescent ions whose lines and bands overlap and extend over the IR (Sm^{2+}) and UV (Ce^{3+}) regions.

These ions are encountered in natural apatites in quite different combinations and thus the fact that characteristic sets of ions for various types of rocks from different regions can be specified [735] is here of importance. Thus one can observe luminiscence of only Eu^{2+} (apatite of Alpine veins, quartz veins), or of only Ce^{3+}, at times with a certain amount of Mn and RE (metasomatic calcite-flogopite veins), of only Mn^{2+} or basically Mn^{2+} (a number of pegmatites, granites, skarns, greisens) of Ce^{3+} and Eu^{2+} (basic rocks, carbonatites, metamorphic rocks), of Eu^{2+} and Mn^{2+} (some alkali rocks, ore deposits, granitoids), of Dy^{3+}, Sm^{3+}, Eu^{2+} (some granitic pegmatites) and of Ce^{3+}, Dy^{3+}, Sm^{3+}, Sm^{2+} (alkali rocks, granitoids, pegmatites), etc. To detect the presence of Nd^{3+} one has to obtain the absorption spectra, and for that of Mn^{2+} the EPR spectra, since both these ions act as sensitizers of the RE^{3+} glow, and in the presence of the RE^{3+} ions they may not show in the luminescence spectra.

In synthetic halophosphates of the apatite composition RE^{3+}, Mn, Sb, ions of mercury-like elements (Ga, Ge, Zn, Tl, Sn, Pb) figure as activators. They find wide use in illumination techniques. Apatite containing Nd^{3+} is a laser material.

Anhydrite. In photoluminescence spectra [831, 834] of natural anhydrites the Sm^{2+} lines (see Fig. 119), Mn^{2+} bands and a characteristic S_2^- spectrum could be observed. In thermoluminescence spectra the presence of Ce^{3+}, Eu^{2+}, Sm^{3+}, Dy^{3+}, Mn^{2+} (see Figs. 122–124) is determined. $CaSO_4$ activated by Mn^{2+} finds application in thermoluminescent dosimetry.

Zircon. In the line spectrum of natural zircons photoluminescence is basically related to Dy^{3+} (see Fig. 125), the Gd^{3+}, Sm^{3+}, Eu^{3+} lines [775, 778, 839] being much less intensive. A broad structureless band in the red-yellow region is related to radiation centers.

5.3.3 Actinides; Absorption and Luminescence Spectra

In more than a century (beginning from observations by Stokes in 1852) of history of optical spectroscopy of the actinide compounds, the most momentous events may be considered as: (1) the discovery of radioactivity by Becquerel, and (2) the use of bright colors and a characteristic luminescence of many uranium minerals for their identification and luminescent field surveys while searching for uranium deposits. More detailed structural-spectroscopic investigations of actinides effected in the years 1950–1960 helped to obtain some characteristics of their electron structure in different crystals, and to interpret optical and luminescent spectra within the framework of the crystal field and molecular orbitals theories [626, 638, 650, 667, 793].

Two sharply different types of the manifestations of actinides are distinguished in their optical spectroscopy: (1) incorporation of uranium (and of other actinides) in compounds as ions U^{6+}, U^{4+}, U^{3+}, U^{2+} and (2) its presence in the form of a complex ion-uranyl UO_2^{2+}. The electron configurations of the U atom and ions may be written down as follows:

U $/Rn/5f^3 6d^1 7s^2$
$U^{2+}/Rn/5f^3 7s^1$
$U^{3+}/Rn/5f^3$
$U^{4+}/Rn/5f^2$
$U^{6+}/Rn/$

With U incorporated in a compound in the form of U^{4+}, U^{3+}, and U^{2+} ions, a complete analogy of their spectra with the spectra of the rare earth ions with the same number of f-electrons is noted. Their absorption spectra are represented by a large number of narrow, weak lines: in the spectrum of U^{4+} in zircon, for instance, one can see over 150 lines [800]. The diagrams of the multiplet levels splitting by the crystal field and the molecular orbitals schemes coincide in principle with the diagrams for the rare earth ions (see Fig. 114).

Optical absorption and luminescence spectra of U^{2+}, U^{3+}, U^{4+} and U^{6+} in synthetic fluorite as well as the EPR spectra [691, 776] along with the spectra of U^{4+} in synthetic zircon [800] have been studied in great detail. There are a number of indications pointing to the presence of uranium in natural minerals [634, 643].

In most minerals uranium is present in the form of uranyl UO_2^{2+} [829, 833, 834]. It is with this that the green, yellow, and orange colors of uranium minerals are associated and from it comes the bright green-yellowish-green luminescence (Figs. 126–128) that forms a basis for luminescent field surveys. The uranyl content lies at the background of the pearl analysis, and gives rise to an artificial lumines-

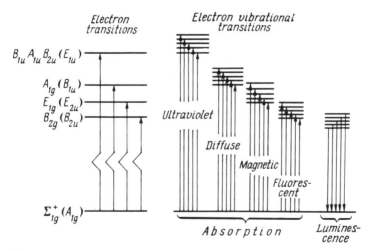

Fig. 126. Diagrams of levels determining transitions in absorption and luminescence spectra of uranyl

Fig. 127. Diagram of absorption and luminescence spectra of uranyl [681]. Absorption series are shown: *1*, fluorescent; *2*, magnetic; *3*, diffuse; *4*, UV

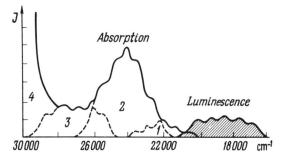

Fig. 128. Luminescence spectrum of uranyl in uranium-containing calcite [829]

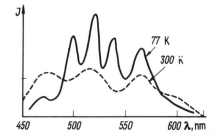

cence of nonluminescent uranium minerals under the effect of acetic or nitric acids.

Uranyl is a linear molecular ion O-U-O which enters the composition of compounds as a structural unit [648, 649, 732, 811, 847]. Consider first its electron structure as an individual grouping of UO_2^{2+}, and then interaction of this grouping as an integral unit with ligand ions of the crystal lattice of whose structure it is a part.

The O-U-O molecule is linear and symmetric relative to the arrangement of two oxygen ions, its symmetry being $D_{\infty h}$. The construction of the UO_2^{2+} molecular orbital diagram is analogous to the construction of molecular orbitals for the linear molecule of CO_2. For this group of symmetry in the case of CO_2 there are four types of transformation: σ and π relative to the molecular axis, and σ_g, σ_u, π_g, π_u, into parity being taken account.

Let us obtain group orbitals of two oxygen atoms O . . . O; in UO_2^{2+} they are quite analogous to group orbitals of CO_2 (see Fig. 150): σ_g, σ_u, π_g, π_u. Then consider transformations of the atomic uranium orbitals: $5f^36d^17s^2$. The atomic orbitals s transform like σ_g; five atomic d-orbitals as σ_g and as π_g and σ_g in twos; seven atomic f-orbitals as σ_u, π_g and σ_u and σ_u in two.

The atomic orbitals U and group orbitals O . . . O transformable according to the same type of symmetry comprise bonding and anti-bonding molecular orbitals, shown very schematically in Figure 129.

UO_2^{2+} has 12 valence electrons (see Fig. 129) (six U $5f^36d^17s^2$ electrons and eight electrons of two O $2p^4$ atoms minus two electrons because of the divalent state of

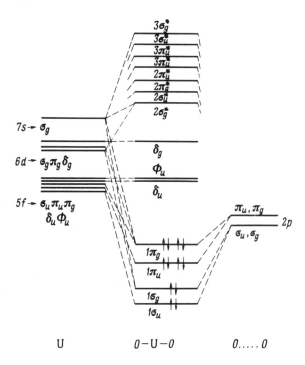

Fig. 129. Schematic molecular orbitals diagram for uranyl

UO_2^{2+}). They are arranged on four bonding MO, forming an electron configuration of UO_2^{2+} $(1\sigma_u^+)^2$ $(1\sigma_g^+)^2$ $(1\pi_u)^4$ $(1\pi_g)^4$. Here four f-atomic orbitals (σ and ϕ_u in twos) and two d-orbitals (σ_g) of uranium are nonbonding.

In passing from an individual UO_2^{2+} grouping to uranyl as a part of the crystal structure the local symmetry goes from $D_{\infty h}$ down to D_{6h} and lower. Accordingly, the atomic and group orbitals transform conformably to the symmetry type of a given group:

$D_{\infty h}$	Σ_g^+	Σ_u^+	Σ_g^-	Σ_u^-	Π_g	Π_u	Δ_g	Δ_u	Φ_g	Φ_u
D_{6h}	A_{1g}	A_{2g}	A_{2g}	A_{1u}	E_{1g}	E_{1u}	E_{2g}	E_{2u}	B_{1g}^+	B_{2g} B_{1u} $^+B_{2u}$.

The MO diagram of uranyl in a crystal is more complicated, for here MO forms with involvement of not only atomic orbitals and group orbitals of axial atoms of O . . . O uranyl, but also of ligand atoms coordinated in an equatorial plane. The MO diagram is not calculated and the levels with which optical transitions are associated are determined ambiguously. The electron configuration of the ground state $(1\sigma_u^+)^2$ $(1\sigma_g^+)^2$ $(1\pi_u)^4$ $(1\pi_g)^4$ corresponds to a singlet term $^1\Sigma_g^+$.

The excited configurations appear with an electron transferred from the last-occupied MO $(1\pi_g)^4$ to the first free MO: σ_u and ϕ_u (see Fig. 129).

A feature distinguishing the molecular absorption and luminescence spectra is the fact that they arise from the transitions between electron-vibrational levels, i.e., the levels of electronic states split into sublevels whose spacing equals the vibrational frequency of the molecule (see Fig. 126). Therefore, each one of the broad electron transitions of the molecules represents a series of vibrational transition bands. In uranyl four series in absorption spectra, and one in the luminescence spectra are distinguished: (1) fluorescent (20,000–22,000 cm^{-1}), (2) magnetic (22,000–26,000 cm^{-1}), (3) diffuse (26,000–29,000 cm^{-1}), (4) UV. Each series has a more or less defined vibrational structure with mean frequency intervals of about 860 cm^{-1} in luminescence spectra and 720–600 cm^{-1} in absorption spectra. At low temperatures and high resolution an additional splitting into vibration lines can be seen originating from the lattice virbations. The region of the fluorescent series of the absorption spectrum appears as a mirror reflection of the luminescence spectrum (see Fig. 127).

Measurements of the uranyl absorption spectra in a large number of minerals [833, 834], such as phosphates, arsenates, vanadates, sulphates, carbonates, and uranium hydroxides, have shown the presence of three bands (series) in the visible with a mean frequency interval from 670 cm^{-1} to 600 cm^{-1}. The different relative intensity of these three series, their mutual disposition, and a partial overlap determine their colors and they depend on the local symmetry, the nature of the cation, and interaction with ligands.

The characteristic luminescence of uranyl in greenish colors is common to many uranium minerals and to minerals containing uranyl as impurity. As in the case of rare earths its quenchers are Fe, Cu, Mn, Pb, and Bi.

5.3.4 Mercury-Like Ions; Pb^{2+} in Feldspars and Calcites

In the activator group next to transition metals, rare earths, and actinides, come isolectronic ions of the Hg^0 type with two outer s-electrons (ns^2): Zn^0, Cd^0, Hg^0, Ca^+, In^+, Tl^+, Ge^{2+}, Sn^{2+}, Pb^{2+}, Bi^{3+}. Of all these ions, well-studied in alkalihalid crystals and zeolite [621, 713, 732, 753, 809, 817, 835], the usual impurity in the minerals is only Pb^{2+}. This type of luminescence has become of interest for mineralogical studies after interpretation of the Pb^{2+} spectra in feldspars and calcites [830].

From the diagram of the Pb^{2+} levels (Fig. 130), the same as in the case of all the mercury-like ions, the assignment of the three bands in the absorption spectrum has been made: $A(^1A_1 \rightarrow {}^3P_1)$ (270 nm in NaCl: Pb), $B({}^1S_0 \rightarrow {}^3P_2)$ (210 nm), $C({}^1S_0 \rightarrow {}^1P_1)$ (190 nm) as well as a broad band in the luminescence spectrum: ${}^3P_1 \rightarrow {}^1S_0$ (621, 830). In like manner the absorption and luminescence spectra of Pb^{2+} in feldspars (see Fig. 130) and calcites are interpreted [830].

A study of the EPR spectra gives ground to associate the green color of amazonites having the highest (up to 0.n%) content of Pb among feldspars, with the electron center Pb^+ [945] (see Chap. 7.3). The fact that the absorption and luminescence spectra helped to interpret in amazonites and other feldspars the Pb^{2+} spectra, whose absorption and luminescence bands fall into the UV region, supports an assumption that the color of amazonites cannot be connected with an ordinary Pb^{2+} ion.

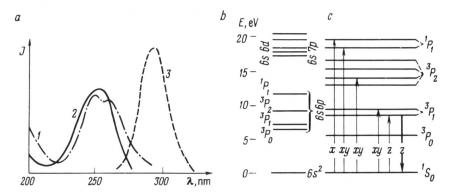

Fig. 130 a–c. Luminescence of Pb^{2+} in amazonite [830]. **a** Absorption (*1*), excitation (*2*) and luminescence (*3*) spectra; **b** schematic diagram of Pb^{2+} free ion energy levels; **c** schematic diagram of Pb^{2+} ion in a tetragonal crystal field

Annealing and bleaching of amazonites leads to reducing the intensity of the EPR spectrum of Pb^+ and to a simultaneous rise in the intensity of the Pn^{2+} absorption and luminescence spectra.

5.3.5 Molecular Ions S_2^-, O_2^-, and F-Centers

The electron-hole centers are unambiguously defined from the EPR spectra (see Chap. 7). The energy absorbed by them is mostly transmitted to activators (luminescent centers) or is spent in making radiationless transitions. In thermoluminescence the electron-hole centers form the most common type of the electrons or holes trapping centers which on heating transmit the energy to luminescent centers (see Chap. 6).

It is only in the case of two types of the electron-hole centers that luminescence proves to be a method as effective for observation as is EPR, namely for molecular ions S_2^-, O_2^- and for the *F*-aggregate centers.

A distinctive feature of the S_2^-, O_2^- luminescence spectra, as well as of other molecular ions, is a quasi-linear vibrational structure of the broad luminescent band (Fig. 131). The S_2^- and O_2^- molecular ions are isoelectronic. From the molecular orbital diagram describing their electron structure the emission transition $^2\Sigma_g \rightarrow {}^2\Pi_{3/2}$ is determined. When observing luminescence spectra at 77 K, a fine structure associated with the frequency of intramolecular vibrations of S_2^- and O_2^- is demonstrable. This frequency depends on the type of the molecular ion, on internuclear distance (S–S or O–O) and upon the particular position of the molecular ion in the structures. For S_2^- the maximum of the emission band lies within the range of 600–700 nm (red-orange glow) with a mean vibrational frequency of 500–650 cm^{-1}, while for O_2^- the respective maximum is 450–550 nm (blue-green glow) with frequency comprising 800–1,200 cm^{-1} [834].

Fig. 131. Luminescence spectra of S_2^- in sulfur-containing aluminosilicates [834]

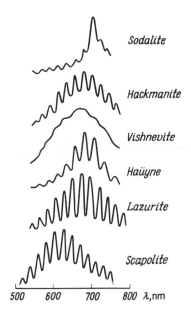

The luminescence of *F*-centers has been studied in great detail in alkaline halides [937], and it has been observed in halite and fluorite [834]. In detection and identification of the F-aggregate center models in alkaline halides polarized luminescence observable at a low temperature in the IR region proved to be quite effective (see Chap. 7.4).

5.3.6 Crystallophosphors of the ZnS Type; Natural Sphalerites and Other Sulphides

Sphalerite is a model system of paramount importance for the theories of luminescence, and a base material for luminophors [591, 599, 610, 623, 660, 664, 676, 685, 709, 757, 770, 813, 814]. The mechanisms of its glow were subject to detailed discussion in the foregoing text (see Figs. 102–111).

Here we shall attempt to systematize the types of luminescent centers in ZnS and to present their models.

A. Centers Whose Luminescence is Due to Transitions Between Levels of the Ion Activator Without Participation in the Glow of the Valence and Conduction Bands, i.e., as in the Case of the Ionic Crystal Luminescence. Hereto belong $ZnS:RE^{3+}$, $ZnS:Mn^{2+}$ [588, 589, 737] which yield the same luminescence spectra as do $CaF_2:RE^{3+}$ or $CaCO_3:Mn^{2+}$.

B. Centers Whose Luminescence Comes as a Result of Transitions Involving Participation of Conduction and Valence Bands, i.e., According to the Semiconductor or Crystallophosphor Schemes. Common to all the centers of this kind are: (1) excitation with ionization of the activator or vacancy, i.e., with formation of an electron or hole center whose lifetime determines the duration of glow (photosensitive centers fixed by EPR with light illumination existing only at the time of luminescence).

(2) Glow resulting from recombination of such an electron center with a hole in the valence band or recombination of the hole center with an electron in the conduction band. The level of an electron or hole center in the forbidden band is an intermediate stage (capture and release of an electron or hole) in "circulation" of the electron from the valence band (or from the activator's level) to the conduction band and back to the valence band [641, 642, 655, 656, 696, 712, 796, 806, 807, 826].

The luminescence centers of the crystallophosphors may be systematized by the way of the activator's charge compensation, this helping to present the model of the electron-hole center (Fig. 132) as well as determine the mechanisms governing excitation and glow. The charge compensation can be nonlocal and local, the local compensation being realized through formation of vacancies (cation or anion vacancies) or by incorporating impurities of a different valence.

1. Centers with Nonlocal Charge Compensation. The luminescence of all the activators entering ZnS without local compensation and with a vacancy or with any other impurity, i.e., with M^+ (Cu^+, Ag^+, Au^+), M^{2+} (transition metals), M^{3+} (Ga^{3+}, In^{3+}, Tl^{3+}), M^{4+} (Si^{4+}, Ge^{4+}, Sn^{4+}, Pb^{4+}) proceeds in two ways:

a) ZnS: Cu^+ (see Fig. 132.1). Excitation: Cu^+ gives off one electron, i.e., captures a hole: $Cu^+ + e^+ \rightarrow Cu^{2+}$ (e^- in the conduction band); at this time an EPR spectrum from Cu^{2+} is observable. Glow: recombination of the conduction band electron with the hole center Cu^{2+}: $Cu^{2+} + e^- \rightarrow Cu^+$.

b) ZnS: Ge^{4+} (see Fig. 132.1). Excitation: the valence band electron is excited into the conduction band, i.e., a free electron appears in the conduction band, and a free hole in the valence band; capture of the free conduction band electron (its localization) by the Ge^{4+} ion attended by the formation of an electron center: $Ge^{4+} + e^- \rightarrow Ge^{3+}$. Glow: recombination of the electron center Ge^{3+} with the conduction band hole: $Ge^{3+} + e^+$ (i.e., S^-) $\rightarrow Ge^{4+}$.

2. Centers with Local Compensation of the Charge Through Formation of Vacancies. (a) Anion vacancy $V_{S^{2-}}$ (see Figs. 132.4, and 105); ZnS: Cu^+ $V_{S^{2-}}$. The vacancy of sulphur is surrounded by three atoms of zinc: Zn_3^{2+} and one atom of copper: Cu^+. Excitation: formation of a free hole in the valence band and of a free electron in the conduction band; capture of free electron by vacancy of sulphur, i.e., formation of an electron center $Cu^+ + Zn_3^{2+} + e^-$ (an analog of the F-center). Glow: recombination of this center's electron with the valence band hole. (b) Cation vacancy V_{Zn}^{2+} (see Figs. 132.2.3 and 104):

ZnS: III = ZnS: Ga^{3+}, Al^{3+} V_{Zn}^{2+} or
ZnS: VII = ZnS: Cl^-, Br^-, I^- V_{Zn}^{2+}.

Here Ga^{3+}, Al^{3+} or Cl^-, Br^-, I^- play a passive role in luminescence, since they merely stimulate the formation of the zinc vacancy. The latter is surrounded by four sulphur anions. Excitation: formation of a free hole in the valence band and of a free electron in the conduction band; capture of holes by the zinc vacancy ,i.e., capture of the hole by one of the sulphur anions surrounding the zinc vacancy, attended by the formation of the hole center $S^{2-} + e^+ \rightarrow S^-$. Glow: recombination of the free electron with the hole center S^-.

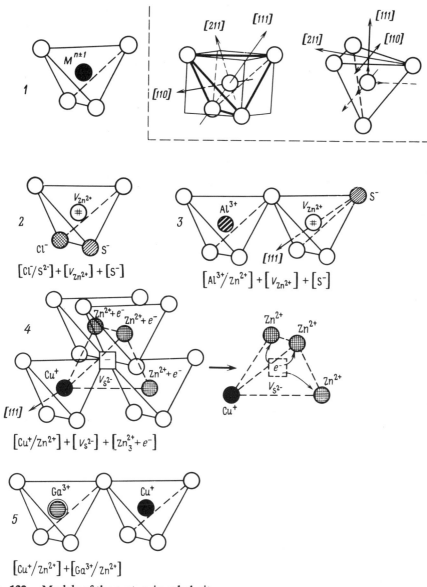

Fig. 132. Models of the centers in sphalerite

3. Centers with Local Compensation of the Charge Through Formation of Associated Pairs: ZnS: M^+, M^{3+} (see Figs. 132.5 and 106). Excitation: excitation of the M^+ ion's electron into the conduction band, i.e., formation of the hole center $Cu^+ + e^+ \rightarrow Cu^{2+}$ and of a free electron in the conduction band, capture of this free electron by the M^{3+} ion attended by the formation of the electron center $M^{3+} + e^- \rightarrow M^{2+}$. Glow: recombination of the electron center $(M^{3+} + e^-)$ with the hole center $(M^+ + e^+)$.

The following features are characteristic of the natural luminescence of sphalerites of: (1) iron content common to them acts as a quencher of luminescence and, therefore, the flow is observable only in the case of low-iron samples (less than 0.1 % Fe), (2) the presence of various impurities compensates for the charge through formation of associated pairs (and not vacancies) or owing to nonlocal compensation. Natural samples were found to contain [675, 785, 786, 788]: Ag^+ (blue glow at 77 K with the band maximum of 470 nm), Cu^+ (green, λ 520 nm), $Cu^+ - Ga^{3+}$ (In^{3+}) and $Ag^+ - Ga^{3+}$ (In^{3+}) (orange-red, λ 620–680 nm), $Cu^+ - Tl^{3+}$ (dark red, λ 825 nm).

For other sulphide minerals, studies have been made [834] for edge luminescence of cinnabar and realgar (Fig. 133) whose mechanism is probably similar to that shown in the Lambet-Klick scheme (see Fig. 103).

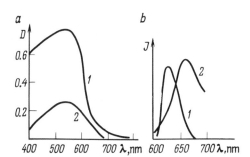

Fig. 133. **a** Excitation and **b** luminescence spectra of natural sulfides. *1* cinnabar; *2* realgar [834]

In the case of many semiconductor sulphide minerals with a small width of the forbidden band, luminescnce in the IR region can be observed [603].

5.3.7 Luminescence of Diamond

In diamonds all the luminescence centers are associated with a nitrogen impurity which substitutes carbon in the structure of diamond. This has nothing to do with some or other peculiarities of nitrogen by comparison with other impurity elements, but is attributable to the fact that its content (up to 0.25 at. %) is at least two orders higher than the content of other impurities in diamond.

The diversity of centers in diamond reflects the diversity of the forms of the nitrogen incorporation in the structure: as single, isolated atoms of N, pairs of N_2 atoms, N atoms with an adjacent vacancy (a vacancy to compensate the N charge with $2s^2 2p^3$-electrons substituting C with $2s^2 2p^2$-electrons), two N_2 atoms and a vacancy, two vacancies and a N atom, a segregation of nitrogen atoms (submicroscopic platelets with a "thickness" of merely a few atomic layers), nitrogen and aluminum, replacing the adjacent atoms of carbon (Table 13).

Different types of these centers are seen in the electron paramagnetic resonance, luminescence, and absorption spectra in the visible and IR regions and are also manifest in the properties of diamonds.

Table 13.   Correlation of the centers models in diamonds with EPR, luminescence and absorption spectra in visible and IR regions and colors (820, 994)

Model	EPR	Luminescence	Absorption	IR	Color
N	+	—	5500 A	—	Yellow-green
N_2	—	—	3200 A	A	—
NV	+	$S1$	$S1$	—	—
N_2V	+	—	—	—	—
NV_2	+	—	—	—	—
VN_2V_2	—	$H3, H4$	$H3, H4$	—	Yellow-green
N_2D	—	—	—	—	—
$N_S(111)$	—	$S2, N9$	$S2, N9$	B_1	—
$N_S(100)$	—	—	—	B_2	—
NA1	+	$N3; A$	$N3$	—	Yellow-green
B	—	—	—	—	Blue

N, Al, B, nitrogen, aluminum, boron; V, vacancy of carbon atoms in the diamond structure; D, dislocation; N_S, segregation of nitrogen atoms (platelets) in planes (111) and (100); $S1$, $S2$, $H3$, $H4$, $N3$, $N9$, usual designations of the luminescence and absorption centers; A, B_1, B_2, usual designations of centers in the infrared spectra of diamonds.

The most detailed models have been obtained for the types of diamond centers which are paramagnetic and, therefore, could be investigated by using EPR [908, 991, 994, 995, 1,005]. The designations of the centers in different domains of research which can be observed by means of corresponding types of spectra are compared with one another and with models of the centers in Table 13.

The diamond luminescence spectra [592, 636, 654, 674, 702, 747, 819, 820] are seen to occur in three regions (Fig. 134): UV (center $N9$, nitrogen segregation in the form of platelets), blue (center $N3$, N and Al atoms) and yellow or yellowish-green (H_3 and H_4 - VN_2V or S_1 and S_2 -NV). Interpretation of these luminescence spectra is based on the band and electron–phonon interaction theories (i.e., on the interactions of electron transitions with lattice vibrations forming the phonon spectrum).

In the diamond luminescence spectra (see Fig. 134) very narrow intensive lines of purely electron transitions (zero-phonon) are distinguished, and broad bands with maxima therein which extend into the longwave direction from the former, and represent a superimposition of the phonon (vibrational) replicas of the electron transitions. Because of the fact that the vibration energy at different points of the Brillouin zone is dissimilar, and also owing to the formation of di- and triphonon repetitions (upon addition of similar and different phonons) 10–15 overlapping lines appear which form a broad band perceived as a blue or yellow glow.

In the diamond absorption spectra, that are approximately symmetric with respect to the luminescence spectra, the phonon repetitions extend into the shortwave region, away from the line of a purely electron transition.

Most natural diamonds display luminescence and the distribution of its different types (blue, yellow, UV) depend on the geological condition of their formation [592, 674].

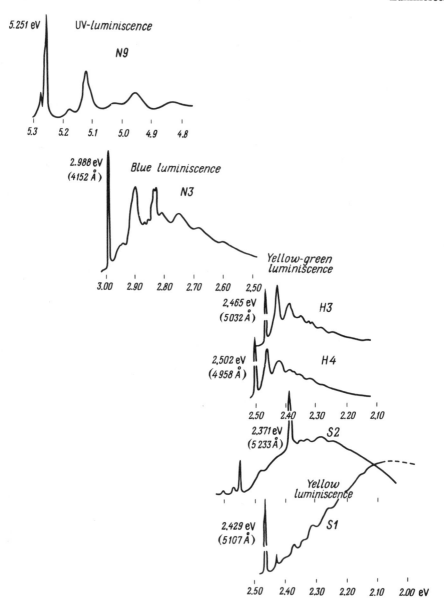

Fig. 134. Luminescence spectra of diamond (after [820, 986])

6. Thermoluminescence

Luminescence of minerals on heating (thermoluminescence), known since the 17th century, has comparatively recently become a method of investigation, after the nature of this phenomenon had been understood (F.Urbach, 1930) as an undulated kindling and waning of the glow brightness under a uniform heating due to electrons escaping from traps. It was, however, not until after development of the theory of the phenomenon (since 1945) that thermoluminescence has found wide application [591, 664, 713, 732].

Models and prerequisites contained in these works have received exhaustive computable formulations. This part of the thermoluminescence theory based on the statistics and kinetics of electronic processes embraced by the band scheme represented until very recently the entire theory. Its important limitation was that the nature of the electron trapping and emission centers remained obscure. Calculations were made with nameless electron trap and emission levels.

New preconditions for further development of the thermoluminescence theory appeared when by the end of the 60s the nature of the capture and emission centers was made clear by models of electron-hole centers (see Chap. 7) and photosensitive luminescence centers in crystallophophors (see Chap. 5.2) were set up with the aid of EPR (see Chap. 3). The kinetic theories of luminescence (thermoluminescence and phosphorescence) have developed largely conformable to investigation of luminophors and thermoluminescent dosimeters, particular attention being focused on the brightness of the glow, its duration, and decay.

From the 50s and onward increasingly numerous investigations into the possible application of thermoluminescence in geology have been under way. These are linked with a whole number of specific theoretical problems that were not sufficiently tackled and elaborated within the framework of kinetic computations in the absence of concrete models of the centers. For this reason, all attempts at determining and estimating the effects of numerous geological factors on the position of peaks and the intensity of thermoluminescence lacked sufficient background, though they furnished an extensive and concrete empirical material. The detection, with the help of EPR, of electron-hole centers widely distributed in minerals (Chap. 7.1) creates a radically novel situation in regard to the understanding of the nature of thermoluminescence in minerals, and is a pre-requisite for a meaningful interpretation of the peaks in the thermoluminescence curves, and their relationships with physicochemical and geological factors.

All the review works on thermoluminescence [591, 664, 713, 732], and on its application in geology, are based on the kinetic theory, and within its framework they offer a complete description of this phenomenon, but fail to include concepts derived as a result of identification of centers in minerals according to the EPR spectra.

However, it is exactly the interpretation of the nature of capture and glow that allows the phenomenon of thermoluminescence to be better understood.

6.1 Mechanism and Parameters of Thermoluminescence

Setups [736, 745] for registration of thermal deexcitation curves (thermoluminescence curves) record changes in the intensity of the glow with uniformly rising temperature. (Fig. 135). The position of the peak depends on the heating rate and, therefore, when indicating its temperature, it is necessary to mention the rate of heating (usually from 0.1–0.5°C s^{-1} to achieve a higher resolution and up to 2–10°C s^{-1} with rapid changes or in the event of a very low intensity of the glow.) An increase of the heating rate from 0.3°C s^{-1} to 1°C s^{-1} results in shifting of the thermal deexcitation peak by 10°C toward higher temperatures. In quantitative estimations the intensity of glow is measured at the rate per unit of the sample volume with an equal area, the uniform degree of comminution, and by taking into account, if necessary, any differences in transparency and color. In selecting geological samples one has to take into consideration the fact that a lengthy illumination and heating by the sun's rays may affect the low-temperature thermoluminescence peaks.

The thermoluminescence peaks carry no information, either on the nature of the corresponding emission centers, or on the centers of the electron trapping. These data are derived from supplementary investigations, largely with the help of EPR.

6.1.1 The Nature of Emission Centers

At each one of the thermal deexcitation peaks (for a sample in the temperature interval fitting the glow maximum) a thermoluminescence spectrum (Fig. 136) in which the assignment of lines and bands enables it to identify an ion activator, a

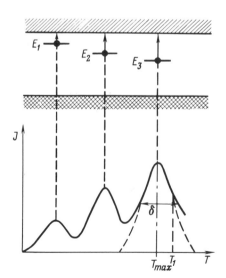

Fig. 135. Thermoluminescence curve and some parameters obtained from it. T_{max} is temperature of the maximum glow of a given peak; I is glow intensity at its maximum; δ is width of the peak at its half-height; T_1 is temperature in the down-sloping part of the curve at half-maximum intensity

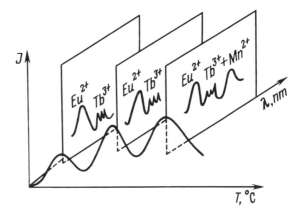

Fig. 136. Thermoluminescence spectra obtained at the temperatures of thermal deexcitation peaks

recombination transition, or radiation of an electron-hole center is obtained. In the event of the emission center being represented by one of the rare earth ions, the characteristic spectrum makes it possible to determine the ion in question unambiguously. If emission comes from a broad structureless band, one must resort to a comparison against synthetized samples containing various activators.

Thus the thermal glow center is an ordinary luminescence center which yields a spectrum similar to that in photocathode-, X-ray and other types of luminescence (with a possible repartition of the lines intensity). In this respect thermoluminescence is only one of the modes of the luminescence excitation. However, here heating does not directly induce luminescence, but imparts the activation energy which liberates electrons from the centers of their trapping. It is these electrons that excite luminescence.

6.1.2 The Nature of Trapping Centers

This is determined by identification of electron and hole centers in a crystal from the EPR spectra (see Chap. 7) and by correlating the thermal stability of the centers established from these spectra with temperatures of the thermoluminescence peaks. In other words, the centers of the electron trapping accounting for the peaks on the thermoluminescence curves are actually the electron or hole centers.

As regards the manner of formation of these centers, one can discern: (1) natural thermoluminescence, i.e., electron-hole centers (trapping centers) already exist in the minerals, having been formed under natural conditions, (2) artificial thermoluminescence: a crystal that is nonthermoluscent owing to the lack of capture centers after irradiation with X- or gamma rays, or with other kinds of emissions producing formation of electron or hole centers, becomes thermoluminescent. Heating of such irradiated samples yields thermoluminescence attended by recombination of the electron-hole centers.

The curves of artificial thermoluminescence arising in consequence of irradiation of a sample at low temperatures or at room temperature may have a different significance.

Irradiation at low temperatures produces formation of short-lived electron-hole centers arising largely due to thermal defects. With the temperature rising from that of liquid nitrogen to room temperature, all these centers recombine one after another or transform one into another, this being accompanied by the thermoluminescence peaks. These centers and the corresponding peaks characterize the real structure of a crystal and its ability to develop thermal defects (in conjunction with the impurity ones) specific to each given structure, but not accumulation of electron centers under natural conditions.

Irradiation at room temperature induces the formation of electron-hole centers largely at the expense of impurity defects that for some reason failed to appear under natural conditions, or did not become fully converted into the ionized state. This irradiation uncovers impurity defects, i.e., the total number of electron traps existing in a crystal and capable of capturing an electron or a hole. The thermoluminescence of these samples yields peaks already at elevated temperatures.

Hence, each thermoluminescence peak has a dual nature, and its assignment must also be dual, since for each peak it becomes necessary to indicate an electron-hole center, which furnishes electrons or holes that recombine on heating, and in this way induce luminescence, and an emission center.

Here deexcitation of energy delivered by the trapping center occurs, which gives a signal of a capture center recombination taking place at a given temperature (Fig. 137).

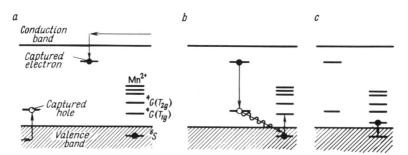

Fig. 137 a-c. Diagrams of levels describing the process of thermoluminescence in calcite [765]. **a** Formation of electron and hole trapping centers; **b** recombination of trapping centers and excitation of the emission center; **c** Mn^{2+} emission

In this respect thermoluminescence means annihilation of electron-hole centers registered by the luminescence of activators excited by these centers or by the luminescence of the centers themselves.

The interrelations of the trapping and emission centers can be understood if one proceeds from the fact of a simultaneous presence of two capture centers, i.e., electron and hole centers. Heating imparts activation energy to an electron center from which an electron becomes released (the center being then annihilated, for instance $Ti^{3+} - e^- \rightarrow Ti^{4+}$) and subsequently migrates to the hole center. Recombination of this electron with the hole center (then the hole center is annihilated, for example as in $O^- + e^- \rightarrow O^{2-}$) delivers energy that is spent in inducing luminescence. In like manner heating can liberate the hole from the hole center (which is then annihilated), the hole then migrating to the elctron center with subsequent recombination.

Recombination results in an energy release spent in exciting luminescence: (1) through transfer of this energy (similar to sensitization) to the ion activator, the closely lying recombination and emission centers being then separated; recombination of different pairs of centers (at different temperatures) can cause a glow of one and the same activator; (2) through formation due to the recombination of a state capable of producing luminescence. For example, recombination of the electron center RE^{2+} in CaF_2 with a hole center F_2^- gives rise to the formation of an excited state $(RE^{3+})^*$, which transforms into ground state with emission:

$$RE^{2+} + F_2^- \rightarrow (RE^{3+})^* + F^- + F^- \rightarrow RE^{3+}.$$

On the thermoluminescence curves of synthetic fluorites with impurities of one of the RE^{3+} ions [597] irradiated at liquid nitrogen temperature, all the peaks of thermal deexcitation of the sample with a single ion are featured by the luminescence spectrum of this ion. In samples with different RE^{3+} the same low-temperature peaks are in evidence (but their glow corresponds to the RE^{3+} ion introduced). This illustrates convincingly the separation of different capture centers and a single emission center, as well as the existence of the same capture centers causing low-temperature peaks in all of these synthetic samples, and the sensitized nature of the energy transfer from the capture center to the emission center. High-temperature peaks in synthetic fluorites stem from centers arising with the participation of the activator, and vary in samples containing impurities of different RE ions. A similar dependence has been established for synthetic anhydrites (see Fig. 145) [831].

To assign a peak in the recombination scheme of thermoluminescence excitation, an indication of two capture centers, the electron and hole ones, is necessary, in addition to determining the emission center.

This scheme presupposes an approximately equal number of electron and hole centers, and their recombination at approximately the same temperature. However, an observation of the centers by means of EPR usually shows prevalence of single type centers. For example, in quartz with Al impurity and without any admixtures of Ti or Ge, an intensive spectrum of the hole center O^-–Al with no electron centers occurs, while in quartz with Ti impurity and without Al a spectrum of a single electron center of Ti^{3+} is in evidence. The sensitivity of thermoluminescence to capture centers is substantially higher than that of EPR but these examples with a controlled set of impurities give no reason to suppose the existence of anticenters with very low contents.

Furthermore, the temperatures providing for the stability of these centers differ. All this complicates the actual picture of thermoluminescence mechanism as compared with the recombination scheme.

6.1.3 Determination of the Thermoluminescence Parameters
[615, 622, 630, 742, 771, 838]

From the thermal deexcitation curve (see Fig. 135) the position of the thermal de-excitation peak (T_{max}) and its half-width at the half-height ($\delta = T_2 - T_{max}$) are measured. Therefrom parameters of the electron trapping center are determined:

E, the "trap depth", i.e., the thermal energy spent in ionizing the electron center, found from the distance between the position of the electron center level in the forbidden band and the conduction band. To calculate this value one has to know the position of the last occupied molecular orbital of the center in the band scheme, and its displacement caused by the thermal energy of the center (owing to which fact the optic energy of the capture center ionization always exceeds the thermal one).

s_0 is the frequency factor that characterizes the capture center from the standpoint of thermal vibrations releasing the electron from the trapping level. Its magnitude is by two to four orders inferior to the frequency of the crystal vibrations ($\sim 10^{12}$ s^{-1}), i.e., s_0 usually equals 10^{10}–10^8 s^{-1}.

Determination of E and s_0 from the thermoluminescence curves is made by using kinetic equations expressing the dependence of changes in the concentration of electrons in the capture centers (variations in the intensity of thermoluminescence) on the temperature and time.

At the root of these calculations lies the expression

$$p = s_0 e^{-E/kT},$$

where p is the probability of the electron escaping from the capture center within a space of 1 s. With it is associated the lifetime of the capture center that determines both the duration of phosphorescence and intensity of thermoluminescence, as well as the existence of the electron center for the space of time measured in terms of the geological scale.

τ is the mean lifetime of an electron in the capture center, inversely proportional to p:

$$\tau = \frac{1}{p} \; ; \; \ln \tau = \frac{E}{kT} - \ln s_0.$$

The intensity of glow, which is proportional to a change in the number of electrons n during time t, is expressed with the aid of differential equations, known as kinetic ones:

$$I = -\, dn/dt = -\, np = -\, n s_0 e^{-E/kT}.$$

Should there occur a retrapping of the electron ("bimolecular" process) then $I = -\, dn/dt = -\, n^2 p$.

Integration of these kinetic equations renders formulas for determining concentration of electrons in the trapping centers, and for estimating the glow intensity at temperature T through the medium of three parameters: the trap depth E, frequency factor s_0 and heating rate β °C s^{-1}.

Over 20 ways of calculating E and s_0 from thermoluminescence curves [622] now exist.

Here we shall dwell upon the results achieved by three such methods:

1. Figures 138–140 show graphic solutions [629] of the Randall-Wilkinson equation for the simplest case of electrons released without retrapping, i.e., for determining the trap depth with reference to the thermoluminescence peak temperature and to the ratio of the frequency factors s_0 to heating rate β, as well as by the temporal stability at a given temperature at the time of its existence versus E and s_0.

2. From among simple expressions for estimating the trap depth, frequent use is made of expressions obtained by Luschik [752] : $E = kT^2/\delta$ with $A_r > A_t$ (where

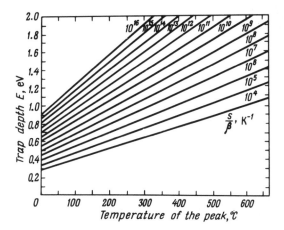

Fig. 138. Temperature of a peak versus energy of the electron or hole capture center ("trap depth") with different s/β (s is frequency factor; β is heating rate). $\beta E/kT^2 = se^{-E/kt}$ [629]

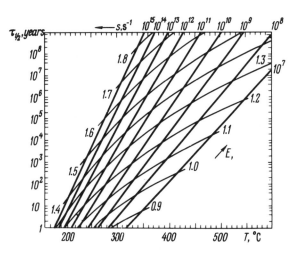

Fig. 139. Lifetime ("decay period") of the trapping center $\tau_{1/2}$ at 20°C versus frequency factor (s) with heating rate $b = 20$°C ($dn/dt = nse^{-E/kT}$) [629]

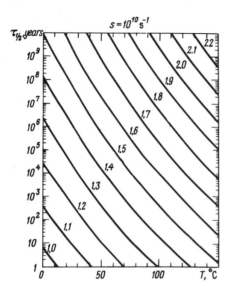

Fig. 140. Lifetime ($\tau_{1/2}$) of trapping centers at different temperatures during existence of the sample; frequency factor $s = 10^{10} c^{-1}$ [$\tau_{1/2} = (\ln 2/s)\, e^{-E/kT}$] [629]

k is the Boltzmann constant equalling $8.62 \cdot 10^{-5}$ eV K^{-1}; δ is the half-width of the peak at half-height; A_t is probability of trapping; A_r is probability of recombination), and $E = 2kT^2/\delta$ with $A_r < A_t$; the expression $s_0 = \beta/\delta \, \exp(T_{\max}/\delta)$ is taken for the frequency factor.

3. Among the most detailed methods worthy of recommending is that proposed by Antonov-Romanovsky [591] in which the thermal deexcitation curves (in I-T coordinates) are remodeled on a log scale ($\lg I$-$\lg T$) and contrasted against calculated curves plotted to the same scale for "linear" and "quadratic" cases. A good fit of the experimental curve to the calculated one specifies the order of reaction and the values for E and s_0.

Instead of measuring the luminescence intensity on heating (thermoluminescence) it is possible to trace changes in the absorption coefficient (thermal bleaching curves), thermostimulated current or exoelectronic emission [743, 804] which in the general case yield the same peaks as does thermal deexcitation.

6.2 Experimental Data, Their Interpretation and Application in Geology

In interpreting the thermoluminescence data, especially as concerns its applications in geology, it is important to realize that this registers the same electron-hole centers which are identified in such great detail by the EPR spectra and described by the molecular orbitals scheme. Peaks on the thermal deexcitation curves actually represent an energy spectrum of the centers.

Elucidation of factors influencing the position and intensity of the thermoluminescence peaks, their geological age, and temperature relationships, is in essence tantamount to determining factors affecting the formation, structural position, the existence and annihilation of the electron-hole centers.

The thermoluminescence parameters are merely a part, and not the most crucial one, of the characteristics featuring these centers. Such parameters do not suffice to determine the nature of the center in concrete crystals. Presently, a matter of extreme importance is to correlate the thermoluminescent data with models of the centers established from the EPR spectra.

An advantage of thermoluminescence is its appreciably greater sensitivity and simplicity of measurements. Following assignment of the thermal glow peak, made by means of EPR, it may be used for identifying this center in a given mineral. The said correlations, however, are ordinarily incomplete, for by using EPR one cannot find centers for all peaks to fit these peaks by their stability, being a single center type is usually in evidence, either electronic or hole. This is possibly due to a different order of EPR and thermoluminescence sensitivity, and to the fact that EPR must be observed at lower temperatures. Similar relations exist between EPR and luminescence: for example, the center S_2^- yielding a characteristic luminiscence spectrum in a number of silicate minerals is not demonstrable in EPR spectra, whereas O_2^- located with EPR in a number of instances is invisible in the luminescence spectra.

Thus, both as regards the theory of the phenomenon and applications of thermoluminescence in geology, combinations of the characteristics of the centers derived from the EPR spectra and from thermoluminescence (as well as from optical absorption spectra) prove most fruitful.

EPR gives concrete expression to the model of a center and enables its concentration and thermal stability to be determined.

6.2.1 Alkali Halide Crystals [599, 713, 732]

In most works the curves of the alkali halide thermal deexcitation have been obtained after irradiation of crystals at low temperatures. In such cases the trapping and emission centers are habitually connected with thermal defects annealed before reaching room temperature, being represented by the F- and F-aggregate and diverse V-centers (see Chap. 7). For LiF and KCl a correlation between hole centers determined from the EPR and low temperature thermoluminescence peaks [837] is established: the H center corresponds to 115 K, V_k to 230 K, V_t to 180 K, V_F to 230 K (designations of centers are shown in Fig. 170), the other low-temperature peaks, however, remaining without assignment.

The thermoluminescence of natural minerals containing diverse sets of impurity defects and vacancies, as well as of electron-hole centers linked with them, and of emission centers, usually displays curves of an intricate shape that changes from one sample to another, but they retain their characteristic features specific to genetic types of the samples.

Quite often a thermal deexcitation curve appears as an envelope curve whose shape is not determined by the actual peaks, but comes as s summation of peaks superposed one upon the other. The displacement of peaks and the appearance of of new peaks may be an apparent one, resulting from re-distribution of intensities from closely lying peaks contributing to the envelope curve. The shape of the thermal deexcitation curve changes when registering different spectral glow regions (with the use of different light filters).

The variability of thermoluminescence curves is determined by the presence of trapping centers on different impurities, liable to be fixed even at very low concentrations, by the formation of various types of centers on a single impurity defect and by means of charge compensation. A change in the shape of the thermal deexictation curve can be caused by a change in the total concentration of the impurity defect, repartition of diverse impurity concentrations, or of the manner of their incorporation in the structure.

Peaks of thermoluminescence of natural minerals usually occur in the region from 100°–150°C up to 500°C and above (the low-temperature maxima mostly fail to sustain).

An immense diversity is noted also in the intensity of thermoluminescence which varies by as much as tens and hundreds of thousands of times.

The complexity of the composition of natural minerals makes it necessary to resort to the thermoluminescence of their synthetic analogs with a controlled composition of impurities. Thermoluminescence in synthetic samples arises as a rule following their irradiation. Then irradiation at low temperatures ionizes defects which differ from those occurring in minerals. Irradiation at a room temperature fills, above all, the centers, yielding lower temperature peaks that fail to persist in minerals. The rate, type, and dose of irradiation, annealing with subsequent irradiation, also produce variations in the formation of centers.

Let us consider features particular to thermoluminescence of some of the best-studied minerals.

6.2.2 Fluorite

The emission centers in thermoluminescence of fluorite are almost always represented by RE^{3+}, while the capture centers may differ (Figs. 141–144). Irradiation

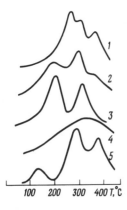

Fig. 141. Thermoluminescence curves for fluorites containing rare earth impurities; *1, 2, 3,* fluorites from pegmatities; *4, 5,* fluorites from hydrothermal deposits

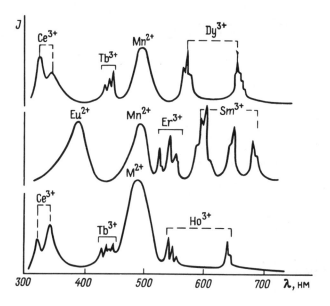

Fig. 142. Thermoluminescence spectra of fluorites with diverse rare earths [834]

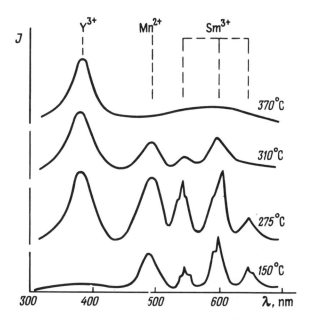

Fig. 143. Thermoluminescencee spectra of natural fluorites at different thermal deexcitation peaks [834]

at low temperatures causes the appearance of five low-temperature thermoluminescence peaks (up to 350 K), independent of the kind of RE, i.e., of peaks associated with thermal defects and with hole centers V, H, V_t, V_F (as in LiF) developing at their expense [593–597, 613].

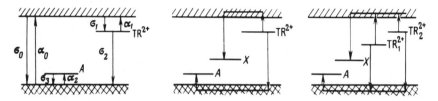

Fig. 144. Band schemes describing thermoluminescence of fluorites[605]. $RE^{2+}_{1,2}$ electron levels at different depth due to for instance, Dy^{2+}, Tm^{2+}, Sm^{2+}, Yb^{2+}, Ce^{2+}; A, X, hole centers created by eletron-donor impurities of RE^{3+} (for example, Ce^{3+}, Pr^{3+}, Tb^{3+}). A, by activators; X, by co-activators; α_0, α_1, α_2, cross-section of emission interaction with centers; δ_1, δ_2, δ_3, δ_4, cross-section of the charge capture by centers. Three band schemes are shown for different correlations of activators and co-activators

The peaks of natural fluorites and the higher-temperature peaks of the irradiated synthetic fluorites depend on the kind of RE present in them. Irradiation (natural or artificial) reduces some of the RE^{3+} ions in RE^{2+}, i.e., it induces formation of electron centers $RE^{2+} \rightleftarrows RE^{3+} + e^-$. Heating contributes to recombination of this center $RE^{2+} + e^+ (RE^{3+})^*$ with the formation of $(RE^{3+})^*$ in an excited state, from which it passes into the ground state by emitting radiation.

High-resolution investigations of the thermal deexcitation spectra allow the existence of centers with differing symmetry to be ascertained. Bearing in mind various modes of compensation (RE^{2+}_{cub}, $RE^{3+} + F^-_{tetr}$, $RE^{3+} + F^-_{trig}$), it becomes possible to trace down the mechanisms of excitation in photo-, X-ray, cathode- and thermoluminescence.

A study of thermoluminescence of synthetic samples of fluorite with RE^{3+} ions introduced into it one by one demonstrated that: (1) introduction of different RE^{3+} produces dissimilar thermoluminescence peaks, which shows that the trapping centers are indeed linked with them in some or other form, (2) with an equal concentration of the RE^{3+} ions and an equal irradiation dose, the thermoluminescence intensity in fluorites containing different RE^{3+} changes by as much as 50,000 times as compared with the weakest glow. This is due both to the probability of trapping and release of an electron (thermal ionization potentials of the RE see [597]) and formation of cubic, tetragonal and trigonal centers by different RE and the oscillator strength of the RE^{3+} on which glow occurs. In the region of up to 300°C the peaks in synthetic samples can be associated (in decreasing order of intensity) with Pr > Nd > Gd > Dy > Tm > Tb > Eu, whereas Sm, Ho, Er, Yb, U are not thermoluminescence activators in this region.

In natural fluorites the thermoluminescence curves are intricate and variable [612, 686, 714, 823, 827, 834]. Two regions are usually distinguished on them: peaks at 260°–320°C, de-excited largely on Pr^{3+} (with presence of Dy^{3+}, Tm^{3+}, Tb^{3+}, Eu^{3+}, Er^{3+}), possibly also on U^{6+} and peaks at 420°–490°C yielding mainly the spectrum of Sm^{3+} or Sm^{3+} and Tb^{3+} with Dy^{3+}. Zonal aggregates of differently colored fluorites proved a suitable object in clarifying factors affecting the geological age dependences of thermoluminescence [714].

6.2.3 Anhydrite

For understanding the nature of the anhydrites thermoluminescence (Figs. 145–147) three groups of observations [761, 831, 834] were of prime importance.

1. From the EPR spectra they were found to display a great variety of electron-hole centers (YO, $O_2^{3-}-Y^{3+}$ etc.) that served as models for the forms of the rare earth incorporation in them (REO, $O_2^{3-}-RE^{3+}$ etc.). The EPR spectra also helped fix the limits of their thermal stability. In the regions where some centers transform into others (retrapping of electrons and holes) the concentration of $O_2^{3-}-Y^{3+}$ increases at 150°–180° and at 300°C, whereas their decomposition occurs at ~ 450°–500°C (see Chap. 7.3).

2. The thermoluminescence spectra display the presence of rare earth ions Ce^{3+}, Sm^{3+}, Dy^{3+}, Eu^{2+}, on which (as well as on Mn^{2+}) bleaching of the natural anhydrite thermoluminescence peaks occurs.

3. The thermoluminescence peaks of the natural samples and synthetic anhydrites under study with one of the RE^{3+} ions (see Figs. 145, 146) practically coincide, in the region of 20°–500°C, merely differing in the relative intensity of individual peaks. Three types of natural anhydrites are distinguished: (1) all the peaks are deexcited on one and the same set of ions: Ce^{3+}, Dy^{3+}, Sm^{3+} (each peak deexciting on all these three ions); (2) Ce^{3+} at 110°C and Eu^{2+} at 180 and 240°C are added to all the peaks with Dy^{3+} and Sm^{3+} in common; (3) all the peaks include only $Ce,^{3+}$ and at 80 and 110°C also Mn^{2+}

The hole centers in anhydrite form due to incorporation in the structure of Y^{3+} and RE^{3+}, mostly not on these ions but on the oxygen ions of the Ca-polyhedron or SO_4-tetrahedron: $O_2^{3-}-RE^{3+}(REO_2)$. This group embraces the trapping center (O_2^{3-}) and the emission center (RE^{3+}). Recombination or transformation of O_2^{3-} induces thermoluminescence of RE^{3+}.

The low-temperature thermoluminescence of anhydrite with Mn^{2+}, Zn^{2+}, Sb^{3+}, Cd^{2+} [761] and properties of $CaSO_4$: Sm, Mn, as one of the most effective dosimeters [810] were also subject to investigation.

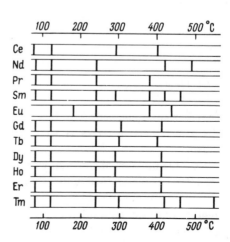

Fig. 145. Thermoluminescence maxima for synthetic rare earth-doped anhydrites [834]

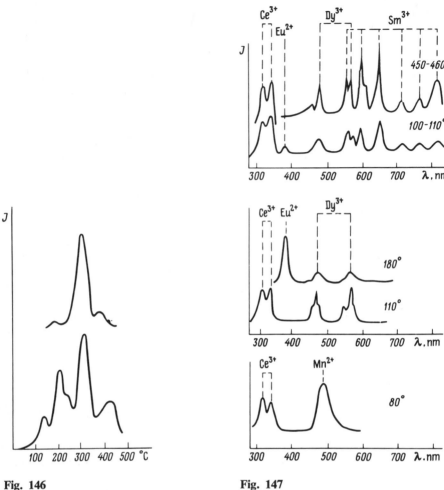

Fig. 146   **Fig. 147**

Fig. 146. Thermoluminescence curves for natural anhydrites [834]

Fig. 147. Thermoluminescence spectra of natural anhydrites [834]

6.2.4 Quartz

The thermoluminescence of the majority of magmatic and metamorphic, and also of many clastic rocks, is determined by the thermoluminescence of quartz and feldspars.

A large number of research works on thermoluminescence of natural quartzes are often of an empirical nature, but have resulted in establishing their general descriptive features and typomorphic potentialities. The thermal deexcitation curves of quartz are of quite an intricate nature and display variability in passing from one genetic type of the sample to another [627, 662, 715, 716, 729, 755, 762, 834].

One observes usually one or two nonelementary peaks in the region of 200°–220°, 280°–300° or 350°–380°C. The number of peaks, their shape and the relative intensity are made use of as typomorphic signs.

Among the emission centers two groups are distinguished: centers in the near UV region of ~ 380 nm and in the blue region of 440–480 nm. Quartz thermoluminescence is known to be of a recombination character. It is associated with the hole centers O^- - Al - M^+, established from EPR, where M^+ are Li, Na, H, with which peaks in the region of 200°–220°C are linked, and electron centers of Ti^{3+}, Ge^{3+} and others, with which peaks in the region of 280°–300°C may be associated. It is possible that the emission centers too are associated with these trapping centers.

6.2.5 Feldspars

In the case of fieldspars (Figs. 148, 149) the thermoluminescence peaks have been observed in approximately the same regions as in that of quartz, being usually in the range of 200°–220° and 280°–300°C [601, 834]. The first correlates with the center $Al-O^-$ - Al, and the second with Ti^{3+}. The blue thermoluminescence may come from these centers (as is presumed to be in the case of quartz) or with the Eu^{2+} impurities to which supposedly the luminescence spectra are assigned. The thermoluminescence is commonly more intensive in microclines than in orthoclases and plagioclases, and in acid plagioclases it is stronger than in the basic ones.

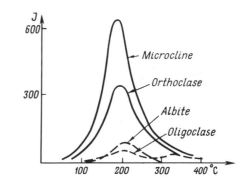

Fig. 148. Thermoluminescence curves for feldspars [834]

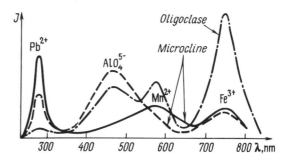

Fig. 149. Thermoluminescence spectra of feldspars [834]

6.2.6 Calcite and Dolomite

In the thermoluminescence of calcite and dolomite, which determine the thermo-luminescence of carbonate rocks, two to three peaks in the region of $220°-240°-260°C$ and $300°-340°-360°C$ are usually noted [607, 617, 744, 760, 763, 764].

In the cases of a low intensity glow or its absence, one irradiates limestones with gamma-rays, which produces the appearance of three or four peaks at $120°-140°$, $220°$ and $320°C$. A correlation with the EPR data makes it possible to contrast the phosphor centers in calcite with the peaks at $160°C$ (PO_2^0), $200°C$ (PO_2^{2-}) and $220°C$ (PO_3^{2-}), but to these temperature ranges other centers may also correspond.

6.2.7 Zircon

In natural zircons, which are concentrators of U and Th and exposed, therefore, to a very high dose of natural α-irradiation by comparison with other minerals, radiation damages in the structure reduce the intensity of thermoluminescence down to its complete disappearance when rated doses exceed $150 \cdot 10^{13} \alpha$ mg^{-1} [734, 844, 845]. The dependence of the glow intensity on the received dose is used in clarifying the mechanism behind the formation of the metamict state. Of five peaks noted for natural zircons ($140°-150°$, $160°-170°$, $240°-250$, $285°-290°$ and $340°-350°C$) most prevalent are the third and the fourth, while in the oldest and high-temperature zircons the presence of fourth and fifth, or only fifth peaks is in evidence.

The thermoluminescent properties of barites [834], clay minerals [816], diamond and other minerals [587, 673, 728, 750, 787, 834] were also studied.

6.2.8 Geological Applications

Eduction of the typomorphical significance of minerals and separation of genera-tion are effected with reference to manifestations of some or other peaks, and to their absolute and relative intensity. Stratigraphic correlations according to the types of thermal deexcitation curve (especially after gamma- and other kinds of irradiation) are based on the persistence of these curves in the samples of rocks and minerals picked up over great distances [627, 803, 841, 854]. In many geological regions the thermoluminescence method is being successfully employed for subdivi-sion of magmatic, metamorphic and sedimentary series and in determining their relative geological age [584, 699, 701]. Changes in thermoluminescence at contacts and in the vicinity of ore bodies and near faults are being traced down [756, 799]. Influence of conditions attending formation, of thermal history, deformations, stresses associated with major tectonic dislocations, meteoritic craters, and under-ground nuclear explosions have been discussed [586, 587, 644, 707, 729, 769]. Attempts to determine the absolute age of rocks, starting with lava streams in historical times and back to older rocks [583, 700, 714, 715, 845, 854], and also of the age of tectites and meteorites [646, 647, 693, 731, 780] are being undertaken.

In all these investigations, largely empirical ones, a more detailed interpretation of the nature and the mechanism governing the dependence of thermoluminescence on various factors is needed. These factors may be classed into three groups.

6.2.9 Crystallochemical Factors

The understanding of their significance for thermoluminescence rests on the fact that the trapping centers are factually electron-hole centers and the emission centers are basically ion activators. Hence, the thermoluminescence intensity depends on (1) the number of precursors, i.e., impurities and vacancies, (2) the possibility of electron-hole centers developing at their expense, the formation kinetics and the lifetime of these centers, (3) the number of formed centers, modes of their compensation, characteristics of the levels' energy and frequency factors, (4) characteristics of the sample respecting the dislocation density and ion disintegrators, (5) concentration of radioactive impurities and related intensity of irradiation.

6.2.10 Physicochemical Factors

These are subdivided into two subgroups:
1. Factors influencing the formation of centers. Three basic genetic types of center exist. (a) Crystallization centers, persisting only in the event of particularly thermostable centers, as, for example, in lazurite. In the overwhelming majority of minerals they become annealed in cooling of the massif or a vein lasting for thousands of years which suffice for decomposition at 400°–200°C of all the ordinary types of centers. The crystallization centers can persevere in sedimentary rocks not subjected to metamorphism; (b) radiation centers which represent the main type of natural centers; (c) centers arising due to strain, stress, and fragmentation.
2. Factors affecting the lifetime and formation of radiation centers after last heating, the period of time and temperature at which the sample exists. Warming up, even of short duration (in the geological time-scale temperatures surpassing 150°–200°C tend to eliminate the centers completely and their accumulation starts anew after the rock has cooled down).

6.2.11 Geological and Geochemical Factors

To understand the effect exercised on thermoluminescence by geochemical aspects of rocks and provinces, by concentration of radioactive elements, radioactive anomalies, relative and absolute age, metamorphism, wallrock alterations, tectonic influences, lithological and facial peculiarities, paleoclimatic conditions, etc., should all be reduced to crystallochemical and physicochemical factors. This is the only way in which the causal relationship of thermoluminescence in minerals can be interpreted.

6.2.12 Geological Age Dependences

In employing thermoluminescence for determining the geologic age two problems have to be solved first.

1. What is the time period necessary for the radiation trapping centers (of electron-hole centers) to form in natural minerals with irradiation doses provided by low contents of radioactive elements in the rocks? If these centers had been formed over a space of thousands and even hundreds of thousands of years, the possibility of using them in determining the geological age is limited to the youngest rocks; if this time extended over millions and scores of millions of years, it proves too short a period in dealing with Mesozoic and Paleozoic rocks and, conversely, if the centers had been accumulating over a period of hundreds of millions and billions of years they could then have been formed in appreciable numbers in the oldest rocks only.

2. What is the duration of existence of these centers (their lifetime), i.e., could they have remained intact since the oldest geological epochs? The time needed for the development of trapping centers is usually determined by comparing a dose of artificial irradiation required for the formation in a preliminarily annealed mineral of the same thermoluminescence peak and of intensity identical to a natural sample, against a dose which this same sample could have received as a result of natural radioactivity coming from small proportions of U, Th, K impurities. However, artificial irradiation can be conveniently effected by means of X-ray and gamma-radiation from an outside source, whereas the natural irradiation is accomplished largely by α-particles of decaying atoms of uranium entering the lattice isomorphically. Therefore, one has recourse to plotting a calibration curve that links a relative equivalent dose, i.e., the dose needed to reestablished the thermoluminescence intensity of the annealed sample and related to the natural α-activity of the sample, with the geological age of the reference sample whose age is known.

The evaluation of the lifetime of trapping centers formed in natural minerals has as its basis the same expression which links the electron's lifetime in a center with trap depth E, frequency factor s_0, and temperature at which this center exists: $\tau = 1/s_0 \exp^{E/kT}$. From the chart in Figure 140 it follows that deep traps can exist for millions of years whereas shallow traps do not persist. Uncertainty creates insufficient confidence as to the values of the frequency factors assessed both by the method of Antonov-Romanovsky [591] and through an experimental measurement of the centers' lifetime at different high temperatures (here extrapolation is made from minutes and hours to millions of years) and by other methods [622].

Experimental measurements of the relationship existing between the number of centers and the dose coming from diverse sources of irradiation, improvements in the calculation methods, direct measurements of the centers' concentration according to the EPR spectra (see Chap. 7) together with more detailed investigations into the nature of the centers and mechanisms responsible for their formation and decay, will all permit the connection between thermoluminescence and geological age to be considered in a more substantial and meaningful fashion.

Empirical observations show positive results to have been achieved by assessing the geological age of limestones according to their relative equivalent thermoluminescence [854] as well as of the age of archeological samples of ancient (up to 7,000 years old) ceramics [583]. In case of a system as complex as fluorite (with set of RE^{3+} and RE^{2+} capable of trapping electrons and holes), there is only a tendency toward a greater glow intensity with age, but no possibility of determining the absolute age, a for relative equivalent dose of the order of 10^5 roentgen characterizes samples dating back to ages from Cenozoic to Precambrian [612].

A comparison of the behavior of $SrSO_4$, $BaSO_4$, $CaSO_4 \cdot 2H_2O$ with a radioactive isotope S^{35} (β-emitter) introduced to them demonstrates: $BaSO_4$ to be much more radiationally stable than is $SrSO_4$ (with the same S^{35} concentration the centers in it develop much later), whereas in $CaSO_4\ 2H_2O$ with S^{35} no formation of centers could not be observed at all.

Observation of the EPR spectra shows that centers induced by artificial irradiation quickly disintegrate with the samples in storage (kept for weeks and months), while the same centers which existed in natural samples remained intact for an unlimited time.

This points to the need for conducting experiments and making calculations for concrete systems with due regard for their actual structure and composition.

7. Radiation Electron-Hole Centers (Free Radicals) in Minerals

7.1 Basic Principles and Methods

7.1.1 Discovery of Free Radicals in Minerals and Their Wide Distribution

The contemporaneous history of research into electron-hole centers in minerals began with applications of the electron-paramagnetic resonance method (EPR, see Chap. 3). Only with its help can one identify and interpret the models of the centers, establish their presence in the crystals, discern the kind of the centers, earlier identified, as well as the whole set of the centers present in a crystal. The very theory of the center is now closely related to the EPR parameters, which are related in their turn to a description of the center within the framework of the molecular orbital theory and position of the center in the crystal structure.

The measurements and interpretations of the EPR spectra of the centers by Bershov, Marfunin, Samoilovich, Vinokurov, Shcherbakova and others, carried out from the middle of the 1960s showed an enormous diversity of electron-hole centers in minerals and their wide distribution.

One can distinguish several branches of the pre-history of these investigations.

1. Since the past century the attention of mineralogists, physicists and chemists has been attracted by the color of some minerals which could not be connected with the impurities of ion chromophors (for example, yellow natural rock salt, amethyst, smoky quartz, etc.). It was noted that this coloration can be bleached by heating and restored through irradiation, electrical charge, or heating in metallic vapors and by other means. Minerals first stimulated systematic studies of the color centers in natural and synthetic crystals, owing to the obscure nature of their color and the tempting possibility to bleach and restore their original color and thus come nearer to understanding the mechanism behind this phenomenon.

2. The ideas about the color centers have been forming, based on materials of alkali halides [969] studied quite extensively since 1920 to 1930. Pohl (1925, 1937) introduced the theory of "color centers" or "F centers" (from German Farbenzentren)[11]. de Boer (1937) suggested the model of the F center as a halogen vacancy trapped electron [889]. At that time the principal method of investigating the centers was the measurement of optical absorption spectra and observation of their changes in the course of irradiation, and heating under the effect of light of different wavelengths. Seitz [983] suggested tentative models of F centers for the observed different absorption bands in alkali halides, these being aggregates of several elementary

---

[11] Presently the F center is referred to only as a center representing anion vacancy with a trapped electron.

F centers and, by analogy with them, the models of the V center (cation vacancy with a trapped hole) and the *V* aggregate centers. Results of these earlier ideas and a survey of the data obtained till the middle of 1950s for the centers in alkali halides, as well as in some minerals, are given by Przibram [791].

The modern period (since the middle of the 1950) in studies of color centers in alkali halides is connected with applications of electron paramagnetic resonance (also polarized absorption spectra and luminescence). These studies resulted in radical changes in understanding this phenomenon. It is with the aid of EPR only that the model and properties of the *F* center have been definitively established and understood in detail.

For the *F* aggregate centers one obtained models different from those suggested earlier, while for the *V* centers the general concept itself was changed. The models of more than 30 centers in alkali halides, interactions of short-lived centers, and the mechanism of their formation have been described by means of EPR and other methods.

At all these stages of the studies of the color centers in alkali halides excellent reviews and monographs were available. Earlier ideas and data (to the 1950s) are summarized by Seitz [983] and Przibram [791]. The latest results for the centers in halides are given in the survey by Schulman and Compton [982], in the special monograph on the *F* center by Markham [947], in the review on the *F* aggregate centers by Compton and Rabin [888]. It should be noted that in all books on solid-state physics published before 1965 the color centers were (1) considered based only on the data for alkali halides and (2) described without taking into account the EPR results which completely changed the models of the color centers in alkali halides.

3. For mineralogists the color centers in minerals, subject to observation because of the color inherent in them, were associated with the *F* centers in alkali halides. However, there is a great difference between centers in minerals and *F* centers in alkali halides: firstly, the centers in minerals form mostly in connection with impurities and other nonthermal defects, and, secondly, the crystal structures of silicate, carbonate and other minerals differ greatly from that of simple binary halides.

Much more similar to the centers in minerals, as was found later, are free radicals studied in chemistry, biology, astrophysics [901, 956] representing fragments of molecules, radicals, and ions trapping or losing an electron. They are called "free" because of the existence of a free valency (unpaired electron) even if they do not occur as isolated ("free") groups but are stabilized by a crystal lattice.

Application of EPR to studies of free radicals in irradiated inorganic crystals [921, 957] gave solid grounds for mineralogists for identifying these radicals in natural minerals. The survey of Atkins and Symons [859] is of special importance for the identification of these radicals from the EPR spectra.

4. In interpreting thermoluminescence data the formation of the glow peaks is related to the electron and hole trapping levels which remained nonassigned to any actual model and only later were attributed to electron and hole centers. All methods of determination of the parameters describing the formation of the centers are valid in considering the electron-hole centers (see Chap. 6).

5. First applications of EPR in mineralogy showed the existence in minerals, besides common transition metal ions (Mn^{2+}, Fe^{3+}, Cr^{3+} etc.; see Chap. 3), also of valence states unusual for geochemistry, VO^{2+}, Nb^{4+}, Ti^{3+}, Zr^{3+}, Hf^{3+}, W^{5+}, Mo^{5+}, etc. At last it was understood that all these paramagnetic centers: free radicals like CO_3^-, SO_2^- etc., molecule ions like O_3^-, S_2^- etc., F centers, ions with unusual oxydation state like Pb^+, Ti^{3+} etc., atomic hydrogen, electron and hole traps in thermoluminescence represent electron-hole centers, i.e., defects (impurities, ions neighboring impurity ions, vacancies, interstitial atoms) trapping electrons or holes. They form according to the general schemes: $AB_m^n \pm e^- \rightarrow AB_m^{n\pm1}$, or $A^n \pm e^- \rightarrow A^{n\pm1}$ or $[V_B] + e^- \rightarrow F$ center. (Here e^- is electron, A is cation, B is anion, and $[V_B]$ is anion vacancy.)

Wide distribution of the centers in nonirradiated minerals as well as, naturally, increasing concentration and formation of new types of centers due to irradiation were established.

For the understanding of the nature of the centers, the methods of their identification and description, one has to resort to the EPR technique (see Chap. 3) as a principal method of their investigation, and also to the molecular orbital theory as the way of describing their electron structure (see [1]). The application of the EPR method and the molecular orbital theory to description of electron-hole centers will be considered below.

7.1.2 Defects and Centers

The defect of a crystal structure (point defect) is any position in the structure representing a violation of the symmetry described by a symmetry space group of the crystal. The space lattice notion means that by putting a node of the lattice onto any element of the crystal structure (atom, interstice, or any other position) one finds the same element of the structure in all the nodes. The presence of a defect means that in one of the nodes a violation of the periodicity has occurred.

Point defects are: (1) anion and cation vacancies, (2) interstitial atoms and molecules, (3) impurity ions (isomorphous or interstitial), (4) atoms in disorder solid solution, (5) vacancy and impurity aggregates.

Larger aggregates of the defect form clusters and then colloidal segregations, while segregations of vacancies gradually form pores of increasing dimensions.

Any impurity ion in isomorphous or interstitial position is a defect of the crystal structure. In discussion the mechanism of defect formation and interaction in alkali halides and other binary compounds a special place is assigned to pairs of defects: (1) two vacancies–those of anion and cation–called Shottky defect and (2) two interstitial ions-anion and cation–called Frenkel defect.

Any defect represents a local deficit of charge electroneutrality, which is why every defect is a center precursor. Thus anion vacancy means the lack of a negative charge. The corresponding structure position is positively charged and thus represents an electron trap. In trapping an electron by the anion vacancy an electron center (F center) forms with subsequent restoration of the electroneutrality.

Another kind of electron center is represented by cations with unusual valence states, formed due to entering the crystal structure of a cation with a charge greater than in the replaced cation, for example:

$$A_{lattice}^+ \longrightarrow M_{impurity}^{2+} \longrightarrow M^{2+} + e^- \longrightarrow M^+$$

substitution   irradiation   electron trapping.

Thus irradiation of $LiF:Ni^{2+}$ gives Ni^+ ions representing the electron centers in which the electron is localized mostly on the cation. These cations have all the properties of the corresponding ions and this makes them different from other electron centers.

In the case of cation vacancy, i.e., positive charge vacancy, or in the case of insertion in the crystal structure of a cation with a lesser positive charge, a deficit of positive charge ensues. This structure position is negatively charged, and thus represents a trap for positive holes. However, the hole trapping process is not "symmetric" with respect to the electron trapping. Let us consider the formation of a hole and of hole centers on two examples.

1. In NaCl each of six Cl^- ions surrounding Na^+ has electron configuration $1s^2 2s^2 2p^6 3s^2\ 3p^6$, i.e., with completely filled electron shells. In the case of Na^+ vacancy one of the electrons in the Cl^- anion shells becomes superfluous.

Irradiation removes this electron and thus one of the negative chlorine anions becomes an atom with the electron configuration [Ne] $3p^5$, i.e., with a "hole" in the outer electron shell. Here the absence of the removed electron in a negative halogen ion is denoted as a hole[12] which behaves like a positively charged particle. The analogy here is similar, for instance to the one existing between Ti^{3+} $(3d^1)$ ion with one $3d$ electron and $Cu^{2+}(3d^9)$ ion in which one electron is lacking to fill the $3d$ shell. In the case of NaCl this hole is localized not on a single ion Cl thus becoming Cl^0, but is distributed among two neighboring ions Cl^-, thus forming the molecule ion: $Cl^0 + Cl^- \to Cl_2^-$, i.e., an electron deficit (a hole) which is distributed here among molecular orbitals of Cl_2^-. An essential difference between the cases of anion and cation vacancies is to be noted. Anion vacancy is not left "empty" in formation of an electron center; it traps an electron forming molecular orbitals with the neighboring Na^+ cation. The cation vacancy remains "empty", while an electron redistribution occurs entirely within the electron shells of anions surrounding the vacancy.

2. In quartz, SiO_2, cation impurities with a lesser charge, for example Al^3, substituting Si^{4+}, creates also a position with a deficit of positive charge, and is thus a trap for the positive hole. However with a surplus electron, removed due to irradiation, the electron deficit (i.e., the hole) localizes not at the cation Al^{3+}, but mostly at a neighboring oxygen anion. Because of this the oxygen ion O^{2-} $(2p^6)$ becomes O^- $(2p^5)$.

Thus Al impurity in quartz leads to the formation of the hole center not at the defect itself, but on the neighboring oxygen ion. Schematically, this may be written as follows:

---

[12] The term "hole" is used conformable to the electron deficit in a ion but not to the lacking ion in a lattice, which is termed vacancy.

$$M^{4+}_{\text{lattice}} + O^{2-} \rightarrow M^{3+}_{\text{impurity}} + O^{2-} \rightarrow M^{3-} + O^{2+} + e^{+} \rightarrow M^{3+} + O^{-}$$

$$\underset{\text{substitution}}{\uparrow} \quad \underset{\text{irradiation, hole trapping.}}{\uparrow}$$

The impurity ion must not necessarily differ by its charge from the replaced ion. Differences in the nuclear charges and their shielding by electron shells can create positions with positive or negative local charge capable of trapping an electron or a hole (see below for instance, Ge and Ti centers in quartz).

The description of centers as point defects trapping electrons or holes is merely a first approximation. Since this description involves the existence of an electron (surplus or lacking electron) a model of a center cannot be confined to its crystallo-chemical position, but must include information concerning its electronic structure. The electron of a center is distributed throughout the group of atoms, and can occur in different molecular orbitals–bonding, antibonding or nonbonding-formed in the general case from atomic orbitals of a number of different atoms. Thus, in the F center of the NaCl type structures the electron trapped by the anion vacancy is distributed among all of the six neighboring Na^{+} cations, and to a lesser degree among the orbitals of atoms of the second and third coordination spheres. In quartz the hole localized at an oxygen ion interacts with impurity Al ion and with alkali ion compensators, occuring in the channels of the crystal structure. In this case the notion of the hole center means the special kind of electron distribution among all these atoms.

In this way, the electron-hole centers represent special electron configurations of atomic groups connected with defect sites in the crystal structure that trapped an electron or a hole. In such cases there is nearly always (except for some aggregate centers with paired electrons) an unpaired electron which gives rise to the appearance of an optical absorption band in a normally transparent region (visible or UV) as well as to the EPR spectrum, paramagnetism, and photoconductivity.

7.1.3 Free Radicals in Crystals

In carbonates, sulphates, phosphates and other oxygen salts the radicals CO_3^{2-}, SO_4^{2-}, PO_4^{3-} etc. can trap holes (i.e., loose electrons) or electrons, just like the simple ions. Compare, for instance

$$Pb^{2+}(5s^2) \underset{\searrow e^{-} \rightarrow Pb^{+}(5s^25p^1)}{\overset{\nearrow e^{+} \rightarrow Pb^{3+}(5s^1)}{}} \qquad CO_3^{2-} \underset{\searrow e^{-} \rightarrow CO_3^{3-}.}{\overset{\nearrow e^{+} \rightarrow CO_3^{-}}{}}$$

By trapping a hole or an electron the radical CO_3^{2-} becomes a free radical: a hole-captured radical CO_3^{-} or an electron-captured radical CO_3^{3-}, just as when the Pb^{2+} ion with two s-electrons in an outer shell, trapping a positive hole, turn into the hole center Pb^{3+} $(5s^1)$, or, by trapping a surplus electron (p-electron) over and above the field $5s^2$ shell, it becomes the electron center Pb^{+} $(5s^25p)$.

Unlike simple ions, however, in radicals the electron or hole is captured not by one oxygen or carbon ion, but by the radical as a whole. Thus, in the CO_3^{2-} radical a hole is trapped not by one of the oxygen ions O^-, but is distributed among all the three oxygens which are here equivalent; this kind of electron distribution leads to axial symmetry of the CO_3^- radical revealed in the EPR spectra.

According to the structure position of a radical in crystals one can distinguish the following principal cases.

1. The hole or electron trapping represents a response to the existence of a defect in the cation position. For example, in $CaCO_3$ the $M^{3+} \rightarrow Ca^{2+}$ substitution stabilizes the CO_3^{3-} center, while the $M^+ \rightarrow Ca^{2+}$ substitution stabilizes the CO_3^- center (just as in substituting $Al^{3+} \rightarrow Si^{4+}$ in quartz a hole is trapped by O^-).

2. Hole or electron trapping is a consequence of heterovalent substitutions in anion (radical) position. For example, in sulphates in substituting PO_4^{3-} for SO_4^{2-} [and in general for $Y^{n-1}O_4^{(m+1)} \rightarrow X^nO_4^m$] a hole is trapped with the formation of the PO_4^{2-} free radical.

3. The existence of radicals AB and AB_2 in compounds of the MAB_3 type (for example, $CaCO_3$) and AB, AB_2, AB_3 in MAB_4 (in sulfates, phosphates etc.) means the existence of oxygen vacancies in these compounds. It should be noted, however, that a radical retains its individuality in structures with vacancies as well as without vacancies (for example SO_3^- in $CaSO_4$ and $CaCO_3$).

4. Radicals can find their way into compounds completely "nonisostructural" with respect to them. For example, SO_2^- enters the KCl structure retaining the same general properties as in $CaSO_4$, $CaCO_3$ and other sulfates and carbonates.

5. Radicals AB_m^n can contain different B ions. For example, it can be SO_2^- and SSO^-, SO_3^- and SSO_2^- with two different types of sulfur.

6. A particular case of AB_m^n radicals are radicals with the same and equivalent A and B ions; O_2^-, S_2^-, Se_4^- etc. Special kinds of center of this type are the Hal_2^- centers, or the hole V centers, in alkali halides (F_2^-, Cl_2^- etc.).

7.1.4 Molecular Orbital Schemes and EPR Parameters

As distinct from ions and impurity cation centers with electron configurations described by atomic orbitals (for example, $5s^25p^1$ for Pb^+), electron configurations of radicals are described by molecular orbitals.

Paramagnetic inorganic radicals are classified according to their molecular orbital types: first, into the types AB, AB_2, AB_3 AB_4 and then, according to the number of outer electrons taken into account in populating the molecular orbitals.

For the most common atoms entering the radicals one usually takes into account the following number of electrons:

$$
\left\{
\begin{array}{l|l|l|l}
B\ 2s^22p^1 = 3 & C\ 2s^22p^2 = 4 & N\ 2s^22p^3 = 5 & O\ 2s^22p^4 = 6 \\
 & & P\ 3s^23p^3 = 5 & S\ 3s^23p^4 = 6 \\
 & & As\ 4s^24p^3 = 5 & Se\ 4s^24p^4 = 6 \,.
\end{array}
\right.
$$

Then the number of electrons in radicals is:

$$\left\{ \begin{array}{c|c|c|c} BO_3^{3-} & CO_3^{2-} & PO_4^{3-} & SO_4^{2-} \\ 3+3\cdot6+3=24 & 4+3\cdot6+2=24 & 5+4\cdot6+3=32 & 6+4\cdot6+2=32 \end{array} \right. .$$

Trapping an electron, for example, by CO_3^{2-} radical with 24 electrons, leads to the formation of a CO_3^{3-} center with 25 electrons, while trapping a hole (a loss of an electron) leads to the formation of a CO_3^- center with 23 electrons.

A similar electron configuration of radicals, as, for example, in the case of SO_4^-, SeO_4^-, NO_4^{2-}, PO_4^{2-}, AsO_4^{2-} all of which have a 31–electron configuration (AB_4^{31}), determines identical for all of them the sequence and the type of molecular orbital, as well as the same general interrelations between the EPR spectra parameters to the same extent as in the case of ions with an identical electron configuration, for example Ti^{3+}, V^{4+}, and Cr^{5+} with the $3d^1$ configuration.

A molecular orbital in which an unpaired electron of a radical lies can be formed mainly from atomic orbitals of oxygens, or from an atomic orbital of carbon, sulfur, phosphorus etc., or essentially from participation of both kinds of atomic orbital.

In the case of an electron being trapped by a radical, this unpaired electron occurs mostly in the atomic orbital of the cation, thus leading to its formal valency of Si^{3+}, P^{4+} etc. and to the formation of the molecular orbitals mainly from s and p orbitals of silicon, phosphorus, etc.

In the case of a hole being trapped by a radical, the remaining unpaired electron mostly lies in the atomic orbitals of the radical oxygens, usually thus causing formation of molecular orbitals from predominantly nonbonding oxygen orbitals.

Therefore, in electron centers the unpaired electron occurs in an antibonding molecular orbital, while in the hole centers it occurs in nonbonding or bonding molecular orbitals.

The type of the molecular orbital formed from carbon (sulfur, phosphorus etc.) orbitals, or from oxygen orbitals, determines the value of superfine interaction of the unpaired electron with the nuclei of the cations and oxygens.

In the case of an electron localization chiefly on a cation (electron center) its strong interaction with this cation's nucleus (and consequently an intensive hyperfine splitting in the EPR spectra) is observed. With the electron localization principally on the oxygen ions (hole center) a weaker interaction of the electron with the cation nucleus is in evidence.

Isotropic component of the hyperfine structure is related to interaction of the s-electron with the nucleus, its anisotropic part is explained by interaction of the p-electron with the nucleus, and then the relation of the isotropic component to the anisotropic one determines the degree of the sp-hybridization and the A-O-A valence angles.

The relative ordering and distances of the outer molecular orbitals determine the spin-orbit interaction of the unpaired electron of a center which in its turn determines the value and the sign of the shift of the g-factor: $\Delta g = g_{exp} - g_e$, where $g_e = 2.0023$ is the free electron g-factor; Δg is usually negative for electron centers and positive for hole centers.

The molecular orbital schemes for the radicals AB_2^{17}, AB_2^{19}, AB_3^{23}, AB_3^{25}, AB_4^{31}, and AB_4^{33} are shown in Figures 150–152.

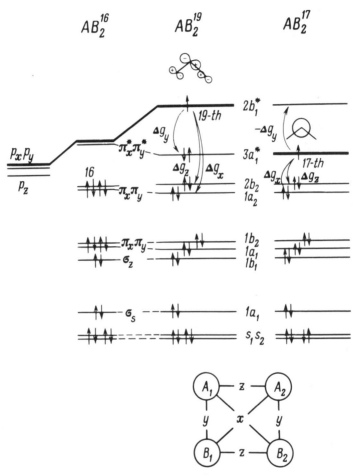

Fig. 150. Molecular orbital schemes for linear molecules AB_2^{16} (for 16-electron CO_2 and other) and for bent radicals (with C_{2v} symmetry) AB_2^{19} (for 19-electron NO_2^{2-}, PO_2^{2-}, SO_2^-, O_3^-, S_3^-) and AB_2^{17} (for 17-electron BO_2^{2-}, CO_2^-, SiO_2^-, NO_2, PO_2, AsO_2, SO_2^+).

AB_2^{19}, ground state: $(s_1 s_2)^4 \, 1a_1^2 1b_2^2 2a_1^2 1b_1^2 1a_2^2 2b_2^2 3a_1^{*2} 2b_1^{*1}$; $^2B_1$; excited states: $\ldots 1a_2^2 2b_1^1 3a_1^{*2} 2b_1^{*2}$; $^2B_2$; $\ldots 1a_2^1 2b_2^2 3a_1^{*2} 2b_1^{*2}$; $^2A_2$; $\ldots 1a_2^2 2b_2^2 3a_1^{*1} 2b_1^{*2}$; $^2A_1$; 19th electron (unpaired) occupies the molecular orbital $2b_1^2$, antibonding, formed mainly from p orbital of the A the atom; strong anisotropic hyperfine structure with small contribution of isotropic component; $+ \Delta g_y > + \Delta g_z > + \Delta g_x \lessgtr 0$.

AB_2^{17}, ground state: $(s_1 s_2)\, ^4 1a_1^2 1b_2^2 2a_1^2 1b_1^2 1a_2^2 2b_2^2 3a_1^{*1}$; $^2A_1$. 17th (unpaired) electron occurs in the $3a_1^*$ molecular orbital, antibonding, made mainly from s and p orbitals of the A atom thus determining superfine structure with large isotropic contribution and considerable anisotropy; $- \Delta g_y > + \Delta g_x > \pm \Delta g_z$

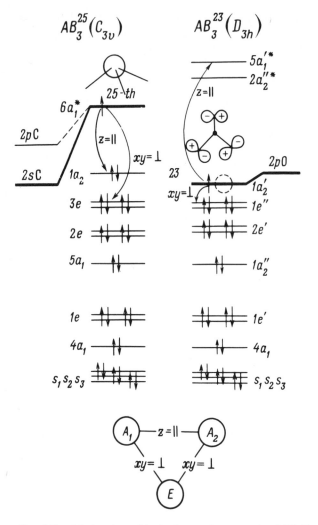

Fig. 151. Molecular orbital schemes for radicals AB_3^{25} (for 25-electron pyramidal CO_3^{3-}, SiO_3^{3-}, NO_3^{2-}, PO_3^{2-}, AsO_3^{2-}, SO_3^-) and AB_3^{23} (for 23 electron plane CO_3^-, SiO_3^-, NO_3, PO_3, AsO_3).

AB_3^{25}, ground state: $(s_1 s_2 s_3)^6 \, 4a_1^2 1e^4 5a_1^2 2e^4 3e^4 1a_2^2 6a_1^{*1}$; $^2A_1$; excited states: . . . $5a_1^2 2e^3 3e^4$ $1a_2^2 6a_1^{*2}$; ^2E; . . . $5a_1^2 2e^4 3e^4 1a_2^1 6a_1^{*2}$; $^2A_2$; 25th (unpaired) electron occurs in the molecular orbital $6a_1^*$ formed mainly from hybrid sp^n orbitals of the central atom A; large and mainly isotropic hyperfine structure with small anisotropic part with maximum along the radical axis; g-factor is close to g_e with axial symmetry, $g_{11} \gtrsim g_\perp \gtrsim g_e$.

AB_3^{23}, ground state: $(s_1 s_2 s_3)^6 (4a_1)^3 (1e')^2 (1a_2'')^2 (2c')^4 (1e'')^4 (1a_2')^1$; $^2A_2$; excited states: . . . $(2e')^4 (1e'')^3 (1a_2')^2$; ^2E; . . . $(2e')^4 (1e'')^4 (1a_2')^2 (2a_2'')^0 (5a_1'^*)^1$; $^2A_1$; 23th (unpaired) electron occurs in the molecular orbital $1a_2'$, nonbonding, formed from atomic $2p$ orbital of the three oxygens, so the interaction with atom A nucleus is very weak, hyperfine structure is small and nearly isotropic; g-factor is axially symmetric, $g_\perp > g_e$, $g_{//} \lesssim g_e$

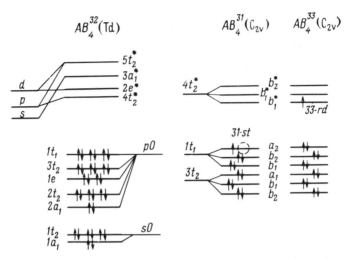

Fig. 152. Molecular orbital schemes for tetrahedral oxyanions AB_4^{32}('SO_4^{2-}, PO_4^{3-}, SiO_4^{4-} . . .) and radicals AB_4^{31} (31–electron SO_4^-, PO_4^{2-}, SiO_4^{3-} in distorted tetrahedron) and AB_4^{33} (33–electron SO_4^{3-}, PO_4^{4-}, SiO_4^{5-}).

AB_4^{32}, ground state: $1a_1^2 1t_2^6 2a_1^2 2t_2^6 1e^4 3t_2^6 1t_1^6$; $^2A_1$.

AB_4^{31} (upper occupied and lower antibonding orbitals only are shown) . . . $b_1^2 b_2^2 a_2^1$; $^2A_2$; 31st (unpaired) electron (trapped hole) occurs in the a_2 molecular orbital made mainly from nonbonding oxygen orbitals, thus hyperfine interaction with the A nucleus is small; deviation of g-factor from isotropic value (in distorted tetrahedron) is determined by splitting of t_1 molecular orbital; the scheme of orbital mixing is the same as for AB_2 (with C_{2v} symmetry), see Figure 150.

AB_4^{33}: 33rd electron occupies the antibonding molecular orbital formed due to splitting of the lower $4t_2^*$ orbital and consisting mainly from sp^n hybrid orbital of the A atom; large hyperfine structure; the scheme of orbital mixing as for AB_2 (see Fig. 150)

7.1.5 The Way to Identify the Electron-Hole Centers from the EPR Spectra

Two stages are distinguished in the course of determining the type of the center: (1) observation of a spectrum with the EPR spectrometer's oscillograph and the measurement of its parameters, (2) interpretation of the measured parameters based on the molecular orbital pattern of the centers, crystal structure, and impurity composition of the mineral.

Unlike other types of spectroscopy of solids, EPR (and NMR) is not merely a registration of a spectrum, but an examination on the oscillograph screen of its reaction to different rotation of the crystal, a procedure similar to universal stage measurements in crystalloptic studies. One can imagine the existence at the site of the center in a crystal structure (for example, SO_4^-, Pb^{3+} and other centers in a crystal) an indicatrix (an ellipsoid) whose principal axes are g-factor components: g_x, g_y, g_z.

By rotating the crystal about two axes (as in the case of the universal stage) relative to the magnetic field generated by the spectrometer's electromagnet, it becomes possible to try out and examine different local structure directions, and to fix the crystal in a position where one of the axes g_x, g_x or g_z coincides with the direction of the magnetic field. On the oscillograph screen the rotation of the crystal is seen to shift the line in the EPR spectrum till it reahces an extreme position ("double extreme position" because of rotation about the two axes), following which the line will start moving in the opposite direction.

A resonance value of the magnetic field H_{res} in this position corresponds ($h\nu = g\beta H_{res}$) to one of the principal values of the g-factor (g_x, g_y, or g_z) and determines its orientation in the crystal.

Hence the radiospectrometer of EPR is a kind of "universal stage", but working in the radiofrequency region and using for interpretation an indicatrix of the g-factors at the site of the atom or center in the crystal structure. By changing H_{res} one can see moving lines with different g-factor values ($h\nu = g\beta H_{res}$) on the oscillograph screen; by changing the orientation of the crystal we observe the line shifting due to its matching the magnetic field direction with different g-factor values of the indicatrix based on the principal values of g_x, g_y, g_z.

One begins with sorting out the lines, singling out all the lines of the same center, thus determining the number of different centers. There are some features of the EPR spectra which are common to all centers: (1) very narrow lines (0.n G) which allows it to discern every type of center, even when there are scores of lines; (2) closeness of observable g-factors to the free electron g-factor ($g_e = 2.0023$); (3) existence in most cases of single fine structure line (in trapping or leaving one electron the spin S of the center is equal to the spin of the electron s, i.e., $S = s = 1/2$); this line is not split by the crystal field, and thus parameters b_m^n (see Chap. 3) are here absent.

As a result of the observations and measurements of the spectrum one obtains: (1) g_x, g_y, g_z values; (2) hyperfine structure (hfs) parameter values A_x, A_y, A_z, as well as a number of hfs lines and their intensity relations; (3) orientation of g_x, g_y, g_z (and A_x, A_y, A_z, which often but not always coincide with the g_x, g_y, g_z directions), a number of nonequivalent positions for each type of center and its orientations. It is convenient to begin analyzing these data by considering the hyperfine structure. The number of hfs lines and their intensity relations in most cases allows determination of which atom nucleus interacts with the unpaired electron of a center. Typical spectra illustrating different cases of hyperfine structure are shown in Figure 153. From the nucleus spin I (see Table 14) and the number of the nuclei n with which the electron interacts, one determines the number of hfs lines equalling $2nI + 1$. (Because of low natural abundance of the isotopes with magnetic momenta, hfs is not observable from oxygen and sulfur atoms, and is not always observable from carbon and silicon atoms).

Molecular orbital schemes of centers are of great importance for determination of the centers. As EPR and optical spectra of transition metal ions are deduced unambiguously from energy levels obtained within the framework of the crystal field theory (see [1]), the EPR parameters of centers are deduced from molecular orbital patterns (see Figs. 150–152). The g-factor values are compared with the free electron g-factor, then relation between g_x, g_y, and g_z values are contrasted against those following from the molecular orbital schemes. Then one considers the same

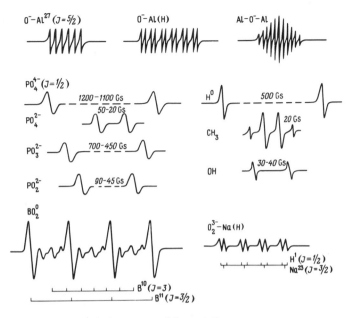

Fig. 153. EPR spectra of the electron-hole centers with superfine structures

Table 14. Nuclear spin (I) and natural abundance (N, %) of isotopes of some elements participating in electron-hole centers formation

Isotope	I	N, %	Isotope	I	N, %	Isotope	I	N, %
H^1	1/2	99.985	Na^{23}	3/2	100	As^{75}	3/2	100
Li^6	1	7.2	Mg^{25}	5/2	10.1	Mo^{95}	5/2	16.5
Li^7	3/2	92.7	Al^{27}	5/2	100	Mo^{97}	5/2	9.4
Be^9	3/2	100	Si^{29}	1/2	4.67	Y^{89}	1/2	100
B^{10}	3	18.83	P^{31}	1/2	100	Zr^{91}	5/2	11.2
B^{11}	3/2	81.17	S^{33}	3/2	0.74	Nb^{93}	9/2	100
C^{13}	1/2	1.1	Cl^{35}	3/2	75.4	Ba^{135}	3/2	6.6
N^{14}	1	99.62	Cl^{37}	3/2	24.6	Ba^{137}	3/2	11.3
N^{15}	1/2	0.38	K^{39}	3/2	93.2	W^{183}	1/2	14.28
O^{17}	5/2	0.04	K^{41}	3/2	6.8	Pb^{207}	1/2	21.1
F^{19}	1/2	100	Ge^{73}	9/2	7.8			

relations for hfs parameters ($A_x \gtrsim A_y \gtrsim A_z$) as well as contributions of the isotropic (A_{iso}) and anisotropic (B_x, B_y, B_z) parts of the hfs parameter.

For many types of center the patterns of molecular orbitals are now available and the relations of the g-factors and hfs parameters can be readily obtained from them.

A center retains its individuality in different structures (for example, SO_2^- in KCl and in $CaSO_4$) though its EPR parameters vary within a certain range (the same range is for a transition metal ion: Mn^{2+} or Cr^{3+} etc. in the case of EPR parameters)

An analysis of the orientations of the g-factors and hfs axes from the viewpoint of the crystal structure (1) helps understand this orientation and the origin of the center formation in the structure, (2) often suggests a possible model for the center, and always provides most convincing evidence for the reliability of the center model.

It is convenient to plot the g-factor orientation data onto a stereographic projection, together with the crystal axes and symmetry elements of the point group. Then the g_x, g_y, and g_z directions are compared with characteristic directions in the crystal structure; for example, with direction sulfur–oxygen in the SO_4^{2-} tetrahedron with an oxygen vacancy, if the center represents radical SO_3^-. There must be four such centers corresponding to the number of possible vacancies of the four oxygens of SO_4^{2-}. The orientation of each of the g-factors must be not only close to the direction of the sulfur–oxygen vacancy, but there must be four directions of the g-factor orientated in a certain manner with respect to crystal axes, and turned to one another at tetrahedral angles. None of the other centers in this structure can have this orientation.

A consideration of the space group of the crystal, multiplicity, point symmetry and the number of geometrically (magnetically) nonequivalent positions (mutually disposed according to the available symmetry elements) allows one to compare with these characteristics the number of nonequivalent positions of the center, determined experimentally from the EPR spectrum, and thus representing also one of the criteria for identifying the model of the center [434, 435, 920].

7.1.6 Systematics of the Electron-Hole Centers in Minerals and Inorganic Compounds

Established with the aid of EPR, center types can be classified proceeding from the two facts; (1) the centers retain their individuality in most different sturctures (the O^- center, for example, retains its properties in quartz, apatite, KCl etc.; O_2^{3-} can be observed in fluorite, anhydrite, sheelite etc); (2) the centers form through trapping by the defects of only one electron (or hole), i.e., usually no stable centers form by trapping two electrons.

Thus one can classify the centers by adding to different defect types of diverse possible types of the electron (or hole). In this way the following types of the centers are obtained (see Table 15):

1. Electron trapping by anion vacancies: F center and F aggregate centers discernable by the number of the aggregated vacancies and the number of the trapped electrons.

2. Electron or hole trapping by impurity cations: impurity cation centers ("ions with unusual valence states"); depending on the electron configuration of the cation, the trapped (or lost) electron can be an s, p, or d electron.

3. Hole trapping by anions in the position of a defect (O^- in KCl) or near it (O^- in tetrahedron with impurity Al ion in quartz): hole anion center with electron configuration p^5.

Table 15. Systematics of electron-hole centers in minerals and inorganic compounds

Anion vacancies trapped n electrons

$n = 1$	F = center	$F^+ = \alpha$	F^-
2	$F_2 = M$	F_2^+	F_2^-
3	$F_3 = R$	F_3^+	F_3^-
4	$F_4 = N$		
	X		

Cations trapped electron or hole

s^1	$H°Li°Na°Mg^+Ca^+Be^{2+}Pb^{3+}$
p^1	$Tl°Pb^+$
d^1	$Ti^{3+}V^{4+}Cr^{5+}Mn^6Y^{2+}Zr^{3+}Nb^{4+}Mo^{5+}Hf^{3+}Ta^{4+}W^{5+}$ ...
$d^{n\pm1}$	$Fe^+ \rightleftarrows Fe^{2+} \rightleftarrows Fe^{3+} \rightleftarrows Fe^4$; $Ni^+ \rightleftarrows Ni^{2+} \rightleftarrows Ni^{3+}$ ...
$d^{10}s^1$	$Cu°Ag°Zn^+$ ...
$f^{n\pm1}$	$TR^+ \rightleftarrows TR^{2+} \rightleftarrows TR^{3+} \rightleftarrows TR^{4+}$

Anions trapped hole

p^5	$O^-S^-Se^-Te^-F°(F^- + F° = F_2^-)Cl°(Cl_2^-)Br^0(Br_2^-)Ir°$

Inorganic radicals trapped electron or hole

AB11	BO^{2-}	CO^-	SiO^-	NO	PO	AsO	SO^+[a]			ClO[d]
13							SO^-	O_2^-	S_2^-[c]	
15								O_2^{2-}	S_2^{2-}	
15								F_2^-	Cl_2^-	
AB₂17	BO_2^{2-}	CO_2^-	SiO_2^-	NO_2	PO_2	AsO_2	SO_2^+			
19				NO_2^{2-}	PO_2^{2-}	AsO_2^{2-}	SO_2^-	O_3^-	S_3^-	ClO_2
AB₃23	BO_3^{2-}	CO_3^-	SiO_3^-	NO_3	PO_3	AsO_3	SO_3^+[b]			
25		CO_3^{3-}	SiO_3^{3-}	NO_3^{2-}	PO_3^{2-}	AsO_3^{2-}	SO_3^-	O_4^-	S_4^-	ClO_3
27								O_4^{3-}	S_4^{3-}	
AB₄31	BO_4^{4-}		SiO_4^{3-}	NO_4^{2-}	PO_4^{2-}	AsO_4^{2-}	SO_4^-			
33			SiO_4^{5-}	NO_4^{4-}	PO_4^{4-}	AsO_4^{4-}	SO_4^{3-}			
MB₄	VO_4^{4-}	CrO_4^{3-}	MnO_4^{2-}	WO_4^{3-} ...						

O^- center in polymeric structures

A—B—A	Si—O^-—Al;	Al—O^-—Al;	B—O^-—B ...

[a] Same for SeO_m^{n-1} etc.   [b] Same for $S(SO_2)^-$ etc.   [c] Same for Se_2^-, Te_2^- etc.
[d] Same for BrO_m^{n-} etc.

4. Electron or hole trapping by normal inorganic radicals: free radicals (i.e., with one free valency); all the mentioned types of defects (vacancies, impurities, anions in defect position) in oxygen salts lead to formation of free radicals: with an electron trapped by oxygen vacancy, for example, in CO_3^{2-} i.e., in $CO_2[V_0]$, the trapped electron is distributed all over $CO_2[V_0]$, i.e., this results in the formation of the free radical CO_2^-; with an electron trapped, for example, by impurity phosphorus in calcite, this electron is then captured by the whole of the radical PO_3^-, substituting CO_3^{2-}, and thus forming free radical PO_3^{2-}; the hole center O^- in compounds with radicals CO_3^{2-}, SiO_4^{4-}, PO_4^{3-}, SiO_3^{2-}, PO_3^- etc. is described as CO_3^-, SiO_4^{3-}, PO_4^{2-}, SiO_3^-, PO_3^0 etc. Further subdivisions in this most important and extensive group of

Table 16. Periodic system and electron-hole centers formation

```
H  He
Li Be
      B   C   N   O   F   Ne
Na Mg
      Al  Si  P   S   Cl  A
K  Ca                        Sc  Ti  V   Cr  Mn  Fe  Co  Ni  Cu  Zn
      Ga  Ge  As  Se  Br  Kr                                        TR
Rb Sr                        Y   Zr  Nb  Mo  Tc  Ru  Rh  Pb  Ag  Cd
      In  Sn  Sb  Te  I   Xe                                        Act
Cs Ba                        Lu  Hf  Ta  W   Re  Os  Ir  Pt  Au  Hg
      Tl  Pb  Bi  Po  At  Rn
Fr Ra
   $s^1$
```

$(H^\circ \quad Be^+ \quad B^{2+} \quad Pb^{3+} \ldots)$

p^1 $d^1(MB_4^{n\pm1})$

$(Tl^\circ Pb^+ \ldots)$ $(Y^{2+}Ti^{3+}V^{4+}W^{5+}Mn^{6+} \ldots)$

$AB_{4,3,2,1}^{n\pm1}$ $d^{n\pm1}$ $d^{10}s^1 \quad f^{n+1}$

$B = O, H, N, F, Cl \ldots$

$A = B, C, N, Al, Si, P, S \ldots$

$X_{2,3,4}^{n\pm1}$

$V_x^- = FF_{2,3,4}$

the centers are conveniently made with reference to (a) stoichiometry type of the radicals: AB, AB_2, AB_3, AB_4; (b) isoelectronicity of the radicals (AB_3^{23}, AB_3^{25} . . .); (c) the type of the atom making part of the radical: sulfur centers, carbon centers etc. It is also very important that all the isoelectronic radicals are described by the same typical molecular orbital pattern, and by the same interrelations of the EPR parameters.

5. In hole trapping by oxygen in polymeric structures varieties of the O^- hole centers form, such as $Si-O^- - Al$, $Al-O^- Al$ etc. One might think that the diversity of polymeric structures such as, for example, silicate structures (insular, chain, ring, sheet, and framework structures) would lead to the diversity of center types in them. However, since oxygen occurs in them mostly in a twofold coordination, the diversity is restricted only to the subdivision of the centers on bridging and nonbridging oxygens.

The position of an element in the periodic system determines the way in which the center linked with this element and electron configuration resulting from the electron or hole trapping by the usual valence states of these elements (Table 16) are formed.

7.2 Description of the Centers

The most systematic way of presenting the characteristics of the centers is by grouping them according to their electron configuration types: for example, the AB_2^{17}

radicals (i.e., all isoelectronic-17-electronic-radicals: BO_3^{2-}, CO_2^-, SiO_2^-, NO_2, NO_2, PO_2, AsO_2), AB_2^{19}, AB_3^{23} etc., then impurity centers with electron configurations d^1 (Ti^{3+}, V^{4+} . . .), s^1, p^1 etc. (see Table 14). All isoelectronic radicals are described by the same typical molecular orbital scheme, and have the same EPR parameters relations.

Let us nevertheless present this section in a different way. Let us adopt as a basis the chemical-mineralogical groupings of the centers, for molecular orbitals and EPR are in fact only means for identifying the centers, while the real significance of the centers for mineralogy and chemistry resides in their specificity in crystals and in features peculiar to their entering into minerals and inorganic compounds.

We shall begin with the description of molecular ions and free radicals prevailing in minerals and most suitable for attacking the problem of the centers on the basis of the molecular orbital, EPR, and crystal structure analysis, and then pass on to considering the impurity and F centers.

The center characteristics include: their chemical features and forms of the entering minerals; the molecular orbital type and the number of electrons; EPR parameters and spectrum features; absorption bands, their assignment within the framework of the molecular orbital and the color associated with them; stability in heating; luminescence, thermoluminescence, and others.

In the following section (see Chap. 7.3) we shall consider the characteristics related to the entering of the centers in the crystal structures of most important minerals, while here we shall consider the centers themselves.

7.2.1 Oxygen Centers: O^-, O_2^-, O_2^{3-}, O_3^- (Table 17)

Hole center O^-. Several types of the O^- center manifestation (Fig. 154) can be distinguished.

1. Oxygen is an impurity ion substituting halogen ion Hal^-, i.e., it represents the defect precursor: (a) in haloides, for example, O^- substituting Cl^- in KCl or F^- in CaF_2 etc. [855, 857, 1,012]; (b) in oxygen compounds where F^-, Cl^- occupies the structure sites, for example, $O^- \rightarrow F^-$ in apatite [970, 921], in synthetic fluorphlogopite [964].

2. Oxygen represents the normal component of a crystal, and occurs in a normal structure position but in the neighborhood of a defect:

a) In nonradical compounds, in simple oxides of the MgO type; just as anion vacancy traps an electron (F center) a cation vacancy traps a hole, i.e., one of the nearest to the cation vacancy oxygen ions loses its electron with the formation of the O^- center.

b) In polymeric structures (especially in silicates) and in the structural types where cation polyhedra (for instance, octahedra) joint to form a chain of bridging oxygen ions (for example, in amblygonite, augelite); when an impurity cation replaces the cation with a larger charge in one of the two neighboring polyhedra (or in the adjacent third polyhedron) the resulting deficit of the positive charge is compensated for by the hole being trapped by the bridging oxygen with concurrent for-

Table 17. EPR parameters of some radicals (A in G)

Radical	Mineral	$g_z(g_\parallel)$	g_y	$g_x(g_\perp)$	$A_z(A_\parallel)$	A_y	$A_x(A_\perp)$	Nucleus
O^-	Apatite(970)	2.0012	—	2.0516	7.0	—	0.3	F^{19}
	Fluorite(877)	2.0016	—	2.0456	53.7	—	15.0	F^{19}
	Microcline(945)	2.0043	2.0070	2.0555	9.2	9.2	4.5	Al^{27}
O_2^-	Zeolite(926)	2.0046	2.0090	2.057	—	—	—	—
O_2^{3-}	Fluorite(877)	2.0042	2.0090	2.0180	—	—	—	—
	Anhydrite(885)	2.0029	2.0262	2.0122	6.3	4.9	4.9	Y^{89}
O_3^-	Anhydrite(875)	2.0056	2.0144	2.0140	—	—	—	—
CO_3^{3-}	Calcite(984)	2.0013	—	2.0031	171.22	—	111.33	C^{13}
CO_3^-	Calcite(883)	2.0055	2.0194	2.0132	—	—	—	—
CO_2^-	Calcite(948)	2.0016	1.9973	2.0032	141	145	182	C^{13}
SO_4^-	Baryte(976)	2.003	2.016	2.071	—	—	—	—
SO_3^-	Celestine(871)	2.003	—	2.005	—	—	—	—
SO_2^-	Anhydrite(871)	2.0058	2.0092	2.0022	—	—	—	—
S_2^-	Baryte(871)	3.972	—	0.42	—	—	—	—
S_3^-	Lazurite(951)	2.005	2.036	2.046	—	—	—	—
SiO_4^{5-}	Zircon(995)	2.0020	—	2.0007	457	—	420	Si^{29}
SiO_4^{3-}	Zircon(995)	2.062	2.004	2.003	(19)	—	—	Si^{29}
SiO_3^{3-}	Zircon(995)	2.004	1.9993	2.0048	218	168	168	Si^{29}
SiO_3^-	Sheelite(902)	2.001	2.011	2.017	(10)	—	—	Si^{29}
SiO_2^-	Zircon(995)	1.9920	1.9960	2.005	202	168	161	Si^{29}
PO_4^{4-}	Phenakite(898)	2.002	—	2.001	1230	—	1095	P^{31}
PO_3^{2-}	Apatite(970)	2.0010	2.0010	2.0010	608	444	459	P^{31}
PO_2^{2-}	Calcite(867)	1.996	2.0007	2.0036	45	38	88	P^{31}
PO_2^0	Calcite(867)	2.0011	2.0036	2.0045	9.4	4.9	6.6	P^{31}

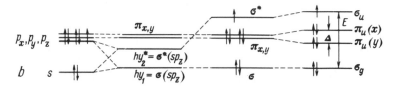

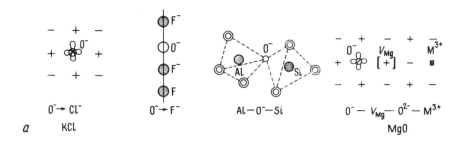

Fig. 154 a and b. Hole center O^-. **a** Structural position; **b** energy level schemes

mation of a O^- center, $M^{n+}-O^--M^{(n-1)+}_{impurity}$. Many important cases belong to this type: the Si-O^--Al center in quartz (due to substitution $Al^{3+} \rightarrow Si^{4+}$), the Al-O-Al in feldspars etc. (see 7.3). Though a hole in these cases is distributed in both tetrahedra: AlO_4 and SiO_4 with bridging oxygen becoming O^- center, but this center is usually approximated as O^- (or AlO_4^{4-}) (see Chap. 7.3)

c) In radical structures O^- becomes free radical XO_m^{n-}, for example, CO_3^-, SO_4^-, PO_4^{2-}, SiO_4^{3-}, WO_4^-, VO_4^{2-}, AsO_4^{2-}. Depending on symmetry, the oxygen ions can be either equivalent and nondiscernible, when the hole is distributed over the whole radical in which no separate oxygen can be distinguished, or, with a lower symmetry, each oxygen ion can trap the hole with formation of O^-, but all the same it is again the radical XO_m^{n-}, though with a nonhomogeneous charge distribution between the oxygen ions. For example, in zircon, $ZrSiO_4$, there are eight nonequivalent positions of O^-, but each of these can be considered as SiO_4^{3-}. On the contrary, in calcite one observes only one CO_3^- in which the hole is distributed over the whole radical and to which a single EPR spectrum corresponds.

There are two ways of approaching the description of the O^- center.

1. In the case of impurity O^- (in halides, apatite etc.) or in oxides of MgO type the O^- center energy levels are considered to be the levels of O^- ion with electron configuration $[He]p^5$ in a weak crystalline field; the unpaired electron occurs in the σ orbital.

2. In the case of O^- in radicals and other compounds where it forms stronger bonds the molecular orbital scheme of the corresponding radical or complex comes under consideration. The hole occurs in the last occupied molecular orbital consisting mainly of the π nonbonding oxygen orbitals. The fact that this molecular orbital is nonbonding allows one to consider the center as O^- and not as radical XO_m^{n-}.

On the other hand the existence in the molecular orbital of a second valence electron of the O^{2-} ion, whose removal leads to the hole formation, explains the possibility of the existence of O^{2-} ion in compounds, while the affinity of isolated O^{2-} ion to the second electron is negative and, in the free state, unstable. In oxygen compounds it is precisely, this electron that is the least strongly bound and in the absence of electrons in antibonding molecular orbitals (as in transition metal complexes), the hole centers always form at the expense of this second electron of oxygen occurring in nonbonding orbitals, these being the last molecular orbitals, thus determining the ionization potential of the compounds as a whole.

From the energy level schemes of O^- or XO_m^{4-} one obtains the EPR parameters: (a) for the σ variety; $g_{||}$ (or g_z) $\gtrless g_e$, $g_\perp$ (or g_x, g_y) $> g_e$; $\Delta g_\perp = \lambda(E_1 - E_0)$ where λ is the spin-orbit interaction parameters, E_0 and E_1 are the energies of the ground and excited state levels; (b) for the π variety: the same expressions, but $g_{||}$ and $g_\perp$ switch places, i.e., $g_\perp(g_x, g_y) \gtrless g_e$, $g_{||}$ $(g_z) > g_e$.

According to the g-factor values the O^- center is axial or pseudoaxial (though the actual symmetry is orthorhombic); $g_{||}$ (g_z) is chosen along the bond direction M-O^-, or X-O^-, or Hal-O^-. For example, in apatite: $g_{||} = 2.0016$ and $g_\perp = 2.0518$ [970]; for O^- in quartz: $g_\perp \approx 2.00$ and $g_{||} = 2.06$ [966].

One often observes hyperfine structure due to interaction of the unpaired electron of O^- with the nuclei of neighboring ions: with F and Cl nuclei in apatite and halides, with Al nucleus in quartz, with two Al nuclei in feldspars, and with Si^{29} in the case of spectra with good resolution.

In the presence of ion compensators (as in quartz with H, Li, Na in structure channels) each line of hyperfine structure (hfs) splits further with emergence of super-hyper-fine structure lines (shfs).

Orientation of the EPR parameters with respect to the crystal structure element depends on local symmetry. Directions of the g-factors, hfs and shfs axes can coincide with one another as well as with characteristic directions M–O, X–O, X–X in the structure, but they can also fail to do so. In quartz, for example, $g_{||}$ closely approaches the direction Si–Al, but does not coincide with it, while the hfs axis runs nearly parllel to the O^- –Al direction, and the shfs axis is orientated approximately along the directions O^-–H, O^-–Li, O^-–Na. Since the hole of the O^- center occurs mainly in the nonbonding oxygen orbital, the hfs and shfs values are usually small. If O^- is shifted towards one of the nieghboring anion (F^- in fluorite) or towards the cation (Mg in MgO) a large hfs value with strong axial symmetry can be observed.

Superoxide Ion O_2^-. The 13-electron $(\pi^*)^3$ radical O_2^- isoelectronic with S_2^-, ClO (see molecular orbital sheme and g-factor axes orientation in Fig. 155) occurs in peroxide compounds NaO_2, KO_2 etc., in alkali halides (grown or heated in oxygen atmosphere or in air) where it replaces halogen anions, in fluorite and apatite ($O_2^- \rightarrow F^-$), in zeolites (adsorbed O_2^- stabilized by hydrogen and interacting with Al), in anhydrite where O_2^- occurs in the position of one oxygen ion, or enters interstitial positions and occurs as HO_2.

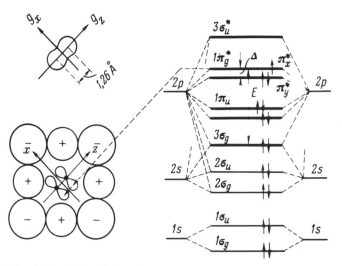

Fig. 155. Molecule ion O_2^-: its position in KCl structure and molecular orbital scheme (after [880]) ·

In all these cases it is a molecule O_2 with a trapped electron (not two structural oxygen ions as in the case of O_2^{3-}). The O_2^- center is always stabilized to a greater or lesser extent by the cations Na, K, Li, by hydrogen, by cation vacancies, by isomorphous or interstitial defects with positive charge deficit up to and including formation of radicals of the MO_2 type.

The g-factor values [954, 1,012, 1,015] are characterized by axial symmetry, relations $g_{||}(g_z) \gg g_e, g_\perp(g_x, g_y) < g_e$ and variability of $g_{||}$ and $g_\perp$ values depending on the degree of stabilization: for free O_2^- $g_{||} = 4.00$, $g_\perp = 0$, and the EPR signal is not observable, but with increasing energy level splitting $g_{||}$ takes on the values from 2.4 in KCl to 2.05–2.03 in zeolites and 2.02 in anhydrite, while $g_\perp$ is 1.95 in KCl, 2.009 in zeolites, and 1.99 in anhydrite.

The orientation of the g-factor axes in crystal structures depends on the position of the stabilizing cation. Hyperfine structure from Al, H and others help determine the structural position of O_2^-.

In alkali halides a broad absorption band near 40,000 cm^{-1} and yellow luminescence appear in connection with O_2^-. In the luminescence spectrum more than 12 equidistant ($\sim 1,000$ cm^{-1}) bands appear between 10,000 and 25,000 cm^{-1} (as in luminescence of S_2^-, where, however, this spectrum covers a much narrower region).

Stability of O_2^- also depends on its structural position; in anhydrite it is stable up to 250°C.

Hole Center O_2^{3-} is a 15-electron oxygen ion pair with a trapped hole, isoelectronic and similar to F_2^- and Cl_2^- centers in alkali halides. As distinct from O_2^-, the precursor of the O_2^{3-} center is not the O_2 molecule, but two oxygen ions each occupying a structural site of its own; trapping of the hole leads to formation of $O_2^{3-} = O^{2-} + O^-$ (as in $Cl_2^- = Cl^- + Cl^0$). In fluorite O_2^{3-} occupies the positions of two F^- anions [877]. In sheelite [995] a hole distributed mainly between two oxygen ions of two neighboring but not contacting WO_4 tetrahedra is described as O_2^{3-}, i.e., $(WO_3)O_2^{3-}$ (WO_3).

In anhydrite [874], following the $Y^{3+} \rightarrow Ca^{2+}$ substitution, each pair of the polyhedron edges can form O_2^{3-} as well as, due to replacement of sulfur with boron in SO_4^{2-}, the O_2^{3-}-B^{3+} center forms a tetrahedron with two oxygen vacancies. In quartz the Si^{4+} cation vacancy in the SiO_4 tetrahedron also leads to O_2^{3-} formation.

Depending on the degree of the other ions' participation in the center formation, the latter may be described either as O_2^{3-} or as MO_2^0, i.e., $O_2^{3-} - Y^{3+} = YO_2^0$, $O_2^{3-} - B^{3+} = BO_2^0$. The unpaired electron occurs in the σ orbital consisting of p_z orbitals of two oxygens; $g_z \gtrless g_e$, g_x and $g_y > g_e$; g_z is directed along the 0–0 bond, but depending on the 0–0 distance the relative position of g_z, g_y, and g_x can change.

Ozonide Ion O_3^-. The 19-electron O_3^-, isoelectronic with ClO_2, SO_2^-, SeO_2^-, NO_2^{2-} (the molecular orbital sheme in Fig. 150) represents the bent molecule AB_2 with the angle 127°, base length 2.24 Å and side length 1.26 Å. It forms upon irradiation of $AgNO_3$, $KClO_3$, K_2SO_4 etc.; following a reaction of the adsorbed O^- with O_2 in MgO, through interaction of alkali metal hydroxides with ozone accompanied by formation of stable ozonide complexes of the NaO_3 type. Among minerals it was observed in anhydrite in the form of captured interstitial O_3^- and MO_3^0 groups, stable up to $150°\,C$ [874].

In all these compounds with O_3^- very narrow ranges of g_1, g_2, and g_3 could be observed with actually always the same $g_{av} = 2.0012$ [875, 1019]. At room temperature one usually observes a rotation of O_3^- and the EPR spectrum becomes isotropic; in cooling down to liquid nitrogen temperature, anisotropic pseudoaxial g factor values with the axis perpendicular to the ion-radical plane are noted.

7.2.2 Carbonate Centers: CO_3^{3-}, CO_3^-, CO_2^-

These usually occur only in carbonates (above all in calcite), but also in carbonate-scapolite, carbonate-cancrinite; isomorphous substitutions $CO_3^{2-} \rightarrow PO_4^{3-}$ or SO_4^{2-} and others are uncommon, and thus one rarely observes CO_m^{n-} centers in non-carbonate compounds (for example, in apatite) (Table 17).

CO_3^{3-}. The 25-electron (C gives $2s^2 2p^2 = 4$ electrons, O_3 $2s^2 2p^4 = 6 \times 3$ and all in all $4 + 18 + 3 = 25$ electrons) AB_3 radical isoelectronic with NO_3^{2-}, PO_3^{2-}, SO_3^-, ClO_3 (the molecular orbital scheme in Fig. 151) form due to electron trapping by the CO_3^{2-} radical. The unpaired trapped electron occupies the antibonding $6a_1^*$ orbital formed mainly from carbon sp^n hybrid orbitals. This determines the emergence of a large isotropic component of hyperfine structure. Usually EPR spectra are axial, but when there is a local compensator (for example Y^{3+} in calcite) the symmetry becomes orthorhombic [949, 984, 985].

CO_3^-. The 23-electron hole AB_3 radical isoelectronic with NO_3 and PO_3 (the molecular orbital scheme in Fig. 151) forms due to the hole trapping (the loss of an electron) by the CO_3^{2-} radical. The unpaired electron occurs in the nonbonding group orbital of oxygen ions and, therefore, the values of the hyperfine structure from carbon (C^{13} 1.1 %), if observed, are very small. Symmetry is axial or—with local compensation—orthorhombic [859].

CO_2^-. Bent (with the bond angle about 136°) 17-electron (as NO_2) AB_2 radical (see Fig. 156); the molecular orbital scheme is shown in Figure 150. It forms due to electron trapping by the vacancy of one of the oxygens of the CO_3^{2-} radical; the trapped electron distributed throughout the whole of the CO_2^- radical occupies the antibonding $3a_1^*$ orbital made mainly of carbon sp^n orbitals; this determines a rather large isotropic component of the hyperfine structure [848, 859, 967].

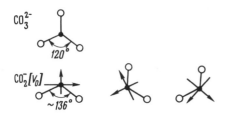

Fig. 156. Nonequivalent positions of CO_2^- center in carbonates

7.2.3 Sulfate and Sulfide Centers: SO_4^-, SO_3^-, SO_2^-, S_2^-, S_3^-

These are studied principally in three groups of compounds: in sulfates, within the framework of aluminosilicates with additional anions (sulfate-cancrinite, sulfate-scapolite, sodalite), in alkali halides (for example, $KCl:SO_2^-$, $KCl:S_3^-$) (Table 17).

Identification of these centers by means of EPR spectra often presents some difficulties because the hyperfine structure from S^{33} (natural abundance only 0.74%) is usually absent. By taking into account the crystal structure data and comparing them against the data derived from samples enriched with SS^{33}, more convincing results are obtained.

SO_4^-. The 31-electron ($S:3s^2 3p^4 = 6$; $O_4:2s^2 2p^4 = 6 \times 4$; $6 + 24 + 1 = 31$ electrons) hole ($SO_4^{2-} + e^+$) radical isoelectronic with PO_4^{2-}, SeO_4^-. Depending on the crystal structure and distortion of the SO_4^{2-} tetrahedron, one distinguishes three varieties of SO_4^- with different molecular orbital schemes (see Fig. 152): (1) SO_4^- in the nondistorted cubic tetrahedron (2) in the trigonal distorted tetrahedron (C_{3v}) and (3) in the orthorhombic distorted tetrahedron (C_{2v}). In the case of cubic local symmetry the EPR spectrum can be observed only at a liquid nitrogen temperature, in the cases of lower symmetries it is observed also at room temperature; the orientation in the case of C_{3v}: $g_z(g_{||})$ along the S–O bond; for C_{2v} one of the g-factors is directed along the L_2 axis, two others along the O–O edges; $g_z > g_e$, $g_x \approx g_y = g_\perp \approx g_e$; $g_{av} = 2.02$–2.03 [854, 963].

SO_3^-. The 25-electron radical isoelectronic with PO_3^{2-}, NO_3^{2-}, AsO_3^{2-}, SeO_3^-, ClO_3 (the molecular orbital scheme in Fig. 151). The unpaired 25th electron occurs in $6a_1^*$ molecular orbital formed mainly from sulfur hybrid sp^n orbitals [859, 874]. This pyramidal radical has axial or pseudoaxial symmetry. In sulfates it can be considered as the SO_4^{2-} tetrahedron with vacancy of one of the oxygens, this vacancy trapped an electron which is distributed throughout the whole of the SO_3^- radical. Typical are nearly isotropic very narrow lines (o.n G) with g-factor values close to the free spin, $g_{||}(g_z)$ is parallel to the direction of oxygen vacancy-sulfur, $g_\perp$ ($g_{x_1} g_y$) occurs in the three oxygen plane, g_y lies along the O–O direction. From the molecular orbital scheme it follows that $g_{||}$ and $g_\perp$ are nearly equal to g_e, but values inferior to the latter are often observed; $g_\perp > g_{||}$; $g_{av} = 2.002$–2.004.

SO_2^-. The 19-electron radical isoelectronic with NO_2^{2-}, ClO_2, SeO_2^-, O_3^- (the molecular orbital scheme in Fig. 150). The unpaired electron occurs in the antibonding $2b_1^*$ orbital. $g_{min}(g_x)$ is perpendicular to the radical plane, g_z coincides with the L_2 axis (if it is there), g_y goes along O–O; g_z and g_y can change places depending on the O–S–O angle; $\Delta g_x \approx 0$; g_y is large and positive; Δg_z is positive; $+\Delta g_y > + \Delta g_z > \Delta g_x \approx 0$; $g_x \approx g_e$ has minimum value, g_y has the maximum value and is equal to about 2.010–2.013, $g_z = 2.006$–2.010 [858, 874, 885, 980].

The SO_2^- center occurs often in sulfates where it enters the SO_4^{2-} tetrahedron, with two oxygen vacancies having trapped an electron.

S_2^-. The 13-electron radical is isoelectronic with respect to O_2^-, and therefore all the features of the molecular orbital scheme (Fig. 155) and the EPR parameters for it are the same as for O_2^-. However a much stronger anisotropy of the EPR spectra ($g_{||} = 0.42$; $g_\perp = 3.972$ for S_2^- in barite) is characteristic of S_2^- and, because of this, it is often difficult to observe it by EPR (low temperatures are necessary) [859, 1,010].

Characteristic luminescence spectra representing band series are in these cases the most convincing method of identifing this center. The EPR spectra of S_2^- are studied in detail in $KCl:S_2^-$ [1,010], the luminescence spectra having been investigated in many minerals: sulfates, framework aluminosilicates with additional anions, and in alkali halides.

S_3^-. The bent 19-electron radical isoelectronic with O_3^-, SO_2^-, S_2O^- (the molecular orbital scheme in Fig. 150). The EPR spectra of S_3^- have been studied in detail in alkali halides, for example, in $KCl : S_3^-$ [980] (see Fig. 170), in natural lazurites and synthetic ultra-marines, where an intensive blue color caused by a 610 nm absorption band is due to S_3^- [978].

In systems containing active sulfur reacting with oxides of four- and six-valent sulfur formation of S_3^- (as well as of S_2^-. SO_2^-, SO_2^-, SSO^-) takes places.

7.2.4 Silicate Centers: SiO_4^{5-}, SiO_4^{3-}, SiO_3^-, SiO_2^-

These are observed not only in insular silicates, but also in polymeric silicate structures (zircon, quartz, etc.) as well as in cases of substition of other anions, for example, $SiO_3^{3-} \rightarrow PO_4^{3-}$ in apatite, $SiO_3^- \rightarrow WO_4^{2-}$ in sheelite (Table 17).

SiO_4^{5-}. The 33-electron (Si: $3s^2 3p^2 = 4$; $O_4 : 2s^2 2p^4 = 6 \times 4 = 24$; $4 + 24 + 5 = 33$ electron) radical isoelectronic with PF_4, PO_4^4 [877]. This is an electron center with an unpaired electron in the silicon antibonding orbital, i.e., it can be considered as Si^{3+} ($Si^{4+} + e^-$). Its formation is probably connected with the presence of interstitial cations, for example, of alkali metals, Most characteristic is a highly intensive (about 400 G) hyperfine interaction with Si^{29}. The anisotropy of the hyperfine structure and of the g-factor depends on the SiO_4 tetrahedron distortion. The presence of the center was ascertained in zircon [856, 994] and quartz [997].

SiO_4^{3-}. The 31-electron radical isoelectronic with well-studied PO_4^{2-}, SO_4^-, SeO_4^- [859] (the molecular orbital scheme in Fig. 152). This radical may be considered to be a hole center O^- in an isolated tetrahedron SiO_4 (or AlO_4) in insular silicates (for example, in zircon) or as a O^- center on nonbridging oxygen in other silicate structures. The hyperfine structure value from Si^{29} is here only about 20 G; the g-factor values are close to those of O^-.

SiO_3^-. The 23-electron hole radical isoelectronic with CO_3^-, NO_3 (the molecular orbital scheme in Fig. 151). It does not form in silicates because the oxygen vacancy in $SiO_3[V_0]$ traps not a hole, but an electron. It is observed when high-charge ions are replaced with silicon, for example, with SiO_3^- replacing WO_4^{2-} in sheelite [995].

SiO_2^-. The 17-electron bent radical with the O–Si–O angle about 127°, isoelectronic with CO_2^- and NO_2 (the molecular orbital scheme in Fig. 150) but according to the way of its formation (at the expense of two oxygen vacancies in the SiO_4 tetrahedron trapping an electron) similar to SO_2^- and PO_2^{2-}. The hyperfine structure from the Si^{29} value is about 200–130 G. Orientation of the g-factor is determined by the configuration of the SiO_2 fragment of the tetrahedron. It has been observed in zircon [902, 994], and quartz [997].

7.2.5 Phosphate Centers: PO_4^{4-}, PO_4^{2-}, PO_3^{2-}, PO_2^{2-}, PO_2^{0}

Apart from phosphates these centers can be observed in carbonates, sulfates, and silicates, when radical PO_m^{n-} are substituted for the CO_3^{2-}, SO_4^{2-}, SiO_4^{4-} anions. The possibility of observing a hyperfine structure from P^{31} ($I = 1/2$, natural abundance 100%) is very helpful in determing the types of phosphate centers.

PO_4^{4-}. The 33-electron (P : $3s^2 3p^3 = 5$; O_4 : $2s^2 2p^4 = 6 \times 4 = 24$; $5 + 24 + 4 = 33$) radical isoelectronic with PF_4, SiO_4^{5-} and similar to SiO_4^{5-} forming through trapping the $PO_4^{3-} + e^-$ electron which may be considered as the P^{4+} ion formation ($P^{5+} + e^-$). The largest hyperfine structure value among phosphorus radicals (1,100–1,200 G in comparison with about 30 G in hole PO_4^{2-} center) and characteristic g-factor values (2.001–2.002) obtained for PO_4^{4-} in phenacite (substitution $SiO_4^{4-} \rightarrow PO_4^{3-} + e^-$) are satisfactorily explained by the scheme of molecular orbitals [920] calculated for this particular case.

PO_4^{2-}. The 31-electron hole radical isoelectronic with SiO_4^{3-}, SO_4^- (the molecular orbital scheme in Fig. 152) represents essentially a O^- center in an isolated tetrahedron [859].

PO_3^{2-}. The 25-electron radical isoelectronic with SO_3^-, NO_3^{2-}, AsO_3^{2-} (the molecular orbital scheme in Fig. 151) the molecular orbital calculation see in [898]; it forms due to the electron being trapped by the oxygen vacancy in PO_4^{3-} tetrahedron (in apatite), or in SiO_4^{4-} (in tremolite), SO_4^{2-} (in anhydrite) tetrahedra, or due to substitution of $PO_3^{2-} \rightarrow CO_3^{2-}$ (in calcite).

The EPR parameters of PO_3^{2-} are very similar in different compounds: the axial hyperfine strcuture of 700–450 G values, the g-factor is less than g_e, isotropic (about 2.001) or weakly anisotropic (1.998–2.008) [859, 874, 917, 964].

PO_2^{2-} and PO_2^{0}. The 19- and 17-electron radicals isoelectronic with SO_2^-, O_3^-, NO_2^{2-} and with CO_2^-, NO_2^{0} were observed in calcites [867].

7.2.6 Impurity Cation Centers

Impurity ions are one of the point defect types and, like vacancies, can capture an electron or a hole (depending on the charge of the defect position). In these cases the impurity ions turn into unusual valent states. The following types of the formation of electron and hole centers on the impurity ions are possible (Table 18).

1. M^{n+1} (impurity) $\rightarrow K^{n+}$ (lattice), i.e., the impurity cation charge is more than or equal to the lattice cation charge and this relation leads to electron trapping: (a) by impurity cation: $[Pb^{2+}/K^+]$ in amazonite $\rightarrow Pb^+$ ($Pb^{2+} + e^-$); (b) by radical: $[Y^{3+}/Ca^{2+}]$ in calcite $\rightarrow Y^{3+} + CO_3^{2-}$ ($CO_3^{2-} + e^-$).

2. M^{n+} (impurity) $\rightarrow K^{n+1}$ (lattice), i.e., the impurity cation charge is less than or equal to the lattice cation charge, this relation leads to hole trapping: (a) by impurity cation: $[Pb^{2+}/Ca^{2+}]$ in calcite $\rightarrow Pb^{3+}$ ($Pb^{2+} + e^+$); (b) by ligand ion: $[Al^{3+}/Si^{4+}]$ in quartz $\rightarrow Al^{3+} + O^-$ ($O^{2-} + e^+$).

In some cases even a difference of effective charges (with equal formal valency) is sufficient for formation of a center. For example, $[Ge^{4+}/Si^{4+}]$ in quartz Ge^{3+} ($Ge^{4+} + e^-$).

Table 18. EPR parameters of some impurity cation centers (A in G)

Center	Mineral	$g_z(g_\parallel)$	g_y	$g_x(g_\perp)$	$A_z(A_\parallel)$	A_y	$A_x(A_\perp)$	Nucleus
			s^1					
Ge^{3+}	Quartz(899)	2.0011	1.9950	1.9939	(228)	—	—	Ge^{73}
					(9)	—	—	Si^{29}
Pb^{3+}	Calcite(942)	2.0045	—	2.0072	13440	—	13540	Pb^{267}
			p^1					
Pb^+	Amazonite(945)	1.390	1.565	1.837	585	164	34	Pb^{207}
			d^1					
Ti^{3+}	Beryl(868)	1.9977	—	1.8391	—	—	22.3	Ti^{47}
							23.2	Ti^{49}
V^{4+}	Sphen(345)	1.9455	1.9492	1.9572	180.6	59.5	55.5	V^{51}
Y^{2+}	Anhydrite(874)	2.0050	2.0072	2.0100	197.3	194.2	190.7	Y^{89}
Nb^{4+}	Zircon(468)	1.862	—	1.908	309	—	138	Nb^{93}
Hf^{3+}	Zircon(468)	1.802	—	1.996			5.8	Hf^{179}
							7.8	Hf^{177}
			$f^{n\pm1}$					
Tb^{4+}	Zircon(873)	2.046	—	7.955	33.1	—	37.3	Tb

Unusual valent states of ions in minerals are nearly always related to electron-hole centers. Normal cation valent states correspond to electron configurations arising due to the loss of all outer electrons, thus leaving completely filled electron shells np^6, ns^2, nd^{10}. The unusual valent states form due to adding to these configurations one additional (trapped) electron, or due to the loss of one electron. In these cases the trapped electron can be the s^1, p^1, d^1, or f^1 electron, depending on the existing configuration of the ion.

In the middle of transition metal groups there remain d^n or f^n electrons in the normal ion's electron configurations, and thus for such ions center formation by electron trapping often results in the emergence of one of the normal valent states ($Fe^{2+} \rightleftarrows Fe^{3+}$) or leads to the configurations $d^{n\pm1}$ or $f^{n\pm1}$. These ions are electron centers, differing genetically from the corresponding usual ions only by their origin.

7.2.7 Hole Center S^-

The significance of the S^- center (i.e., $S^{2+} + e^-$) in sulfides is the same as that of the O^- center in oxygen compounds and Hal_2^- (for example, Cl_2^-) in halides. At present the available data S^- covers only simple sulfides ZnS and CdS where S^- forms in the same way as O^- in simple oxides MgO and CaO.

Its formation is linked with entering of heterovalent impurities while replacing cation ($M^{3+} \rightarrow Zn^{2+}$) or anion ($X^- \rightarrow S^{2-}$).

The structural position of S^- is shown together with other centers in ZnS in Figure 130.

7.2.8 Atomic Hydrogen in Crystals

Atomic hydrogen evolving in irradiated crystals containing hydride ions, hydroxyls, or water molecules may be held for an electron center formed, for example, according to the following schemes:

a) H^- (isomorphous) $+ [V_{an}] \rightarrow H^0_{int} + F$ center, i.e., isomorphous ion H^- and anion vacancy $[V_{an}]$ produce after irradiation a neutral interstitial hydrogen atom H^0 and a F center; the latter forms through the anion vacancy trapping the electron lost by the hydride ion H^-.

b) $[OH]^-$ (isomorphous) $\rightarrow O^-$ (isomorphous) $+ H^0$ (interstitial). The evolving atomic hydrogen easily diffuses through the lattice, and because of this the H^0 centers are either unstable at a rising temperature (in KCl already up to 100 K) or need to be stabilized by "traps".

Because of its small size and electroneutrality, atomic hydrogen interacts weakly with the surrounding ions, and thus can be considered with good approximation to be a nonperturbed free atom in its spherically symmetric ground state. This explains the features of the EPR spectrum of atomic hydrogen in crystals: its isotropy, closeness of the g-factor to the free electron value, closeness of hyperfine splitting in crystals (about 500 G) to that of the free hydrogen atom (506 G), similarity of its EPR spectra in different compounds.

Strong interaction of the electron of atomic hydrogen with proton having the nuclear spin $I = 1/2$ and very large magnetic moment yields two lines of hypefine structure ($2mI + 1 = 2 \cdot 1 \cdot 1/2 + 1 = 2$) with the splitting about 500 G. Additional splitting of each of these two lines is caused by the weak interaction with nuclei of the surrounding ions:

— in KCl and NaCl the weak interaction with Cl^{35} and Cl^{37} nuclei (both having $I = 3/2$) gives $2mI + 1 = 2 \cdot 4 \cdot 3/2 + 1 = 13$ lines [859, 890],
— in CaF_2 from eight F^{19} ($I = 1/2$) nine lines [859, 887],
— in beryl from one Na^{23} ($I = 3/2$) four lines [858, 932].

In alkali halides KCl, NaCl, KBr [915, 975, 1008] optical spectra, EPR and double resonance of H^0 center have been studied [859]. The U, U_1, and U_2 optical absorption bands are interpreted as follows:

— U: hydride ion H^- substituting Cl^- or Br^-,
— U_1: hydride ion H^- in interstitial position (with anyon vacancy yielding the α absorption band),
— U_2: atomic hydrogen in interstitial position.

All these absorption bands occur in the UV region, contiguous to the absorption edge, and are related to the charge transfer. Of these only the H^- and OH^- absorption bands are observed at room temperature (in alkali halides) while others bleach at above 100 K. Hyperfine structure of EPR spectra and double resonance data distinctly establish the interstitial position of H^0 in the tetrahedral interstice surrounded by four chlorine anions. In CaF_2, SrF_2, and BaF_2 crystal structures the cubes with F^- in the eight vertices, occupied by Ca^{2+}, or Sr^{2+}, or Ba^{2+} alternate with similar but empty cubes and the atomic hydrogen enters the empty cubes [859, 1019]. Because of interatomic distances in CaF_2 equalling $R = \sqrt{3} \cdot a \cdot \sqrt{2} = 2.36$ Å and F^- ionic radius 1.36 Å a space with the radius of 1.00 Å is left there for H^0. An estimation of charge density distribution from the EPR hyperfine structure shows that about 92% of the charge density occurs within the limits of a sphere with a radius of 1.00 Å.

Atomic hydrogen has been observed in natural fluorites [988], in quartz [1014] and calcite [953]. In natural "alkali" beryls (vorobyevite) irradiated by X-rays at room temperature, the H^0 centers form at the expense of water molecules in structural channels [946]. Observation of secondary hyperfine structure (splitting of each line of hydrogen doublet into four lines) shows that water molecules occur in channels contacting sodium ions.

7.3 Models of Centers in Minerals

7.3.1 Prevalence and Significance of Centers in Minerals

Prior to application of the EPR the presence of "color centers" has been observed in some minerals without, however, any more precise definition of their models. Later, the spectra of "strange" valence states of the atoms quite uncommon in geochemistry and previously unsuspected were identified, along with free radicals and molecular ions. Finally, observations of many samples taken from scores of the most widespread minerals stemming from different areas, types of rock, and ore deposits proved the electron-hole centers to be widely distributed in nature and to be of essential importance for geology and prospecting.

Since they were discovered in such major rock-forming minerals as quartz, feldspars, calcite, anhydrite, and such accessory minerals as zircon and apatite, a new opportunity was provided for characterizing these minerals and the rocks enclosing them. Especially important are the centers for the minerals (like quartz and barite) with rather restricted isomorphous substitutions, and without structural state changes which could be used for discerning these minerals in rocks of different origin. However, even for feldspars with a vast set of variable structural, chemical, and optical characteristics, the centers present the possibility for further more detailed subdivisions. The diversity of the center types, the existence of many centers in each mineral, variations in their relative and absolute concentrations, their presence in nonirradiated samples or their appearance after artificial irradiation, the genetic meaning, their very existence in minerals, the accumulation of trapped electrons over geological ages, thermal stability of different centers, features specific for the real structure and geochemical conditions, deficit or excess of sulfur or oxygen, etc., all this makes them characteristic tracers for the same minerals, formed in different condition. Many of these minerals with the centers represent widely distributed minerals of gold, tungsten, tin, rare metal and other deposits (quartz, feldspars, calcite, barite, anhydrite, beryl etc.). The electron-hole centers can be used in discerning genetic types of ore deposits for their evaluation, in investigating their zonality and the stages of mineralization. These applications are essentially similar to the applications of thermoluminescence in geology. However, the center characteristics as obtained by means of EPR have the following advantages:

1. The EPR spectra make it possible to determine directly the concentrations of the centers (to observe each one of them in a mineral) and to do so rather quickly, provided a prior identification of the center's model is available. At this stage the entire assembly of the centers becomes much more numerous than can be judged by broad overlapping thermoluminiscence peaks.

2. The centers observed by means of EPR spectra are identified unambiguously in contrast to "anonymous" thermoluminescence peaks.

3. With the help of EPR one can obtain all the structural and chemical information about the centers.

4. This information includes not only the data on the trapped electrons and holes, but also on the center precursors: impurites, vacancies, interstitial ions and molecules which can be used for the characterization of condition attending the mineral formation.

5. The parameters and models of the uncovered set of the centers represent a precise and nontransient characteristic of the real structure of a mineral (it is only the number and concentration of the centers that vary), while the thermoluminescence peak positions do not represent a constant property of the mineral, varying in many cases from one sample to another, and depending on factors that are still outside control.

The electron-hole centers represent a new characteristic of minerals the same as isomorphous substitutions, order–disorder relations, and cation distribution between nonequivalent sites. Their geochemical significance consists first of all in enormous expansion of a list of molecules, radicals and ions observable, and earlier unknown in natural processes (free radicals SO_3^-, SO_2^-, CO_2^- . . . , molecule ions S_3^-, S_2^-, superoxide ion O_2^-, ozonide O_3^- . . ., ions Pb^{3+}, Pb^+, Hf^{3+} . . ., atomic hydrogen). The existence of electron-hole centers was established in lunar samples [473], and their presence presumed in the interstellar dust [1016].

Center formation is a general property of the mineral (natural inorganic) matter of the earth, moon and cosmic space. Three conditions are necessary for their formation: the real structure with defects, weak natural radioactive irradiation, and geological and cosmic scales of time.

Free radicals, earlier well known in chemistry, biology and astrophysics, have thus now become a subject of studies also in geochemistry and mineralogy.

7.3.2 Features Specific to the Structural Type and Models of the Centers in Minerals

In the above text the general description of the electron-hole centers and, above all, of the EPR parameters and molecular orbital schemes of the centers was given. Now let us see what undergoes a change during of a formation of a center type in different crystal structure types.

First of all, measurements showed that general characteristics of a center (g_x, g_y, g_z and A_x, A_y, A_z values and interrelations deduced from the molecular orbital scheme) are retained in different crystal structures, but some variations of the EPR parameters are observed. What is more, the structure position of a center is of greatest importance for its identification. The crystalline field axes at the site of the center are reflected in the g-factor and hyperfine structure axes orientation. The superhyperfine structure axes orientations are related to the position of neighboring ions with magnetic nuclei. The number of lines of the superhyperfine structure (shfs) depends on the number and position of the ligand ions with magnetic nuclei. The number of symmetry related structurally nonequivalent positions of the same center is also closely related to the crystal structure. The center axes orientation can be measured with great precision by means of EPR spectra, while the number of center positions and the number of shfs lines are observed immediately on the oscillograph screen.

These experimental data are compared with the symmetry space group of the crystal structure and with the cation–anion, anion–anion, cation–cation directions in the crystal structure.

The space group data are also of spectroscopic significance in controlling the number of the center positions in the structure observed from EPR spectra, the point symmetry of the center positions, and their orientation with respect to crystal axes. Tabulated results of such an analysis of space groups made from the viewpoint of the EPR are now available and help establish the structural position of a center [425, 434, 435].

An analysis of the directions in a crystal structure involves calculations of their orientation, plotting them in stereographic projection, their multiplication by means of the symmetry elements and their comparison with experimentally determined orientations of the g-factor and hyperfine structure axes. In many cases these data already allow identification of the center type and its structural position (see below descriptions of the centers in quartz, anhydrite and others). Moreover, in such a way one can distinguish the centers formed on each of the nonequivalent atoms of the crystal structure (see below description of centers in feldspars).

The identification of a center in a crystal as well as the understanding of its model and its relation to the structural type, its geochemical features which caused insertion of a set of impurities and vacancies in the structure can be considered as complete only after determination of the center position in the structure (together with an analysis of the EPR parameters, based on the molecular orbital schemes).

Now we shall consider the centers studied in detail in some of the most important minerals.

7.3.3 Quartz

One can distinguish the following center types in natural and synthetic quartz: (1) impurity centers, (a) electron centers (titanium, germanium and in general $M^{4+} + e^-$), (b) hole centers (aluminum, with iron and in general with M^{3+}, M^{2+} ions, as well as atomic hydrogen); (2) vacancy centers, (a) electron centers (with oxygen vacancies), (b) hole centers (silicon vacancies) [857, 861, 899, 908, 910, 914, 918 936–939, 953, 955, 966, 996, 997, 1013].

According to the precursor type, the electron (e^-) and hole (e^+) centers in quartz can be grouped as follows:

e^-	e^+
Ti	Al
Ge	Fe
V_0	V_{Si}
H^0	—

Each one of these types has subtypes and varieties, depending on the kind of the ion-charge compensator in structural channels (H^+, Li^+, Na^+) as well as on the position of the compensator. There are also geometrically nonequivalent positions, i.e., some positions of the same center related to one another through the symmetry elements exist.

The electron centers Ti^{3+} and Ge^{3+} are stable in the presence of the compensators M^+ and are not stable without them. On the contrary, the hole centers O^-–Al are stable without an ion compensator near the center, and form unstable varieties in the presence of M^+.

Other impurites can be inserted into synthetic quartz crystals (W, Co, Ga and others), and the similar center types can be formed with these impurities. Most widely distributed centers in natural quartz are O^- — Al, Ti^{3+} and O_2^{3-}.

Let us consider first the EPR parameters and varieties of the centers and then their structural positions (the center orientation types in the quartz structure).

Titanium Centers: Ti^{3+}, i.e, $Ti^{4+} + e^-$ substituting silicon; the existence of their varieties is caused by the presence of H^+, Li^+, or Na^+ ions in channels near some or other edge of the silicon-oxygen tetrahedron (see Fig. 159). Each of the center varieties with very close EPR parameters (Table 19) has three nonequivalent positions.

Hyperfine structure from two titanium isotopes is observed: Ti^{47} ($I = 5/2$, natural abundance 7.75%) and Ti^{49} ($I = 7/2$; 5.51%) as well as superhyperfine structure caused by additional interaction with the nuclei H^1 ($I = 1/2$; 99.98%; each hfs line is split into shfs doublet), Li^7 ($I = 3/2$; 92.57%; each hfs line is split into three shfs lines with relative intensities 1:2:1), Na^{23} ($I = 3/2$; 100%, shfs triplets)

The Ti^{3+} centers produce an absorption band of 496 nm with a pink color. The center is stable up to 200–300° C; X-ray or gamma irradiations restore the bleached center.

Germanium Centers: Ge^{3+} ($Ge^{4+} + e^-$) have the same type of orientation and the same varieties as Ti^{3+} (Table 19). Hyperfine structure is observed from Ge^{73} ($I = 9/2$; 7.8%).

Table 19.   EPR parameters of some varieties of centers in quartz (A in G) [899, 944, 1017]

Center	n^*	g_1	g_2	g_3	A_1	A_2	A_3	Nucleus
Ti^{3+}—Li^+	3	1.912	1.930	1.979	1.16	—	1.35	Li^7
							27.1	Ti
Ti^{3+}—H^+	3	1.916	1.936	1.984	4.43	—	5.0	H^1
							22.8	Ti
O^-—Al	6	2.0045	2.0045	2.0590	5.0	6.3	6.3	Al^{27}
O^-—Al/H^+	6	2.0034	2.0093	2.0581	7.9	9.2	9.8	Al^{27}
					1.3	0.49	0.7	H^1
O^-—Al/Li^+	6	2.0033	2.0083	2.0618	6.5	8.0	8.1	Al^{27}
O^-—Al/Na^+	6	2.0021	2.0096	2.0437	7.1	8.5	8.8	Al^{72}
$SiO_3^{3-}(E_1)$	6	2.0017	2.0005	2.0003	457	—	389	Si^{29}
					9	—	7	Si^{29}
$SiO_3^{3-}(E_2)$	6	2.0020	2.0007	2.0005	—	(412)	—	Si^{29}
					—	(0.4)	—	H^1

n is number of the each center positions in quartz structure.

Aluminum Centers (Table 19): O^- — Al, or AlO_4^{4-} is the particular case of hole O^- center with all its characteristics. The syperhyperfine structure arising due to the interaction of unpaired O^- electron with the Al^{27} nucleus ($I = 5/2$; 100%) in the defective tetrahedron yields a distinctive six-line EPR spectrum. There are six nonequivalent positions of the center. The substitution of aluminum for silicon ($Si^{4+} \rightarrow Al^{3+}$) is accompanied by a charge compensation through incorporation of H^+, Li^+, Na^+ ions in the channels near the Al tetrahedron. Irradiation causes formation of the O^-–Al center with simultaneous diffusion of the ion compensators.

The absorption bands 450–460 nm (2.7 eV) and 620 nm (2 eV) connected with the O^-–Al center make quartz assume a smoky tint. After irradiation practically all natural quartz samples show O^-–Al centers.

The center is stable up to 300° C, and after heating can be restored by the X-ray or gamma irradiation. The O^-–Al/H^+, O^-–Al/Li^+, O^-–Al/Na^+ centers are unstable at room temperature, but the O^-–Al/Li, H center (with simultaneous presence in the channels near O^-–Al of both Li and H compensators) is stable up to ~180° C; one of the radiation citrine color corresponds to this center variety (absorption bands 390–410 nm and 620 nm).

Fe^{3+} Iron Ion EPR Spectra are observed in all natural and synthetic amethysts, in brown and green quartz, nonradiation citrines. Three types of the spectra are distinguished:

1. $Fe^{3+} \rightarrow Si^{4+}$ with alkali metal ions in the channels; there are three positions of the iron; quartet hyperfine structure due to interaction with Li, Na (both have $I = 3/2$); spin-Hamiltonian parameters: $D = 9.34$ GHz, $E = 1.72$ GHz, $c = 0.012$ GHz.

2. $Fe^{3+} \rightarrow Si^{4+}$ with proton in the channels; the same three positions; doublet hyperfine structure ($I = 1/2$ for proton).

3. Fe^{3+} in interstitial pseudotetrahedral positions; $D = 7.9$ GHz, $E = 1.7$ GHz, $c = 0.002$ GHz; six positions. All these cases represent the spectra of usual Fe^{3+} ions and not those of the centers. They are, however, the precursors of the amethyst centers. Irradiation causes a decrease of the Fe^{3+} concentration; the amethyst color grows in intensity with weakening of the Fe^{3+} spectra.

Both states which can be produced from Fe^{3+} by irradiation: Fe^{2+} and Fe^{4+} are not investigated by means of EPR (their spectra, at least that of Fe^{2+} can be observed only at the liquid helium temperature). Thus their existence is supported principally by optical absorption spectra, which in this case are less convincing.

Note should be taken of the difference between Fe^{3+} and Al^{3+} insertion consequences. The analogy between them is obvious ($Fe^{3+} \rightarrow Si^{4+}$ and $Al^{3+} \rightarrow Si^{4+}$). However, if in the case of Al^{3+} insertion in quartz structure the hole is trapped by oxygen with the formation of O^{-}–Al (or AlO_4^{4-}), in the case of Fe^{3+} the scheme of center formation changes. The hole in the nonbonding oxygen orbital would be in this case unstable because of the presence of Fe^{3+} ion $3d^5$ electrons in antibonding orbitals. It is probably at the expense of these $3d^5$ electrons that the hole center $Fe^{4+}(3d^4) = Fe^{3+}(3d^5) + e^+$, or FeO_4^{4-} forms.

Centers with Oxygen Vacancies. As in other structures with tetrahedral oxyanions these are oxygen vacancies in the SiO_4^{4-} radical which have trapped an electron that is being distributed over the whole of the radical forming electron centers SiO_3^{3-} (or $SiO_3 V_0 e^-$, or Si^{3+} in the tetrahedron with oxygen vacancy). The fact that quartz has a framework polymeric crystal structure and each oxygen here is a bridging one, i.e., it belongs simultaneously to two SiO_4 tetrahedra, does not essentially complicate the center model because the electron trapped by the vacancy is drawn off towards one of the tetrahedra $SiO_3 V_0$.

Two varieties of the SiO_3^{3-} center arise because of the existence of vacancies of two types of oxygen: (1) oxygen ions forming Si-O bonds with the length of 1.617 Å, directed at an angle of 44° to the L_3 axis, belonging to shorter (2.604 Å) oxygen–oxygen edge of the tetrahedron (Table 20), (2) oxygen ions with Si–O lengths 1.597 Å at an angle of 66° to L_3, belonging to the longer O–O edge (2.640 Å). Each of these center types has six positions.

The first variety is often designated as $[Si(E_1) e^-]$ and the second one as $[Si(E_2) e^-]$.

Similar centers exist also when an electron is trapped by an oxygen vacancy in the tetrahedron with Ge instead of with Si, i.e., GeO_3^{3-} or $[Ge(E_1)e^-]$ and $[Ge(E_2) e^-]$. A large hyperfine structure from Si^{29} ($I = 1/2$; 4.7%), equal to about 400 G or from Ge^{73} ($I = 9/2$; 7.8%), equal to 220–260 G, indicates that these centers are indeed essentially Si^{3+} and Ge^{3+} centers. In the case of GeO_3^{3-} centers additional weak hyperfine structure from Si^{29} of the first (8–10 G) and second (0.4–1.0 G) coordination spheres is also observed. The varieties of these centers owe their existence to the presence or absence of alkali ions or protons in channels, their kind, and position.

The SiO_3^{3-} centers are rarely observed in natural quartz, but appear in considerable quantities in neutron or gamma-irradiated samples. The absorption bands assigned to these centers are found in the far UV region (207–218 nm, i.e., about 5.7 eV):

Table 20.   Directions Si-O, Si-S, O-O in quartz structure (left-handed quartz)

	φ	θ	n	r_{A-B}
Si—O_1	267.5	44	6	1.617
Si—O_4	147.5	44		
Si—O_2	357.5	66	6	1.597
Si—O_5	117.5	66		
$Si^1/_3$—Si_0	246	54	6	3.057
O_1—O_4	210	28.5	3	2.604
O_2—O_5	30	61.5	3	2.640
O_1—O_5	254	79	6	2.614
O_2—O_4	46	79		
O_1—O_2	316	47	6	2.637
O_4—O_5	106	47		

The φ angle is counted from $a_1 = +x = [2\bar{1}\bar{1}0]$, the angle θ from $c = z = L_3$ axis. Designations Si-O_1, O_2-O_5 and other correspond to Figures 157 and 158; n is number of nonequivalent positions of a given type; r_{A-B} is interatomic distances (Å). From the six directions of the same type the three (Si-O_2 or Si-O_4 or O_1-O_5 . . .) are connected by L_3 symmetry axis; the directions Si-O_2 and Si-O_4 are connected in pairs by L_2 axis (as well as Si-O_2 and Si-O_3, O_1-O_5 and O_2-O_4 and so on).

The formation of SiO_2^- centers, i.e., with two oxygen vacancies, is also probable.

Centers with Silicon Vacancies. In the quartz crystal structure the existence of the silicon vacancy leads to hole trapping by two oxygen ions with formation of a O_2^{3-} center (i.e., $O^{2-} + O^- \rightarrow O_2^{3-}$), which can only form on two of the four O–O edges of the SiO_4 tetrahedron, namely at those perpendicular to the L_2 axis. On two different edges (O_1–O_4 and O_2–O_5; see Table 20 and Fig. 157) two center varieties form, each of them having three positions.

Hyperfine structures from Si^{29} nuclei of the neighboring tetrahedra and from protons in channels are observed. The center is stable up to 300–350°C.

Structural Positions of the Centers in Quartz. The projection of the quartz crystal structure is shown in Figure 157, and to this the following comments should be made.

In quartz there is only one kind of silicon (located at the L_2 axis) and only one kind of oxygen (in the common position). Each oxygen ion belongs to two unequal edges (2.604 and 2.640 Å), each is bonded with two silicons: with the first Si at an angle of 44° to the L_3 axis and Si–O distance 1.617 Å, with the second Si at an angle 66° and Si–O distance 1.597 Å, each located identically with respect to the channels. However, the edges in SiO_4 tetrahedron perpendicular to L_2 are different: one O–O edge is shorter and its oxygens form the bonds with Si at an angle of 44°, while other O–O edge is longer and the angle is 66°. Alkali metal and hydrogen ions in the channels located on the L_2 axis can occur either near shorter or near longer edges. These directions are described in Table 20 and shown in a stereographic projection in Figure 158.

Consideration of the Si–O, O–O, Si–Si directions shown in Figures 157 and 158 allows to distinguish three types of the center orientations in quartz as well as to determine their structural position.

Fig. 157. Crystal structure of quartz (left-handed); projection on the plane perpendicular to $z = L_3$ axis. The oxygen atoms are designated 1–6 according to the Table 20 (cf. also Fig. 158); number in the circles indicates heights above the projection plane in units of hundredth of the lattice parameter c

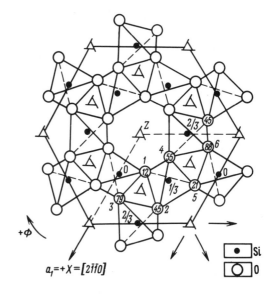

$a_1 = +X = [2\bar{1}\bar{1}0]$

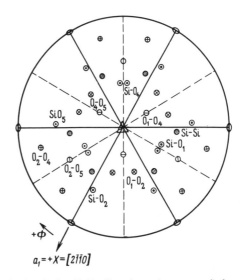

$a_1 = +X = [2\bar{1}\bar{1}0]$

Fig. 158. Stereographic projection of the Si-O, O-O, Si-Si directions in quartz (left-handed) crystal structure (see Table 20 and Fig. 157). The $+\theta$ angle is counted from $z = L_3$ axis. Only one of the directions of each type is labeled (Si-Si, O_2-O_5 etc., corresponding to the labeled atoms in Fig. 157); other directions of the given type are obtained with the action of the L_3 and L_2 symmetry axes. The points of the intersection of the directions with upper hemisphere only are indicated; this explains the positions of the O_1-O_2 O_2-O_4 and other directions connected by L_2 axis. *Note:* Figures 157 and 158 show the structure of left-handed quartz: $+x$ = electronegative $[2\bar{1}\bar{1}0]$ axis = L_2; $+y$ = mechanical $[01\bar{1}0]$ axis = $\perp L_3$; $+z$ = optical $[0001]$ axis = L_3. The $+\phi$ is counted from $+x$ clockwise, the $+\theta$ angle from $+z$. In order to pass to right-handed quartz it is necessary to rotate the projection at 180° around the y–y axis. Then the angles indicated for left-handed quartz correspond without change also to righthanded

The first type (Fig. 159): One of the g-factor axes coincides with the L_2 axis in the quartz crystal, which at the same time is the L_2 axis in the SiO_4 tetrahedron and intersects the centers of the O_1–O_4, O_2–O_5 edges; the two other g-factor axes are approximately parallel to the O_1–O_4 and O_2–O_5 edges. An exact coincidence of one of the g-factor axes with L_2 is unambiguously established by the existence of only three positions of the center; this type of orientation alone gives in quartz three positions of the center, whereas all other orientations give six positions for each center, any slightest deviation of the g-factor orientation from L_2 would have immediately manifested itself in splitting the three spectra into six spectra of the center.

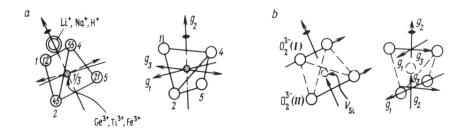

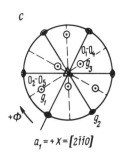

Fig. 159 a–c. First type of orientation of the electron-hole center axes (g-factor axes) in quartz. **a** Electron centers Ti^{3+}, Ge^{3+} and Fe^{3+} ion (with the ion compensators Li^+, Na^+, H^+); *on the left,* a portion of the crystal structure in projection on (0001) plane as in Figure 157 (the designations of the atoms are the same); *on the right,* the same tetrahedron in a perspective view (In the (0001) projection the g_1 and g_3 axes are projected in one line and their bifurcation in the figure is shown for clearness). **b** Hole centers O_2^{3-} (with Si vacancy). **c** Stereographic projection of the axes of the centers shown above and the centers obtained by the symmetry elements; there are three center positions (cf. with Figs. 157, 158)

This orientation type includes the electron center Ti^{3+} with any ion compensators, Fe^{3+} ions substituting Si^{4+} and hole centers O_2^{3-}.

The second type (Fig. 160): one of the g-factor axes runs approximately parallel to the Si–O direction, two other g-factor axes are in the oxygen plane of the tetrahedron, perpendicular to the Si–O direction. One of the latter axes runs parallel to one of the O–O edges, an other is directed approximately along the bissectrix of the angle between the two other edges.

The SiO_3^{3-} centers, i.e., $[SiO_3 V_0 e^-]^{3-}$ with vacancy of one of the oxygen ions, toward which the g-factor axis points, belong to this orientation type.

Two different inclinations of the Si–O directions ot the z axis of the crytal (Table 20)—at the 44° or 66° angles–give two varieties of the SiO_3^{3-} center: (1) $[Si(E_1)e^-]$, $[Ge(E_1)e^-]$ and (2) $[Si(E_2)e^-]$, $(Ge(E_2)e^-]$.

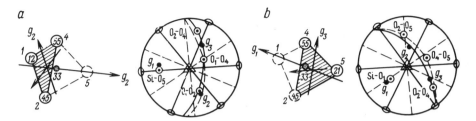

Fig. 160 a and b. Second type of orientation of the electron-hole centers axes (g-factor axes) in quartz. **a** Electron centers SiO_3^{3-}, i.e., oxygen O_5 (or O_2) vacancy trapped electron distributed over whole radical $[SiO_3\,V_{05}]^{3-}$, i.e., it is the center $[Si(E_1)e^-]$ or $[Ge(E_1)e^-]$; the same orientation type arises in the case of O_2 vacancy; **b** electron centers SiO_3^{3-}, i.e., oxygen O_1 (or O_4) vacancy trapped electron: $[SiO_3V_{01}]^{3-}$, i.e. it is the center $[Ge(E_2)e^-H^+]$; the same orientation type arises in the case of O_4 vacancy. *On the left*, projections on (0001) plane (parts of Fig. 157); *on the right*, stereographic projections (parts of Fig. 158)

The third type (Fig. 161): one of the g-factor axes runs approximately parallel to the Si–Si direction, the second axis is approximately in the Si–O–Si plane, being

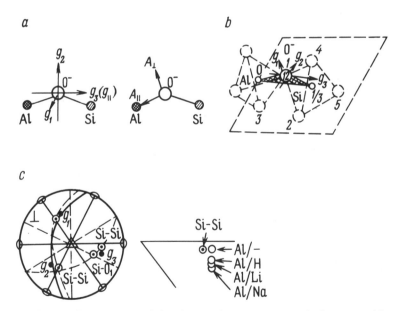

Fig. 161 a-c. Third type of orientation of the electron-hole centers axes (g-factor axes) in quartz. **a** g-factor axes directions of the Al-O$^-$-Si center (on the left) and hyperfine structure from Al^{27} axes directions (on the right). **b** Al-O$^-$-Si center axes; projection on (0001) plane (part of Fig. 157); **c** stereographic projection of the Si-Si, Si-O$_1$-Si (part of Fig. 158), as well as that of g-factor axes of the Al-O$^-$-Si center; on the right in a greater scale the changes of the g_3 position of the Al-O$^-$-Si centers without compensation and with different ion compensator: H$^+$, Li$^+$, Na$^+$

perpendicular to Si–Si, and the third axis is perpendicular to the Si–O–Si plane. The Si–O$^-$–Al center belongs to this orientation type.

The perturbation influence of H$^+$, Li$^+$, Na$^+$ ions located in the channels near the SiO$_4$ tetrahedron is of two kinds: (1) depending on the situation of these ion compensators on the L_2 axis near the shorter O$_1$–O$_4$ or longer O$_2$–O$_5$ edges (see Table 20; Fig. 157) the emerging varieties of each of the centers differ only slightly in their EPR parameters, (2) the presence of these ions results in only small fluctuations in the center axes orientation within the limits of the given type.

For example, in the case of Ti^{3+} with H, Li, or Na compensator the coincidence of one of the g-factor axes with the L_2 axis remains, but the inclination of the two other axes changes slightly. In unstable varieties of the O$^-$–Al center with different compensators (H$^+$. Li$^+$, Na$^+$) the g-axis position changes regularly with the increasing charge of the ion-compensator (see Fig. 161c).

Quartz Color and Electron-Hole Centers. The existence of colored varieties of quartz is mostly due to the presence of the electron-hole centers. The color of a smoky quartz is associated with the O$^-$–Al center, that of morion is attributable to the same O$^-$–Al plus intensification by the centers with oxygen and silicon vacancies, the pink color of quartz is linked with Ti^{3+}. Among yellow-green and honey colors of citrines one distinguishes three genetic types: (1) radiation ("real") citrines whose color is due to the same O$^-$–Al, but with two compensators in channels: Li and H and with one more variety of O$^-$–Al which shows a stronger anisotropy of the EPR spectrum; (2) "secondary morion" color of citrines appearing after heating of morion at 300°–320C;° (3) "secondary amethyst" (iron) color of citrines similar to the color of brown quartz appearing after heating amethyst at 450°–500° C, which transforms the Fe^{3+} and amethyst centers into submicroscopical segregations of iron oxides. The precursor of the violet color in amethyst is Fe^{3+} which after irradiation turns into the hole center Fe^{4+} (FeO$_4^{4-}$).

7.3.4 Feldspars

As in quartz, the most widely distributed centers in feldspars are O$^-$ and Ti^{3+} (as well as Fe^{3+} impurity [921, 945]. They occur most frequently in microclines, then in orthoclase, sanidine, plagioclases and in the latter their frequency diminishes from albite to anorthite. After irradiation of the samples the O$^-$ center develops in all feldspars.

As distinct from quartz, in which this center is designated as O$^-$–Al, and is characterized by its hyperfine structure with six lines (because for Al^{27} $I = 5/2$ and $2I + 1 = 6$), in feldspars the Al–O$^-$–Al center forms with two Al atoms, one of which may be called "structural Al" and the other "impurity Al"; the hyperfine structure is represented here by 11 lines ($2nI + 1 = 2 \cdot 2 \cdot 5/2 + 1 = 11$). In microcline each "structural Al" tetrahedron is surrounded by four silicon tetrahedra; insertion into them of impurity Al leads to the formation of four Al (structural) O$^-$–Al (impurity) centers at oxygens O(A_1), O(B), O(C_0), O(D_0); at oxygens O(A_2), O(C_m), O(C_D) no centers form in microclines.

Let us calculate the number of nonequivalent types of the Al–O⁻–Al centers in microcline. Since in the EPR spectra the positions mutually related to the symmetry center are indiscernible and Figure 162 shows that the tetrahedral rings T_{10}–T_{2m}–T_{10}–T_{2m} are mutually related by the center of symmetry, it suffices to consider only one such tetrahedral ring. Since two Al tetrahedra T_{10} in the ring are also mutually related to the center of symmetry, it is sufficient to consider only one of them T_{10} tetrahedron in the ring. At the expense of the impurity Al which substitutes Si in four tetrahedra surrounding T_{10} (with structural Al), four Al–O⁻–Al centers of equal intensity develop: $O^-(A_1)$, $O^-(B)$, $O^-(C)$, $O^-(D_0)$. Cross-hatched twinning common in microclines doubles the number of observed spectra.

In orthoclase the T_{1m} and T_{10} tetrahedra change into simple T_1 tetrahedra, statistically 50% of each of them being occupied by aluminum. Thus, there arise eight nonequivalent types of the Al–O⁻–Al centers, pair-wise linked through the plane of symmetry here, while the $O^-(A)$ centers coincide. In sanidine the formation of the $O^-(A_2)$ centers is also probable.

Microcline and orthoclase can be distinguished by the presence of the O^- centers (1) through precise measurements of the g_x, g_y, and g_z positions (which, however, differ only little therein; besides overlapping of the spectra from different O^- centers make accurate measurement difficult), (2) through measurements of relative intensities of the nonequivalent types of center.

The g-factor axes running approximately parallel to the Al–Al direction and the hyperfine structure axes oriented in the O^-–Al direction fail to coincide, and this complicates the picture of the spectrum to a still greater extent. In all amazonites, and only in amazonites, the spectrum of Pb^+ occupying the K^+ position (Fig. 163) and the spectrum O^-–Pb on the oxygen, making part of the K-polyhedron, are observed [945].

7.3.5 Framework Aluminosilicates with Additional Anions: Scapolite, Cancrinite, Sodalite, Ussingite Groups

Unlike feldspars, where the centers form mostly due to substitutions of Al and Si, all the diversity of the centers in these groups is linked not with the aluminosilicon framework, but with the "additional anions" Cl^-, CO_3^{2-}, SO_4^{2-} entering large voids in the framework and coordinated by Na and Ca cations specific for these groups. The additional anions and their vacancies are the precursors of diverse free radicals, molecule ions and F centers [875, 952, 963, 978].

The radicals and centers are of great importance for these groups. All kinds of colors (the characteristic blue color of lazurite; yellow, bleaching pink, violet of other minerals in these groups) are linked only with the centers. Such mineral varieties as hackmannite (pink bleaching solalite), glaucolite (blue scapolite), and lazurite owe their existence to the presence of definite types of center. The presence of centers explains also the luminiscence common to most minerals of these groups. The centers in sodalite have acquired a special importance when this mineral was found to display photochromism, i.e., a reversible process of acquiring coloration following UV irradiation and bleaching under the action of visible light. Because of

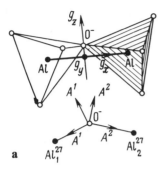

Fig. 162 a and b. Position of the Al-O⁻-Al center in feldspar structure. **a** Model of the ▶ center: oxygen belonging to two AlO_4 tetrahedron traps hole with formation of O^-; the g_x axis is approximately parallel to the Al–Al direction, g_z is perpendicular to g_x in Al–O–Al plane; the hyperfine structure axes are directed from O^- towards Al^{27} nuclei; the shaded tetrahedron contains "structural" Al, unshaded, impurity Al; **b** The crystal structure of orthoclase in projection on (201) plane; alumino-silicon tetrahedra are shown only. Symmetry centers are designated by small crosses; $A_1B_1C_1D$ labeled different oxygen ions; T_1 and T_2 are different (Al, Si)O_4 tetrahedra. Subscripts m and o appear in the microcline structure in which pass the shown in this figure orthoclase structure in the case of small distortion. *Heavy lines with perpendicular arrows* at oxygen ions indicate the g_z directions of the O^- centers formed at corresponding oxygen ions (as in Fig. 162a)

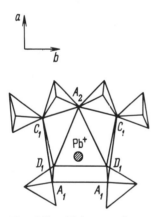

Fig. 163. Pb⁺ center in amazonite. A_1C_1D are labeled nonequivalent oxygen ion in the coordination poly-hedron around Pb⁺

this the mineral has found application in redioelectronics and television. Earlier, mineralogists observed this phenomenon in hackmannite, which bleached in the light, and regained its color in the dark.

Conformable to the composition of additional anions and impurity anions replacing them, one can distinguish sulfide (S_2^- and S_3^-), sulfate (SO_4^-, SO_3^-), and carbonate (CO_3^{3-}, CO_3^-) center groups as well, as other center groups (with halogen vacancies, with adsorbed O_2^- etc.) Of great importance for their identification in the silicate minerals under review is the fact that their properties, especially EPR

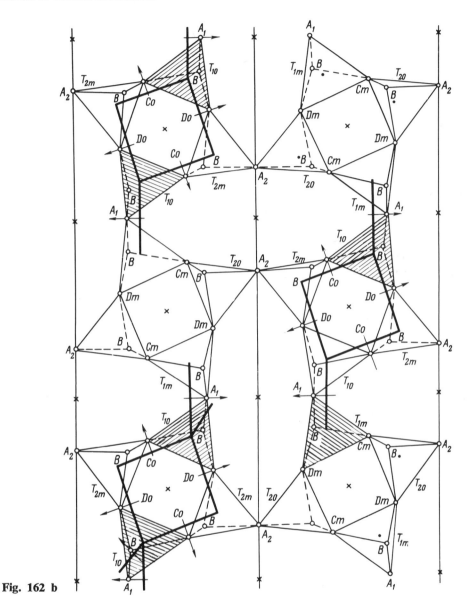

Fig. 162 b

parameters, are similar to those of the same centers in carbonates, sulfates, or alkali halides with radicals and molecules (KCl:S$_3^-$, KCl:SO$_2^-$ etc.) captured in the course of their growth or treatment.

A systematic comparison of the center types established from the EPR spectra, the color, absorption spectra and luminescence with isomorphous substitutions in the cation (Na$^+$ → Ca^{2+}) and anion (Cl → SO$_4^{2-}$, CO$_3^{2-}$) parts and with variations in the size of voids and channels occupied by the centers formed on the additional anions allows it to form a general picture of the electron-hole centers formation in these groups of minerals

In sodalite (Fig. 164) one observes the EPR spectra of SO_4^- and O_2^- centers and luminescence spectra of the S_2^- center. The 600 nm absorption band which determines the blue color of sodalite comes from the SO_4^- center. Hackmannite, a bleaching pink variety of sodalite, and photochromic synthetic sodalites contain as precursor Cl^- vacancies and S_2^{2-} molecules. Following UV irradiation (or other kind of irradiation) two centers emerge: the hole center S_2^- (due to the loss of an electron, i.e., hole trapping) and the electron F center (representing here Cl^- vacancy trapped electron interacting with four neighboring Na cations, i.e., it is Na_4^{3+}). Absorption bands 400 nm (S_2^-) and 530 nm (F center) emerge which account for the pink color. The effect of visible light or of heating removes the electron from the F center, and it is captured by the S_2^- center, resulting in bleaching.

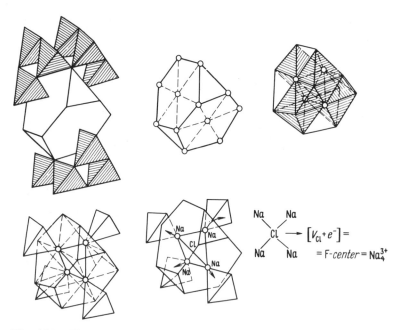

Fig. 164. Electron-hole centers in sodalite and lazurite: Cl^- vacancy $+ e^- \rightarrow F$ center; substitution $Cl^- \rightarrow SO_3 + e^- \rightarrow SO_3^-$; substitution $Cl^- \rightarrow S_3 + e^- \rightarrow S_3^-$; Cl^- vacancy and substitution of Cl^- by $S_2^{2-} \rightarrow F$ center and S_2^-

The deep blue color of lazurite and synthetic ultramarine come from the S_3^- center (absorption band 600 nm); the blue-green color of lazurite appears with increasing contents of the S_2^- center (absorption band 400 nm). Cancrinites have the following centers: in yellow carbonate-cancrinites–CO_3^-, in pink cancrinites–CO_3^{3-} and in blue sulfate–cancrinite (vishnevite)–SO_4^-, SO_3^-, S_2^-. In scapolites one can observe a number of sulfate (SO_3^-, SO_2^-) and carbonate (CO_3^-, CO_2^-) centers and probably an F center on the Cl vacancies.

The absence of hyperfine structure in the EPR spectra of sulfate and carbonate centers, as well as the complicated nature of their crystal structures, the unavailability in many cases of single crystals (lazurite, ussingite), and the overlapping of absorption bands and of the EPR lines of different centers make it difficult to identity the centers in these groups.

7.3.6 Zeolites

In connection with studies of electrostatic fields acting upon exchange cations in the structures of synthetic zeolites—molecular sieves, the superoxide molecular ions O_2^-, adsorbed in large voids of an aluminosilicon-oxygen framework, and F centers (or Na_4^{3+}), formed in decationized zeolites in a vacuum have been described [926, 973, 1000].

7.3.7 Zircon

Numerous impurities in zircons (Nb, Hf, Ti, Mn, Y, Gd, Er, Dy, Tb, Tm, Al, P, M^+) lead to formation of diverse centers both on the impurity ions themselves and in their surroundings [873, 902, 994]. The crystal structure of zircon is similar to that of anhydrite, but, because of different ion charges and chemical bond states, the manner of the center formation in these closely related structural types is different. The following schemes of the center formation in zircon can be distinguished.

1. Substitutions of $M^{5+} \rightarrow Zr^{4+}$ and $M^{4+} \rightarrow Zr^{4+}$ lead to the formation of electron centers $M^{5+} + e^-$ (Nb^{4+}) and $M^{4+} + e^-$ (Hf^{3+}, Ti^{3+}).

2. Substitutions of $M^{3+} \rightarrow Zr^{4+}$ lead to the formation of hole centers either on the impurity ion itself ($Tb^{3+} + e^+ \rightarrow Tb^{4+}$) or on the oxygen (due to substitution of $Y^{3+} \rightarrow Zr^{4+}$ there forms an O^- center which can be described here as SiO_4^{3-}).

3. Due to the existence of the pairs Y^{3+}–Y^{3+} in the case of nonrandom distribution of Y^{3+} impurity one can observe either the O^- center interacting with the two Y^{3+} or a complex of two SiO_4 tetrahedra with a trapped hole.

4. Three varieties of the center Zr^{3+} ($Zr^{4+} + e^-$) form at the expense (a) of an adjacent oxygen vacancy, (b) interstitial alkali metal ion, (c) due to the presence of the Y^{3+} ion pair at the neighboring cation sites.

5. Substitution of $Al^{3+} \rightarrow Si^{4+}$, as in quartz, leads to the formation of the O^-–Al center (AlO_4^{4-}).

6. Phosphorus impurities cause formation of a PO_4^{2-} center in the SiO_4^{4-} position.

7. Oxygen vacancies after irradiation yield the SiO_3^{3-} and SiO_2^- centers.

7.3.8 Beryl, Topaz, Phenakite, Euclase, Kyanite, Danburite, Datolite

Beryl. Besides isomorphous impurities Cr^{3+}, Fe^{3+} (including $Fe^{3+} \rightarrow Si^{4+}$), Mn^{2+}, established by means of EPR, beryl usually contains electron center Ti^{3+} ($Ti^{4+} + e^-$) replacing Al [873]. Several types of natural beryl can capture atomic hydrogen, methyl (CH_3) radical (932, 946) in the channels. After irradiation there form also the radicals $(H_2O)^+$, OH [977] and "Maxixe" center CO_3^- [897].

The H^0 and CH_3 centers have been observed also in clinohumite and enstatite and the H^0 center in tourmaline (elbaite) [946, 965].

Topaz, Phenakite, Euclase, Kyanite. For all these minerals similar to beryl, the presence of isomorphous impurities of Cr^{3+} and Fe^{3+} and electron centers Ti^{3+} is characteristic [873]. In topaz and euclase also the presence of V^{4+} has been ascertained. Three complexes with Ti^{3+} are distinguished in topaz [976]: $Ti\ O_4F_2$, $Ti\ O_4(OH)_2$, $Ti\ O_4(OH,F)_2$. In all these minerals the O^- center also occurs (AlO_4^{4-} or SiO_4^{3-}). In phenakite the PO_4^{2-} radical has been studied in detail [898].

Danburite, Datolite. The O^- center exists in donburite and datolite as BO_4^{-4}.

7.3.9 Calcite

The presence of numerous impurities substituting calcium were established by means of the EPR and the luminescence spectra: Mn^{2+}, Fe^{2+}, Ni^{2+}, Co^{2+}, Ag^{2+}, Cu^{2+}, Pb^{2+}, UO_2^{2+}, V^{3+}, Fe^{3+}, Y^{3+}, Gd^{3+} and other rare earths. Hole centers formed on the cation are Pb^{3+} and Ni^{3+}. However, the diversity of the centers in calcite is due to anion substitutions: CO_3^{2-} is replaced by the free radicals PO_3^{2-}, PO_3^0, PO_2^{2-}, PO_2^0, AsO_3^{2-} AsO_2^{2-}, SO_3^-, and BO_2^0; there are also the centers CO_3^{3-}, CO_3^-, and CO_2^- [868, 933, 934, 948, 949, 967, 984, 985].

7.3.10 Anhydrite

About forty center types have been observed in anhydrites [874, 875].

In the case of substitution $Y^{3+} \rightarrow Ca^{2+}$ or rare earths $\rightarrow Ca^{2+}$ in $Ca-O_8$ polyhedron, the O_2^{3-}, or $(M^{3+}O_2^{3-})^0$, centers form at each pair of the oxygen edges lying in planes of symmetry, i.e., the hole is statistically distributed among the O–O edges of the $Ca-O_8$ polyhedron (Fig. 165). Another group of the centers is linked with anion substitutions: PO_2^{2-}, BO_2^0, with vacancy formations: SO_3^-, SO_2^-; with the presence of thiosulfate radicals SSO_3^-, SSO_2^-. The third group of the centers is represented by interstitial ozonide ions O_3^-, superoxide ions O_2^-, molecular ions S_2^-. There are also hydrogen-containing varieties of the O_2^{3-} center: $O_2^{3-}-M^{3+}/H$, and the centers $O_2^{3-}-M^{3+}/M^+$ with hyperfine structures from sodium and potassium.

The existence of the varieties of anhydrite, for example, yttrium-containing, rare earth, phosphorus- and boron-containing, with ozonide, superoxide, and S_2^-, with thiosulfate radicals containing hydrogen and alkali metal allows the anhydrites coming from different types of mineral deposits to be discerned.

7.3.11 Barite and Celestine

In the crystal structures of these minerals with large 12–vertex polyhedra around barium and strontium, usually containing only small quantities of impurities (less than $0.0n$ %), detectable neither by EPR nor luminescence spectra, the diversity

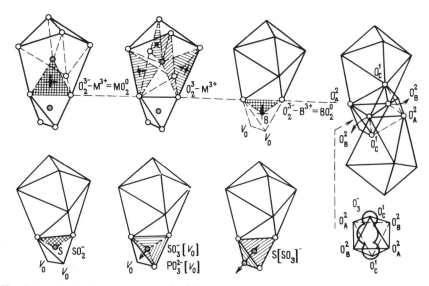

Fig. 165. Electron-hole centers in anhydrite

of the centers is due mainly to the prevalence of sulfate centers SO_4^-, SO_3^-, SO_2^- and (in barite) to that of S_2^- (861). Barite, like anhydrite, has also centers formed through substitution of yttrium for barium. The formation of centers in barite is controlled by features peculiar to the crystal structure: one of the *g*-factor axes of all the centers coincides with the crystal axis *b* (coinciding also with the symmetry axis L_2 of the Ba polyhedron), while the other two axes lie in the (010) plane at different angles to the c axis of diverse centers (Fig. 166).

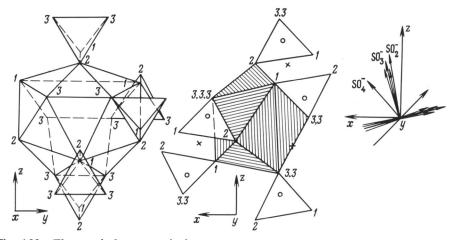

Fig. 166. Electron-hole centers in baryte

7.3.12 Apatite

Three groups of centers and of impurity ions are distinguished in apatite due to the existence of the following three structure elements: tetrahedral oxyanion PO_4^{3-}, a chain of F (or Cl) ions along the c axis, and the Ca^{2+} cations at two nonequivalent sites [904, 923, 970, 971, 1005].

1. Substituting PO_4^{3-} leads to the formation of the centers (a) CO_3^{3-} and CO_3^-, and (b) SiO_3^{3-}.

2. Many varieties come up due to the formation of two centers originating from the fluorine (or chlorine) ions linear: (a) the F center representing here a fluorine vacancy with a trapped electron and (b) the O^- center formed due to the substitution of $O^{2-} \rightarrow F^-$. The existence of the varieties of these centers in the chain is determined by the presence of different ions in the neighborhood of the center: F–O^-–F, F–O^-–Cl, Cl–O^-–Cl, O^2–O^-–V_O, or F–[F center]– O^{2-}–V_F–F, F–[F center]–F.

Substitutions in the cation part ($Ca^{2+} \rightarrow Mn^{2+}$, Fe^{3+}, RE^{2+}, RE^{3+}) does not directly lead to the center formation. The pink color of apatite comes from the F center, the blue color from the O^- center.

The observation of the centers in apatites has served (as in the case of quartz and feldspars) to study various geological problems, such as that of accessory apatites in magmatic rocks; the electron-hole centers can be helpful in differentiating apatites of continental and marine phosphorites, etc.

7.3.13 Sheelite

The most characteristic precursors in sheelite are Ca vacancies, as well as oxygen vacancies and isomorphous substitutions W $\rightarrow$ Mo and Si, O $\rightarrow$ F, Ca $\rightarrow$ Pb.

In most cases here both electron and hole centers (Fig. 167) form on the WO_4^{2-} complex (except for $Pb^{3+} = Pb^{2+} + e^+ \rightarrow Ca^{2+}$ observed in synthetic sheelite) [950, 995, 1020].

Electron centers in sheelite are usually W^{5+} (or $WO_4^{3-} = WO_4^{2-} + e^-$); two varieties exist because of the presence of calcium vacancies in two different structural positions; their existence lowers the center symmetry to the orthorhombic one and determines the orientations of the centers in the structure.

The hole centers on WO_4^{2-} in sheelite owe their origin to the fact that the WO_4^- ($WO_4^{2-} + e^+$) center does not remain individual but the unpaired electron is divided between those two WO_4 complexes which are neighbors through the Ca vacancy. As a result the following varieties of WO_4^- exist: (a) WO_4^-–V_{Ca}–WO_3F^-, i.e., unpaired electron (trapped hole $WO_4^{2-} + e^+$) interacts across the calcium vacancy with the fluorine replacing an oxygen in the neighboring tungsten-oxygen tetrahedron; three different locations of fluorine in WO_3F determine the existence of three varieties of the center. Substitution $F^- \rightarrow O^{2-}$ in sheelite is confirmed also by the EPR observations of the $[MnF_4]^{2-}$ complex replacing $[WO_4]^{2-}$ [346]; (b) WO_4^-–V_{Ca}–WO_4^-, i.e., two WO_4^- centers interact across the calcium vacancy; as a result the two-electron center with spin $S = 1$ and with hyperfine structure from two W^{183} forms;

Fig. 167. Electron-hole centers in sheelite

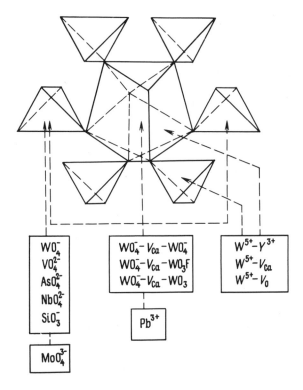

(c) $WO_4^- - V_{Ca} - WO_3$, i.e., after irradiation the calcium vacancy leads here to the WO_4^- formation, and is accompanied by an oxygen vacancy in the neighboring tungsten-oxygen tetrahedron.

Replacing tungsten by silicon leads to the emergence of the hole center SiO_3^-, i.e., $SiO_3^{2-} + e^+$. As in anhydrite, where substitution $BO_4^{5-} \rightarrow SO_4^{2-}$ forms the BO_2^{2-} center with two oxygen vacancies ($BO_2^- + e^-$), in sheelite the substitution of an ion with a lower charge in the oxanion radical leads to the formation of an oxygen vacancy.

Similar centers arise in compounds with radicals CrO_4^{2-}, MoO_4^{2-}.

7.3.14 Fluorite

The crystal structure of fluorite can be represented by cubes whose vertices are occupied by F^- ions with alternating empty and Ca^{2+} occupied cubes (Fig. 168).

The following types of impurity ions and formation of the centers in fluorite are distinguished [855, 857, 877, 884, 900, 906, 913, 941, 988, 1019]:

a) nonlocal charge compensation of rare earth: $RE^{3+} \rightarrow Ca^{2+}$; with retained cubic symmetry;

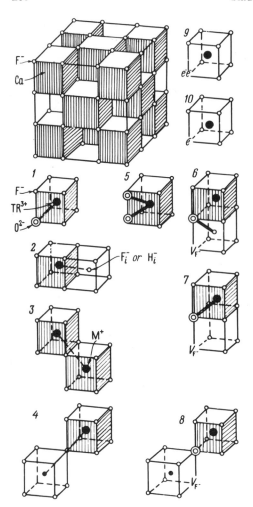

Fig. 168. Structural position of rare earths and Y^{3+} in fluorite and electron-hole centers formation (*shaded cubes are occupied by Ca^{2+} and unshaded cubes are empty*)

b) local charge compensation: (1) by oxygen: $O^{2-} \to F$; trigonal symmetry (see Fig. 168.1), (2) by interstitial fluorine; tetragonal symmetry (Fig. 168.2) or trigonal symmetry (Fig. 168.4), (3) by hydride ion H^-: insterstitial H^-, tetragonal symmetry (Fig. 168.2), or trigonal symmetry (Fig. 168.4); isomorphous $H^- \to F^-$, trigonal symmetry, (4) by univalent cation: $RE^{3+} \to Ca^{2+}$ and $M^+ \to Ca^{2+}$, orthorhombic symmeiry (Fig. 168.3);

c) electron-hole centers formation.

Yttrium centers:

YO_2^0 ($O_2^{3-} - Y^{3+}$) (Fig. 168.5);

$(OF_2)^{3-} - Y^{3+} - V_F$ or $(OF) - Y^{3+} - V_F$ (Fig. 168.6);

$(OF) - Y^{3+} - V_F = (YOV_F)^+$ (Fig. 168.7).

$(OF_{int} - Y^{3+} - V_F) = (YOF_{int}V_F)$ (Fig. 168.8).

Photochromic centers:

$(RE^{3+} + F \text{ center}) = (RE^{3+} + V_F + e^-e^-)(\text{Fig. 168.9})$;

$(RE^{3+} + F \text{ center}) = (RE^{3+} + V_F + e^-)$ (Fig. 168.10);

(here RE^{3+} are represented by La, Ce, Gd, Tb, Lu).

7.4 Electron-Hole Centers in Alkali Halide Crystals

7.4.1 F Center

Alkali halides crystals as test models in the study of ionic crystals and the F center in alkali halides, as an elementary color center, represent the simplest systems which, as in the case of the hydrogen atom in quantum mechanics, are fully understandable and readily quantifiable.

This is above all due to the simplicity of the crystal structure, an easy growth of large homogeneous single crystals, to the possibility of introducing into them a great variety of impurity ions, to their transparence within a wide spectral interval from UV to IR, as well as to their "transparency" in the superhigh frequency and radiofrequency regions.

The significance of alkali halide crystals increased rapidly thanks to the accumulation of diverse experimental and theoretical data, and to a greater understanding, first, of the simplest phenomena observed in them, then to quantitative calculations made with an ever closer degree of approximation and by taking different factors into account.

Many basic concepts describing the nature of color centers, of luminescence, of optical obsorption in UV and visible regions, interrelated processes in dielectrics and semiconductors, have been formulated on the ground of the material derived from alkali halides, and thus carry specific marks inherent in them, though also applicable to other quite different types of compounds.

The special position of the F center among all other color center types is determined by the fact that because of its elementariness, the properties of the center can be quantified proceeding from the first principles. It is also a prototype of the F aggregate centers. For the center vast experimental data have been accumulated, empirical relations, with special theories explaining variations of several parameters. A survey of these data was presented by Markham [947].

At present the F center model has been confirmed by means of electron paramagnetic resonance, and double resonance showed that the electron of the F center trapped by the anion vacancy interacts equally with six neighboring cations. This description, however, gives only the structural position of the F center electron. Since the electron is not a material point, and its behavior is described by wave functions, the F center description involves also molecular orbitals represented by linear combinations of atomic orbitals of the cations surrounding the vacancy.

The F center wave functions (Fig. 169) are written as follows:

$$\phi_{F \text{ center}} = \frac{1}{\sqrt{6}} \sum_{\alpha=1}^{6} [\varphi^{1/2}\psi_{\alpha s} + (1 - \varphi^{1/2}{}_{\alpha p\sigma})],$$

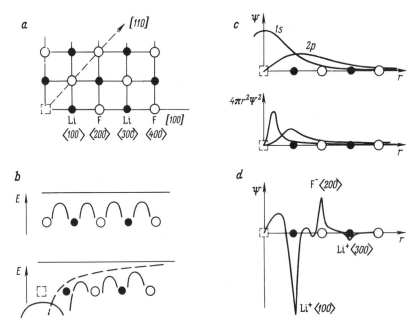

Fig. 169 a-d. Potential energy and wave functions of F center electron (after [947]). **a** Vacancy and nearest ions positions along [100] and [110] directions; **b** potential energy of electron in normal lattice and in field of vacant lattice node; **c** wave function and electron density distribution of F center electron; **d** wave function of F center electron in "large molecule" approximation

where α is the cation, σ is the orbital directed along the axis connecting the vacancy center with the α cation, φ is the parameter describing a contribution of s and p states in the molecular orbital.

Quantitative calculations of the F center, based on different approximations, have been considered on numerous occasions (see [947]).

F centers give the most intense absorption bands in optical absorption spectra of alkali halides, causing their coloration. The absorption band arising after irradiation or additional coloration has a characteristic bell-like shape, its half-width being about 0.30–0.35 eV at room temperature and 0.15–0.20 eV at liquid helium temperature. The F band shifts its position depending on the composition of alkali halides toward a longer wavelength in the series Li–Na–K–Rb–Cs and the F–Cl–Br–I is accompanied by color changes. The maximum absorption position is related to lattice parameters by an empirical equation of the type

$$\lambda d^n = k,$$

where λ is absorption maximum of the F band (in Å; or energy of the transition in eV), d is interatomic distance, n and k are empirical parameters.

Numerical expressions for this general equation have been obtained by Mollwo [(see [947]):

$\lambda(Å) = 600\ d^2(Å)$ or $\varepsilon(eV) = 20.7/d^2$.

Better coincidence gives the numeral expression suggested by Ivey (see [947]):

$\lambda(Å) = 703^{1.84}\ d\ (Å)$ or $\varepsilon(eV) = 17.6/d^{1.84}$.

Least square treatment of the experimental data gives $\varepsilon = 17.4/d^{1.83}$ [947].

Similar equations have been deduced also for other types of centers in alkali halides, allowing correlation of absorption bands observed in them. However, these equations are purely empirical and are of a limited significance: they are valid only for the NaCl structure type, being already inadequate for the series of alkali halide solid solutions.

Attempts at theoretical substantiation of these equations showed them to be a simple empirical approximation of a more complicated relationship. The transition energy depends on different powers of the interatomic distance (from d^{-1} to d^{-7}) and may be written in first approximation as

$$\Delta \varepsilon_F = (c_1 - c_1)d^{-1} + (c_2 - c_2)d^{-2},$$

where parameters c have a complicated dependence on kinetic energy, Madelung energy, and interatomic distances.

Concentration of the F centers may be determined by using absorption coefficient of the F band by means of the Smakula formula:

$$N_F f = \frac{9\ mc}{2e^2}\ \frac{n}{(n^2 + 2)^2}\ \alpha_{max}\ w,$$

where N_F is the number of F centers in 1 cm^3, f is oscillator strength being constant for the F center in a given alkali halide crystal; m and e are the electron mass and charge, c is light velocity, n is index of refraction for the wave length corresponding to the absorption band maximum, α_{max} is the maximum absorption coefficient in cm^1, w is half-width in eV.

The formula was originally deduced from the classical dispersion theory, but the same results can be obtained on the basis of quantum-mechanical calculations though approximating different models. By substituting numerical values for the physical constant, the formula may be written

$$N_F f = 1.29 \cdot 10^{17}\ \frac{n}{(n^2 + 2)^2}\ \alpha_{max}\ w.$$

For a given alkali halide index of refraction n, line width w (it is independent of concentration but varies with temperature) and oscillator strength f are constant. Thus, the F center concentration is determined from this formula by measuring only the absorption coefficient. However, if n and w are easy to measure, to determine f it is necessary to find separately for each alkali halide the F center concentration and then, by using the Smakula formula, calculate the oscillator strength f.

The F center concentration can be determined by chemical analysis, or by electron paramagnetic resonance, or by magnetic susceptibility measurements. For KCl $f = 0.66$–0.85, for NaCl $f = 0.7$–0.9.

Having determined n, w, and f, one can obtain the F center concentration by means of the absorption coefficient. For KCl $n = 1.4902$, $n/(n^2 + 2)^2 = 0.084$, then

$$N_F = 1.06 \cdot 10^{16} \frac{w}{f} \alpha_{max},$$

when $w = 0.35$ eV and $f = 0.81$ we obtain $N_F = 4.6 \cdot 10^{15} \alpha_{max}$. For instance, for one of the KCl crystal with $\alpha_{max} = 310$ cm^{-1} and $w = 0.35$ and assuming $f = 1$ we get $N_F = 1.06\,{}^{\prime}10^{16} w/f \alpha_{max} = 1.15 \cdot 10^{18}$ F centers in 1 cm^3. For NaCl with $n = 1.5442$, $n/(n^2 + 2)^2 = 0.081$, $w = 0.469$eV, $f = 1$ one obtains $N_F = 4.9 \cdot 10^{15}$ α_{max}, if $f = 0.87$ (as determined from comparison with chemical analysis data) $N_F = 5.6 \cdot 10^{15} \alpha_{max}$.

In order to form an idea as to the order of magnitude of $f = 0.8$–0.9 let us compare this value with the oscillator strength f for a different optical transition of the iron group ions: the f value for d–d transitions, imparting color to most minerals, is of the order of 10^{-4}, i.e., three orders less than f for the F center in alkali halides. For this reason such extremely low concentrations of the F centers produce a marked coloration. For $N_F = 10^{16}$ F centers in 1 cm^3 α_{max} is about 2 cm^{-1} for KCl and NaCl; for $N_F = 10^{18}$ the α_{max} reaches very high values: up to about 200 cm^{-1}, and for $N_F = 10^{19} - 10^{20}$ alkali halides become opaque in the visible region.

X-ray irradiation of KCl for 3 h gives the absorption coefficient value of about 500 cm^{-1} and after 20 h of irradiation it reaches the value of about 2,000 cm^{-1}.

In the short-wave part of the F center spectrum besides the F band one can observe a much weaker K band (as a shoulder at the short-wave side of the F band) and then still weaker L_1, L_2, and L_3 bands. The intensity relation of the absorption bands $F:K:L_1:L_2:L_3$ is about 100:10:1:1, 5:2. In some halides the L_1, L_2, and L_3 bands appear as a weak short-wave tail of the F band.

These bands do not signify the formation of new centers, but occur as a result of transitions from the ground state of the same F center to upper excited states.

The model of the F center is similar to the hydrogen atom: a single electron interacts with a positive charge arising in place of the anion vacancy. Because of this, there exists an analogy also in optical absorption spectra of the F center and the hydrogen atom. The ground state of the F center is similar to the $1s$ state of the hydrogen atom, and so are the excited states to $2s$, $2p$, $3s$, $3p$ etc. The transitions $ns \rightarrow ms$ are forbidden. Thus the first and most intense transition is $1s \rightarrow 2p$; the $1s \rightarrow 3p$ and other similar transitions are also allowed.

Since the sum of the oscillator strengths of all transitions of the one-electron system is equal to unity, and the F band oscillator strength is equal to 0.7–0.8, then only 0.2–0.3 falls to the share of all other transitions; that explains the low intensity of absorption bands related to them.

The F center $1s \rightarrow 2p$ transition energy has been calculated theoretically more often than any other parameters of the alkali halides. Recent calculations give for the $E_{2p} \rightarrow E_{1s}$ energy the value 2.616 eV, close to the experimental value of 2.65 eV.

The absorption band β in the UV, superimposed on the tail of the first exciton absorption band, is assigned to the exciton linked with the F center.

The anion vacancy designated as α in work on alkali halides may be considered as an F center with a lost electron, i.e., as a positively ionized F center ($\alpha = F^+$ center). An exciton absorption band α connected with this F^+ center exists.

The F absorption band is related to a new center representing an F center with one more trapped electron, i.e., it is an anion vacancy with two trapped electrons of a negatively ionized F center (F^- center).

Electron Paramagnetic Resonance of the F Center provides most important information for understanding the nature of the F center: the F center model has been definitively established by EPR, the electron interaction with the six nearest cations was observed, as well as that with anions and cations of the next coordination spheres up to the 16th, and the amount of the s state of the F center electron estimated.

The electron density distribution over the "large molecule" of the F center is a new source of information on the wave function of the F center electron.

EPR studies of the F center as such (hyperfine structure, g-factor theories and others) have become a complex and ramified branch of the F center theory.

The EPR signal of the F center unpaired electron in alkali halides differs essentially from EPR signals from other center types and from those of the F centers in other classes of compounds. Instead of a single narrow line or several narrow lines (about $0.n-0.0n$ G) one can observe in alkali halides an unusually broad band (scores and hundreds G). The width of this band ranges from 58 G in KCl, 65.5 G in LiCl, 168 G in NaCl to 450 G in RbBr and about 800 G in CsCl [1006].

The width of the line is caused by the unresolved hyperfine strcutre arising from the interaction of the F center electron with the nuclei of the nearest-neighboring ions: (a) the nuclei of all the cations of alkali halides have magnetic isotopes with a high natural abundance (Li^6, Na^{23}, K^{39}, Rb^{87}, Cs^{133}), (b) the interaction, though much weaker, occurs with anions of the next nearest sphere, and the nuclei of all anions of alkali halides also have magnetic isotopes with high natural abundance (F^{19}, Cl^{35}, $Br^{79,81}$), (c) the spin of all these nuclei is $I = 3/2$ (except of Cs^{133} with $I = 7/2$), (d) the number of the nearest cations N is equal to 6 in the NaCl strcutrue type and 8 in the CsCl structure type (while in most other center types the hyperfine structure arises due to interaction with the nuclei of one or two nearest atoms).

The number of hyperfine structure lines is $(2NI + 1)$, and the number of spin states, whose densities determine the relative intensities of these lines, is $(2I + 1)^N$. Taking into account only the interaction with the nearest cations ($N = 6$, $I = 3/2$), one obtains $(2NI + 1) = 19$ lines of the hyperfine structure and $(2I + 1)^N = 4096$ spin states. Calculation of the distribution of these 4096 spin states between energy levels giving 19 lines allows obtaining the envelope curve nearly coinciding with that observed experimentally. If in addition to this, account is taken of the interaction with the anion, a still closer coincidence is then achieved. In some alkali halides, such as LiF, NaF, RbCl, and CsCl, one can observe a partially resolved hyperfine structure. The number of lines exceeds 19, because of anisotropy of the hyperfine structure and of some other factors. The broadening of the each line is related to a weaker interaction with the nuclei of anions and cations of the next coordination spheres.

More detailed information concerning paramagnetic resonance of the F centers in alkali halides is obtained by means of electron-nuclear double resonance (EN DOR), which makes it possible to measure direct transitions between nuclear spin levels near the paramagnetic center. The double resonance raises the resolution by several orders, enables it to measure isotropic and anisotropic parts of the hyperfine interaction between the electron and nuclei of the cations and anions of different coordination spheres.

In Table 21 the complete pattern of the hyperfine interaction of the F center electron with the nuclei of the LiF crystal [947] is offered. In this table it is seen that though electron density of the F center electron extends beyond the eighth sphere (in NaF the hyperfine structure is measured from the nuclei of the 16th sphere), the interaction with remote nuclei is insignificant. This measurement also confirms the existence of anisotropic hyperfine interactions with axial symmetry with respect to the F center–nucleus direction.

Table 21.   Hyperfine interaction of F center electron in LiF with neighboring nuclei Li^7 and F^{19} in different coordination spheres

Sphere	Atom position	Nucleus	Distance, Å	A, MHz	B, MHz
1	<100>	Li^7	2.01	39.06	3.20
2	<110>	F^{19}	2.85	105.94	14.96
3	<111>	Li^7	3.50	0.5	0.68
4	<200>	F^{19}	4.02	0.48	1.12
5	<210>	Li^7	4.50	0.27	0.28
6	<211>	F^{19}	4.93	0.88	0.69
8	<220>	F^{19}	5.69	1.34	0.56

A, isotropic part of hyperfine structure; B, anisotropic.

The g-factor of the F center in alkali halides is characterized (1) by its closeness to free electron g-factor value ($g_e = 2.0023$), (2) by an always observed small negative shift relative to the g_e, (3) isotropy. The g-factor shift is related to a weak spin-orbit interaction.

7.4.2 F Center in Compounds of Other Types

The F center occurs not only in alkali halides, but also in other compounds in which halogen, oxygen, sulfur vacancies trapping electrons arise. However, the F center is far from being able to form in all classes of compounds. In crystals with radical groups and negative ion vacancy, the electron is captured not by this individual vacancy but is distributed in the molecular orbitals of the whole radical. For example, the CO_3^{3-} center emerges in carbonates, SO_3^- in sulfates, and so on. Thus the formation of the F center proper is possible in two types of heteropolar compounds: (a) binary compounds (without radicals): haloides, oxides, sulfides; (b) more complex compounds in cases when they have a negative ion which does not enter the radical part of the structure (occurring in structural channels or voids, for example, in sodalite, zeolite, and so on). While the general concept of the F center (anion vancancy with a trapped electron) is retained in diverse compounds, its peculiarities come from the difference in the composition of cations surrounding the vacancy and from the different structural type of crystals.

The F center has been observed in the alkaline-earth metal compounds: (1) fluorides with fluorite structure (CaF_2, SrF_2, BaF_2) or with rutile structure (MgF_2), (2) oxides, sulfides, and selenides (with NaCl structure). The existence of F centers could be established in these compounds by EPR, and their correlation with the optical absorption F band effected. The hyperfine structure of the EPR spectra, so important for understanding and calculating F centers in every compound, has been studied in detail for CaF_2 and BaF_2 as well as for all alkaline-earth oxides: MgO, CaO, SrO, BaO. We shall dwell upon some peculiarities of the EPR spectra in these compounds.

Explanation of the g-Factor Shift. The g-factor values in alkaline metal compounds in the first place have a greater shift relative to the free electron g-factor (for BaO $g = 1.936$) than in the case of alkali halides, and in the second, the shift can be negative (as in alkali halides and generally in electron centers as distinct from the hole centers) and positive (up to $g = 2.0062$) in MgS.

Should the F center electron occur only in the vacancy and in the s orbits of the nearest cations surrounding the vacancy, then the corresponding g-factor would be equal to the free electron g-factor (recall that the s orbit means a state with a zero orbital moment: $l = s = 0$).

However, since the electron occurs partially in the p-orbitals of the nearest cations a spin-orbit interaction takes place. Then the g-factor is determined by the expression:

$$g = 2.0023 - k \frac{\lambda}{\Delta} ,$$

where λ is the spin-orbit coupling constant of the nearest ion where the F center electron occurs, Δ is the distance between s and p states of the F center, k is a constant that takes account of the "polarizability" of the atom.

In this expression the sign of the g-factor shift is determined by the sign of the spin-orbit coupling constant λ: if λ is positive the g factor shift is negative and vice versa.

In compounds of divalent ions, the F center has an important distinctive feature: the divalent anion vacancy with a trapped electron does not become neutral as it is in alkali halides, but retains a single positive charge. Thus, the F center in oxides, sulfides, selenides, as distinct from halides, "attracts" electrons of the nearest divalent cations and anions. For example, for MgO:

$$Mg^{2+} - e^- \to Mg^{3+},$$
$$O^{2-} + e^- \to O^-,$$

i.e., one has to take for the spin-orbit coupling constant λ the values of λ for Mg^{3+} and O^- (the cation and anion without an electron). O^- is isoelectronic with F° ($2p^5$) having a negative and rather high spin-orbit coupling constant $\lambda = -270$ cm^{-1}, this explaining a positive shift of the g-factor values.

In alkaline-earth compounds a rather great increase of λ in heavy metals is of essential significance. For Ba ion it is so large ($\lambda = 1128$ cm^{-1} as compared with $\lambda = 150$ cm^{-1} for Ca) that its contribution exceeds the contribution from anions and causes a negative shift of the g-factor in Ba compounds.

Explanation of Hyperfine Structure in the EPR Spectra. The features of the EPR spectra of the F center of alkaline-earth compounds are determined by the isotope composition of the nearest cations and the next-nearest anions surrounding the F center, for due to the F center electron's interaction with magnetic nuclei of these cations and anions, there emerges a hyperfine structure which determines the diversity of the F center EPR spectra [1006, 1009].

The natural abundance (C) and spin values (I) for the magnetic isotopes of the cations and anions in alkaline-earth compounds are as follows:

Cation	C, %	I	Anion	C, %	I
Mg^{25}	10.1	5/2	O^{17}	0.04	5/2
Ca^{43}	0.13	7/2	F^{19}	100	1/2
Sr^{87}	7.02	9/2	S^{33}	0.74	3/2
Ba^{135}	6.6	3/2	Se^{77}	8.3	1/2
Ba^{137}	11.3	3/2			

In contrast to alkali halides, even nonmagnetic nuclei prevail in the alkaline-earth compounds (except fluorides). In CaO both magnetic isotopes Ca^{43} and O^{17} are found in negligible amounts, and thus the EPR spectrum of the F center in CaO is represented by a single narrow line (0.1 G instead of 60–800 G in alkali halides). Measurements of the hyperfine structure from Ca^{43} are possible only by using samples enriched by this isotope.

For Mg, Sr, Ba oxides the EPR spectra consist of an intensive narrow central line and weak, but quite distinct hyperfine structure lines. The number of these lines equals $(2mI + 1)$, where I is the nuclear spin of the nearest cation, m is the number of the equivalent nuclei with which the electron interacts.

The number of nuclei in the case of a 100% or near content of the magnetic isotope is determined by the vacancy coordination ($m = 6$ for crystals with NaCl structure and 8 for crystals with CsCl structure). With a low content of magnetic nuclei their number is determined by the statistical probability of the magnetic nuclei lying near the vacancy (1006):

$$P_m = \binom{2}{m}(f)^m(1-f)^{6-m},$$

where P_m is the probability of finding the nuclei of a given isotope near the vacancy, f is the content of the isotope (natural abundance for nonenriched samples), $\binom{6}{m}$ is binomial coefficient.

For example, in a CaO sample enriched with a Ca^{43} isotope, its content being brought up to $f = 2\%$, one obtains: 88.6% F centers with no neighboring Ca^{43} nuclei (i.e., $m = 0$), 10.8% F centers with a single neighboring Ca^{43} nucleus ($m = 1$), 0.6% F centers with two and more neighboring Ca^{43} nuclei. The EPR spectrum of such a sample represents a superimposition of an intensive narrow central line (arising from the fact that 88.6% of the F center have no neighboring magnetic nuclei) on the hyperfine structure of $(2mI + 1) = [2 \cdot 1 \cdot 7(2 + 1)] = 8$ lines from Ca^{43} isotope with $I = 7/2$.

In Ca, Mg, Sr, Ba fluorides each line (i.e., the central one arising from non-magnetic isotopes and hyperfine structure lines) splits additionally into seven lines of hyperfine structure emerging due to a weaker interaction of the F center electron with the fluorine anions nearest to the cations. Since the natural abundance of F^{19} is 100%, the number of interacting F^{19} nuclei depends on the coordination. In fluorite-type structure each Ca^{2+} is surrounded by eight F^- ions, but each F^- has four nearest Ca^{2+} and six nearest F^-. Then $(2mI + 1) = (2.6 \cdot 1/2 + 1) = 7$ lines, i.e., the EPR spectrum of the F center in CaF_2 represents seven lines of the hyperfine structure from F^{19} with intensities relation $1:2:3:4:3:2:1$.

In ThO_2 single crystals with fluorite-type structure irradiated by electrons, gamma and X-ray, one can observe the EPR spectrum [961] assigned to an electron trapped by oxygen vacancy with $g_{\parallel} = 1.9739$ and $g_{\perp} = 1.9644$.

F Center in Minerals. In complex structures with radical ions, the F center formation is possible in the structures where some anions occur in solitary (detached from the radicals) positions. In silicates F centers have been found in cases when an F center electron is trapped by large voids occurring in some silicate structures. In sodalite (see above, Chap. 7.3) the F center designated here as Na_4^{3+} represents an electron trapped by chlorine vacancy (occurring in structure voids) and divided among four nearest Na^+ ions. In the EPR spectrum of this center one observes 13 lines of hyperfine structure arising due to interaction with four nuclei of the Na^+ ions (nuclear spin for Na^{23} $I = 3/2$): $2mI + 1 = 13$ lines with the intensities relation $1:4:10:20:31:44:31:20:10:4:1$. The similar EPR spectrum consisting of 13 lines and connected with the Na_4^{3+} center has been observed in synthetic zeolite (NaY type) after heating in sodium vapors or after gamma- or X-ray irradiation in vacuum.

The EPR spectrum of a synthetic zeolite NaX type consists of 19 lines and is linked with the Na_6^{5+} center, i.e., electron trapped by a void in a zeolite framework, and distributed between the six nearest Na^+ ions: $(2mI + 1 = 2 \cdot 6 \cdot 3/2 + 1 = 19$ lines).

In synthetic apatite some varieties of the F center form due to an electron being trapped by fluorine and chlorine vacancies (see above Chap. 7.3).

S Center—Surface F Center. After gamma-ray or neutron irradiation of MgO, CaO, and SrO powders, anion vacancies form at the crystal surfaces, these vacancies trapping an electron [962]. Oxygen adsorbtion by the surfaces is correlated with the formation of these centers. Because of the absence near this surface vacancy of one of the cations surrounding it, the polarization of other cations increases, leading to a somewhat greater g-factor shift. The g-factor values of the S centers stand close to those of the F centers in the same compounds, but have a slighly larger negative shift.

7.4.3 F Aggregate Centers

Besides the elementary F center described above, in alkali halides F aggregate centers form, representing individual centers consisting of aggregates of two, three and four F centers adjacent to one another [868, 969, 982]. They are designated correspondingly as F_2, F_3, and F_4 centers (Fig. 170). The optical absorption bands related to these centers are on the longwave side of the F band and are designated as M (F_2 center), R_1 and R_2 (F_3 center), N (F_4 center) bands. Before the nature of these centers (as F_2, F_3, F_4) has been ascertained the designations of these absorption bands were taken also as the designations of the centers themselves. Together with the ionized varieties (F_2^+, F_3^+, F_2^-), as well as with the modifications arising in associations with impurity cations, they form the F aggregate centers group.

Systematics of F Aggregate Centers. Extensive studies have been made to establish the models of the F aggregate centers giving rise to optical absorption bands, M, R_1, R_2, N in alkali halides.

Seitz [983] suggested for the first time a hypothesis in which anion and cation vacancies were regarded as aggregates with trapped electrons (Seitz's models correspond to the following modern interpretations: the M center of Seitz corresponds to the F_2^+ plus cation vacancy, $R_1 = F_2^+$, $R_2 = F_2$). Before the 1950s these models were generally accepted. Polarized bleaching and polarized luminescence studies of anisotropic F aggregate centers rendered important information about their properties but did not allow choosing between different models. It was only in conjunction with additional studies on thermal equilibria of the F and M centers that a present-day model for the M center as a double F center (i.e., F_2), and of the R and N centers as F_3 and F_4 were suggested [888]. However, only the EPR and ENDOR studies confirmed definitively the models of $M = F_2$ and $R = F_3$ centers.

As a result, the following simple and systematized scheme of the aggregate center models has taken shape:

$F_2 = M$: two conjugated F centers (two anion vacancies with two trapped elec-

Elementary center	F	$F^+ = \alpha$	$F^- = F'$	F_A
F aggregate center	$F_2 = M$	F_2^+	$F_2^- = F_2'$	M_A
	$F_3 = R$	F_3^+		
	$F_4 = N$			
		ionized centers		with impurity cation

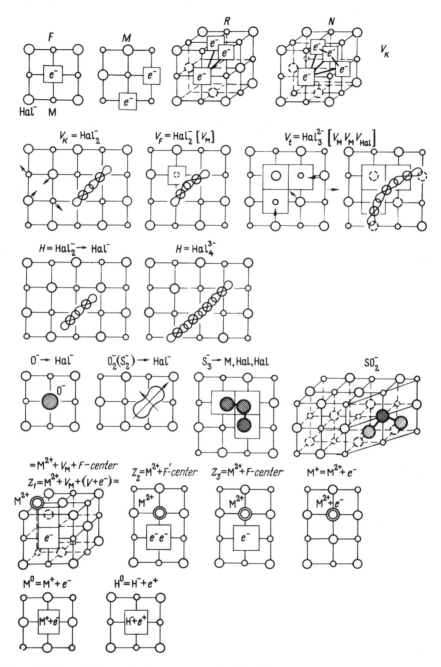

Fig. 170. Models of electron-hole centers in alkali halides

trons); D_{2h} symmetry; an analoge of the hydrogen molecule H_2; $F_3 = R$: three conjugated F centers (three anion vacancies with three trapped electrons); C_{3v} symmetry; R_1 and R_2 absorption bands; $F_4 = N$: four conjugated F centers (four anion vacancies with four trapped electrons).

The orientation and structural position of these centers are determined by the peculiarities of the crystal structure type (in alkali halides it is NaCl or CsCl types). Along [100] directions anions alternate with cations. Along [110] only anions or only cations occur. Thus the F_2 center is orientated along [110]; the F_3 center consists of three F centers lying in the (111) plane and orientated in pairs along [110], [011], [101]; the F_4 center has two varieties: N_1 represents four F centers in the (111) plane, N_2 represents four F centers located at the tetrahedron vertices. Among ionized F aggregate centers one can distinguish: F_2^+, i.e., two anion vacancies that trapped electron or F center with an anion vacancy; F_3^+, i.e., three anion vacancies that trapped electron; F_2^-, i.e., two anion vacancies that trapped three electrons. Together with impurities the centers form with a monovalent cation neighboring impurity: F_A and M_A.

Optical absorption bands of the F aggregate centers (the Molnar-Petrof series) were originally designated in order of their appearance in bleaching of KCl: A, B, C, D, E, G, but later on some of them were given other designations: M, R_1, R_2, N and others are compared in Table 22. (These designations were proposed in honor of the scientists who had observed them for the first time: M—Molnar, O—Okamoti, the F center's bands: K—Kleinschrod, L—Lüty; the R band after red).

Table 22. Optical absorption bands of F aggregate centers in some alkali halides (eV)

Petroff nomenclature	A	B	C	E	D	G	—
Other designations	A	B	M	R_1	R_2	N	O
Center type	F_A	—	F_2	F_3		F_4	—
Absorption bands							
NaCl	—	—	1.73	2.27	2.08	1.50	—
KCl	2.1	($\sim F$)	1.57	1.9	1.7	1.25	0.84
LiF	—	—	2.77	3.87	3.28	2.25	—

A definitive confirmation of the electron and hole centers is always obtained with the aid of the EPR spectra. In the case of F_2 and F_3 centers, to overcome certain difficulties it was necessary to arrange special conditions for measurements. The F_2 center is diamagnetic (there are two paired electrons with spin S equal to zero), while an F_3 center with a spin $S = 1/2$ has too short a relaxation time and is not observable in the EPR spectra even at lowest temperatures. Moreover, an EPR signal from these centers, if it could be observed, would be overlapped by a stronger signal coming from a usually present F center.

Earlier observations (parallel optical and EPR) of F center transformation into an F_2 center only confirmed the diamagnetic behavior of the latter. Therefore, further EPR studies of F_2 (M) and F_3 (R) centers were made in the following manner [888, 990]. In the case of the F_2 (M) center an EPR signal was obtained during optical illumination with the light of 365, 435, 540, or 580 nm. By means of this illumination the center was brought into an excited state, with both electrons transferred from the state with antiparallel spins ($S = 0$) into the state with parallel spins ($S = 1$), i.e., into the triplet state ($2S + 1 = 3$). As a result, three lines in each of the six nonequivalent positions of the center could be observed in the EPR spectrum.

In the case of the F_3 center the EPR spectrum was obtained during illumination which would bring the center into the lower metastable quartet state $^4A_1$ with sumultaneous action of an orientated pressure and at very low (helium) temperature, to suppress relaxation processes [990].

7.4.4 V Centers and Molecule Ions Hal_2^-, Hal_2^{3-}

The color centers in alkali halides are usually divided into (a) electron centers: F and F aggregate and (b) hole centers: V centers. A parallelism exists between them (but restricted to certain limits). Thus heating in metal vapors gives rise to the formation of F centers, while heating in halogen vapors leads to the formation of V centers (with excess of halogen). In this way (i.e., through additive coloration) one can obtain in crystals F or V centers separately. Upon irradiation of alkali halides one can obtain F and V centers simultaneously, these being to a certain extent complementary.

In optical absorption spectra of alkali halides one can observe, besides a most intense F band and F aggregate centers bands, occurring mainly in visible region also V absorption bands in the UV region. (Note that because these V bands occur in the UV, the color of alkali halides is not linked with the hole V centers). There are three absorption bands in the NaCl spectrum (V_1, V_2, V_3) and seven bands in the KCl spectrum (from V_1 to V_7). The corresponding hole centers have been designated by symbols V_1, V_2, V_3, and so on.

The models of these centers were originally suggested by Seitz [983] by analogy with his models for F centers: V_1 center (hole trapped by cation cacancy) is "antimorph" of the F center, V_2 center (two holes trapped by cation vacancies) is "antimorph" of his R_2 center model, etc. (see also other earlier models in [982]).

However, the EPR spectra studies of the V centers in alkali halides [890, 924, 925, 981] showed that V centers are not "symmetric" with respect to the F center, and the very concept of the hole center in halides has to be changed.

The specificity of hole centers in halides (in comparison with hole centers O^-, S^- etc.) is linked with monovalency of the ions composing them. When an anion, for example, Cl^-, traps a hole, the arising atom Cl^0 is stabilized through its being displaced towards the neighboring Cl^- ion with formation of a molecule ion Cl_2^-. Hole trapped by cations vacancy (see below) leads also to the same type of the molecule ion Hal_2^- (V_F center).

Thus, the elementary hole center in alkali halides is the molecule ion Hal_2^-. This molecule ion retains its individuality also in all compounds. The molecule ions

Hal_2^-, Hal_3^- form a basis of different hole centers in halides located in positions of different aggregates of vacancies and interstitials.

Molecule Ions Hal_2^- occur in three varieties (three different centers), depending on defects with which they are linked (Fig. 170).

1. V_K Center (see Fig. 170) is a molecule ion Hal_2^- taking up the position of two usual ions, when one of them loses its electron. This center is linked neither with vacancies nor interstials, i.e,. it forms in a lattice that does not contain any defects except for this hole [924, 925]. It is called a self-trapped hole (or autolocalized hole), being a defectless Hal_2^- center, that is a hole analogous to polaron. Molecule ions Hal_2^- are orientated along [110] directions in the NaCl type structures, i.e., along the cube face diagonals. There are six such directions, and hence six nonequivalent positions of the center.

In the fluorite V_K center $= F_2^-$ (see Fig. 168) is orientated along [100] and, accordingly, has three nonequivalent positions; it disappears at 138 K. Similar $V_K = F_2^-$ centers have been observed in SrF_2, BaF_2, and $SrCl_2$.

2. H Center (see Fig. 170) is a molecule ion Hal_2^- replacing the position of one Hal^- ion. Chemically, it is equivalent to an interstitial Hal^0 atom, but is better described as Hal_2^- ($Hal^0 + Hal^- = Hal_2^-$). It is an interstitial Hal_2^- center. Establishing of the model of this center by means of EPR [924, 925] was the first precise and unambiguous proof for the existence of interstitial atoms in crystals, enabling appraisement of the paramount part they play in processes of radiation damage formations. A hole localized at the site of two halogen ions Hal_2^- has an additional weaker interaction with two adjacent Cl^- ions, thus the H center can be represented as the Hal_4^{3-} complex consisting of four colinear halogens that trapped a hole and are elongated along the [110] direction.

3. V_F Center (see Fig. 170) is a molecule ion Hal_2^- replacing the position of the cation M^+ vacancy, i.e., it is a hole center antimorph to the F center; i.e., a hole trapped by the cation vacancy, but divided not among six anions surrounding it, but only between two of them. This model of an elementary hole center with cation vacancy in crystals of monovalent ions was also originally established by EPR [988]. The V_F center is also orientated along [110], but is slightly bent because of the presence of a cation vacancy.

In natural fluorites after their X-ray irradiation at 77 K and heating above 138 K one can observe V_K centers representing F_2^- molecule ions taking up the Ca^{2+} vacancy position [988]. Besides hyperfine structure from two F^{19} nuclei there is weak hyperfine splitting from two other fluorine nuclei.

XY^- Centers are mixed H centers consisting of two different halogen ions: FCl^-, FBr^-, $BrCl^-$, etc. [981].

Electron Structure, EPR and Optical Spectra of Hal_2^- Molecule Ions. All the Hal_2^- ions have the same 11 electron configuration (for example, $Cl^- 3p^6$, $Cl^0 3p^5$ lead to 11 outer electrons of the Cl_2^- molecule ion). These eleven electrons are distributed all over the molecular orbitals: . . . $(\sigma_g)^2 (\pi_u)^4 (\pi_g)^4 (\sigma_u^*)^1$. The unpaired electron occupies the antibonding σ orbital. This oribtal may be also considered to have a hole in it.

If one looks at the Hal_2^- center as a molecule with an axial and not an orthorhombic (as in crystals) symmetry then the ground state is to be designated as

$^2\Sigma_u^+$ and the excited states as $^2\Pi_g$, $^2\Pi_u$, $^2\Sigma_g^+$ (see Chap. 7.2). The most intense absorption band in the UV region corresponds to the transition from the ground state $^2\Sigma_u^+$ to the $^2\Sigma_g$ state (an electron transition corresponding to it from the σ_g orbital into the σ_u^* orbital is equivalent to the hole transition from the σ_u^* to σ_g orbital and is sometimes designated as $\sigma_u^* \to \sigma_g$).

The similarity of the molecular orbital scheme for Hal_2^- molecule ions is a factor safeguarding the individuality of the latter in different crystals, in chemically different varieties of F_2^-, Cl_2^-, Br_2^-, I_2^-, FCl^-, etc., and in different center types V_K, H, V_F, XY^-. This is manifest in the closeness of optical absorption bands which practically also coincide with those for solutions, in similarity of g-factors and hyperfine structures.

Molecule ion Hal_3^{2-} (V_t center; Fig. 170) is a more complicated center. Here, a hole is trapped by three halogen ions replacing the position of two cation and one anion vacancy, and forming an isosceles triangle (hence the subscript t in the designation of this center V_t).

References

1. Marfunin A. S.: Physics of Minerals and Inorganic Materials. An Introduction. Berlin-Heidelberg - New York: Springer, 1979. (This book contains references to crystal field theory, molecular orbital theory, energy band theory, and chemical bonds in minerals and inorganic materials.)

Mössbauer Spectroscopy

2. Amthauer G., Annersten H., Hafner S. S.: The Mössbauer spectrum of ^{57}Fe in titanium-bearing andradites. Phys. Chem. Mineral. *1*, (1977)
3. Amthauer G., Annersten H., Hafner S. S.: The Mössbauer spectrnm of ^{57}Fe in silicate garnets. Z. Kristallogr. *143*, 14–55 (1976)
4. Andersen E. B., Fenger J., Rose-Hansen J.: Determination of Fe^{2+}/Fe^{3+}-ratios in arfvedsonite by Mössbauer spectroscopy. Lithos *8*, 237–246 (1975)
5. Annersten H.: Mössbauer studies of natural biotites. Am. Mineral. *59*, 143–151 (1974)
6. Annersten H.: A Mössbauer characteristic of ordered glauconite. N. Jahrb. Mineral. Monatsh. *8*, 378–184 (1975)
7. Annersten H. (1976): New Mössbauer data on iron in potash feldspar. N. Jahrb. Mineral. Monatsh. *8*, 337–343 (1976)
8. Astakhov A. V., Viotkovski Yu. B., Generalov O. N., Sidorov S. V.: Studies of some sheet and boron-containing silicates by NGR method. Kristallographia *20*, 769–774 (1975)
9. Baguin V. I., Gendler T. S., Kuzmin R. N., Rybak R. S., Urasaeva T. K.: Weak ferromagnetism of natural hydrogeothite. Physica zemli *5*, 71–82 (1976)
10. Bancroft G. M.: Mössbauer Spectroscopy. An Introduction for Inorganic Chemists and Geochemists. New York: McGraw Hill, 1973
11. Bancroft G. M., Maddock A. G., Burns R. G., Stone A. J.: Applications of the Mössbauer effect to silicate mineralogy. Geochim. Cosmochim. Acta *31*, 2219–2246; *32*, 547–559 (1967)
12. Bancroft G. M., Burns R. G.: Mössbauer and absorption spectral study of alkali amphiboles. Mineral. Soc. Am. Spec. Pap. N 2, 137–148 (1969)
13. Bancroft G. M.: Quantitative site population in silicate minerals by the Mössbauer effect. Chem. Geol. *5*, 255–258 (1970)
14. Bancroft G. M., Brown J. R.: A Mössbauer study of coexisting hornblendes and biotetes: quantitative Fe^{3+}/Fe^{2+} ratios. Am. Mineral. *60*, 265–272 (1975)
15. Banerjee S. K., Johnson C. E., Krs M.: Mössbauer study to find the origin of weak ferromagnetism in cassiterite. Nature (Londen) *225*, N 5228, 173–175 (1970)
16. Barabanov A. V., Tomilov S. B.: Mössbauer study of the isomorphous series anthophyllite-gedrite and cummingtonite-grunerite. Geochem. Int. 1973, 1240–1247 (1973)
17. Baranowskii V. I.: Concerning quadrupole splitting values in Mössbauer spectra of iron compounds. Zhurn. Strukt. Khim. *7*, 141–143 (1966)
18. Barros F. de S., Hajemeister D., Viccaro P. Mössbauer Study of Co^{57} implanted in diamond. J. Chem. Phys. *52*, 2865–2867 (1970)
19. Barsukov V. L., Durasova N. A., Malysheva T. V., Bobr-Sergeev A. A.: Mössbauer studies of tin insertion in biotite and aluminosilicate glass. Geochimia *6*, 758–766 (1970)

20. Bashkirov Sh. Sh., Kurbatov G. D., Manapov R. A., Penkov I. N., Sadykov E. K., Chistiakov V. A.: NGR study of ilmenite. Dokl. Adak. Nauk 173, 407–409 (1967)
21. Belov V. F., Devisheva M. N., Zheludev I. S., Makarov E. F., Stukan A. R., Trukshanov B. N.: Mössbauer effect in Mn and Mg-Mn ferrites. Physica tverd tela 6, 3435–3438 (1964)
22. Belov V. F., Khimich T. A., Shipko M. N., Voskresenskaja I. E., Okulov E. N.: NGR studies of iron tourmalines. Kristallographia 18, 192–194 (1973)
23. Belov V. F., Korovushkin V. V., Belov A. F., Korneev E. V., Zheludev I. S.: Nonequivalent positions of iron and electron-nuclear interactions in tourmaline. Physica tverd tela 16, 2410–2411 (1974)
24. Borshagovskii B. V., Marfunin A. S., Mkrtchyan A. R., Nagyaryan Stukan R. A.: Mössbauer study of isomorphous substitution in ilvaite. Phys. Status. Solidi. (b) 43, 479–482 (1974)
25. Bowen L. H., Weed S. B., Stevens J. G.: Mössbauer study of micas and their potassium-depleted products. Am. Mineral. 54, 72–84 (1969)
26. Brown F. F., Pritchard A. M.: The Mössbauer spectrum of iron orthoclase. Earth Planet. Sci. Lett 5, 259–260 (1968)
27. Buckley A. N., Wilkins R. W. T.: Mössbauer and infrared study of a volcanic amphibole. Am. Mineral. 56, 90–100 (1974)
28. Burnham C. W., Ohashi Y., Hafner S. S., Virgo D.: Cation distribution and atomic thermal vibrations in an iron-rich orthopyroxene. Am. Mineral. 56, 850–876 (1971)
29. Burns R. G., Prentice F. J.: Distribution of iron cations in the crocidolite structure. Am. Mineral. 53, 770–776 (1968)
30. Burns R. G.: Crystal field spectra and evidence of cation ordering in olivine minerals. Am. Mineral. 55, 1608–1632 (1970)
31. Burns R. G., Greaves C.: Correlations of infrared and Mössbauer site population measurements of actinolites. Am. Mineral. 56, 2010–2033 (1971)
32. Burns R. G., Huggins F. E.: Cation determinative curves for Mg-Fe-Mn olivines from vibrational spectra. Am. Mineral. 57, 967–985 (1971)
33. Burns R. G., Tossell J. A., Vaughan D. J.: Pressure-induced reduction of a ferric amphibole. Nature (London) 240, N 5375, 33–35 (1972)
34. Burns R. G.: Mixed valencies and site occupancies of iron in silicate minerals from Mössbauer spectroscopy. Can. J. Spectrosc. 17, 51–59 (1972)
35. Bush W. R., Hafner S. S., Virgo D.: Some ordering of iron and magnesium at the octahedrally coordinated sites in a magnesium - rich olivine. Nature (London.) 227, N 5265, 1339–1341 (1970)
36. Goldanskii, V. I. Herber R. H. (eds.): Chemical Applications of Mössbauer Spectroscopy. New-York - London: Academic Press 1968
37. Clark A. M., Fejer E. E., Donaldson J. D., Silver J.: The ^{119}Sn Mössbauer spectra, cell dimensions, and minor elements contents of some cassiterites. Mineral. Mag. 40, 895–898 (1976)
38. Cordey-Hayes M.: An interpretation of the Mössbauer spectra of some tin compounds in terms of the valence state of the tin atom. Rev. Mod. Phys. 36, H. 1, N 2 (1964)
39. Coster M. de, Pollak H., Amelinckx S.: A study of Mössbauer absorption in iron silicates. Phys. Status Solidi 3, 282–288 (1963)
40. Cruset A., Friedt J. M. (1971): Stabilization of aliovalent ions observed by Mössbauer emission spectroscopy in ^{57}Co - doped simple ligand compounds. Phys. Status Solidi. (b) 47, 655–662 (1971)
41. Cruset A., Friedt J. M.: Mössbauer study of the valence state of ^{57}Fe after ^{57}Co decay in $CoFe_2O_4$. Phys. Status. Solidi. (b) 45, 189–193 (1971)
42. Danon J.: Chemical Applications of Mössbauer Spectroscopy. New York - London: Academic Press, 1968
43. Da Silva E. G., Abras A.: Mössbauer effect study of cation distribution in natural chromites. J. Phys. (France) 12, Suppl., 783–785 (1976)

44. Dézsi J., Fodor M.: On the antiferromagnetism of α-FeOOH. Phys. Status. Solidi. *15* 247–254 (1966)

45. Dowty E., Ross M., Cutitta F.: Fe^{2+}-Mg site distribution in Apollo 12021 pyroxenes: Evidence for bias in Mössbauer measurements and relation of ordering to exsolution. Proc. 3rd Lunar Sci. Conf., Geochim. Cosmochim. Acta, Suppl. *3*, 1, 481–492 (1972)

46. Dowty E., Lindsley D. H.: Mössbauer spectra of synthetic hedenbergite-ferrosilite pyroxenes. Am. Mineral. *58*, 850–868 (1973)

47. Drickamer H. G., Frank C. W.: Electronic Transitions and the High Pressure Chemistry and Physics of Solids. London: Chapman and Hall, 1973

48. Dundon R. W., Walter L. S.: Ferrous ion order-disorder in meteoritic pyroxenes and the metamorphic history of chondrites. Earth Planet. Sci. Lett. *2*, 372–376 (1967)

49. Eibschütz M., Hermon E., Shtrikman S.: Determination of cation valencies in Cu_2^{57} $Fe^{119}SnS_4$ by Mössbauer effect and magnetic susceptibility measurements. J. Phys. Chem. Solids *28*, 1633–1636 (1967)

50. Eirish M. V., Dvorechenskaya A. A.: NGR studies of the position and role of Fe^{3+} ions in clay minerals structure (relation to crystal chemistry of the minerals). Geochimia *5*, 748–757 (1976)

51. Eissa N. A., Sallam H. A., Ashi B. A., Hassan M. Y., Saleh S. A.: A Mössbauer effect study of plant and animal fossils. J. Phys. D. Appl. Phys. *9*, 1391–1397 (1976)

52. Ermakov A. N., Alexandrov S. M., Kurash V. V., Malysheva T. V.: NGR studies of Mg-Fe substitutions in borates of ludwigite-vonsenite series. Geochemia *10*, 1217–1223 (1969)

53. Ernst W. G., Wai C. M.: Mössbauer, infrared, X-ray and optical study of cation ordering and dehydrogenation in natural and heat-treated solid amphiboles. Am. Mineral. *55*, (1970)

54. Ershova S. P., Babeshkin A. M., Perfiliev Yu. D.: Mössbauer study of iron oxidation in cummingtonite. Geochemia *2*, 252–256 (1970)

55. Evans B. J., Ghose S., Hafner S. S.: Hyperfine splitting of Fe^{57} and Fe-Mg order-disorder in orthopyroxenes ($MgSiO_4$-$FeSiO_3$). J. Geol. 75 (1967)

56. Evans B. J., Hafner S. S.: ^{57}Fe hyperfine fields in magnetite (Fe_3O_4). J. Appl. Phys. *40*, 1411–1413 (1967)

57. Evans B. J., Sergent E. W., Jr.: ^{57}Fe NGR of Fe phases in "magnetic cassiterites", I. Crystal chemistry of dodecahedral Fe^{2+} in pyralspite garnets. Contrib. Mineral. Petrol. *53*, 183–194 (1975)

58. Fatseas G. A., Dormann J. L., Blanchard H.: Study of the Fe^{3+}/Fe^{2+} ration in natural chromites (Fe_xMb_{1-x}) (Cr_{1-y-x} $Fe_yAl_z)O_4$. J. Phys. *12*, Suppl. 787–792 (1976)

59. Forester D. W.: Mössbauer search for ferric oxide phases in lunar materials and simulated lunar materials. Proc. 4th Lunar Sci. Conf., Houston, Texas, *3*, 2697–2707, (1973)

60. Forsyth J. B., Hedley I. G., Johnson C. E.: The magnetic structure and hyperfine field of goethite (FeOOH). J. Phys. (Proc. Phys. Soc.) C *1*, 179–188 (1968)

61. Frank C. W., Drickamer H. G.: High pressure chemistry and physics of iron compounds. In: The Physics and Chemistry of Minerals and Rocks. Strens, R. G. (ed.) London: Wiley 1976

62. Gabriel J. R.: Computation of Mössbauer spectra. Mössbauer Effect Methodology *1*, 121–132 (1965)

63. Gakiel U., Malamud M.: On the valence of iron in tripuhyite: a Mössbauer study. Am. Mineral. *54*, 299–301 (1969)

64. Gakiel U., Shtrikman S.: Theory of magnetically induced electric field gradients in cubic Fe^{2+}. Phys. Rev. *167*, 258–270 (1968)

65. Gangas N. H., Simopoulos A., Kostikas A., Yassoglou N. J., Filippakis S.: Mössbauer studies of small particles of iron oxides in soil. Clays Clay Mineral. *21*, 151–160 (1972)

66. Gapeev A. K., Gendler T. S., Kuzmin R. N., Novakova A. A., Pokrovskii B. I.: Study of Mössbauer spectra of $Fe_{3-x}Mg_xO_4$ ferrites - spinels. Kristallographia 17, (1972)

67. Gendler T. S., Kuzmin R. N., Urasaeva T. K.: Mössbauer effect study of hydrogoethite. Kristallographia 21, 774–781 (1976)

68. Gelberg A.: On point-charge calculations of the quadrupole splitting in Mössbauer spectra. Rev. Roumaine Phys. 14, 183–188 (1969)

69. Ghose S., Hafner S. S.: $Mg^{2+}-Fe^{2+}$ distribution in metamorphic and volcanic orthopyroxenes. Z. Kristallogr. 125 (1967)

70. Ghose S., Weidner J. R.: $Mg^{2+}Fe^{2+}$ order - disorder in cummingtonite (Mg, Fe)$_7Si_8$ $O_{22}(OH)_2$: a new geothermometer. Earth Planet. Sci. Lett. 16, 346–354 (1972)

71. Ghose S., Tsang T.: Structural dependence of quadrupole coupling constant e^2qQ/h for ^{27} Al and crystal field parameter D for Fe^{3+} in aluminosilicates. Am. Mineral. 58, 748–755 (1973)

73. Gibb T. C., Greenwood N. N., Twist W.: The Mössbauer spectra of natural ilmenites. J. Inorg. Nucl. Chem. 31, 947–954 (1969)

74. Goldansky V. I., Makarov E. F., Stukan R. A.: Relation of chemical shift and quadrupole splitting in NGR spectra to electronic structure of iron atom. Theor. Exp. Chim. 2, 504–511 (1966)

75. Goldansky V. I., Dolenko A. V., Egiasarov B. G., Zhaporozhets V. M., Ratnikov V. M.: Gamma-Resonance Methods and Instruments for Phase Anlysis of Mineral Raw Materials. Moscow: Atomisdat, 1974

76. Goncharov G. N., Ostanevich Ya. M., Tomilov S. B. Mössbauer study of iron sulfides. Izv. Akad. Nauk, Ser. Geol. 8, 79–88 (1970)

77. Goncharov G. N., Ostanevich Yu. M., Tomilov S. B., Cser L.: Mössbauer effect in the FeS_{1-x} system. Phys. Status. Solidi. 37, 141–150 (1970)

78. Goncharov G. N., Rozhkova L. V.: NGR study of magnesium-containing magnetites. Geochimia 8, 1260–1265 (1973)

79. Goncharov G. N., Kalyamin A. V., Lury B. G.: NGR study of iron-manganese nodules from Pacific. Dokl. Akad. Nauk SSSR. 212 (1973)

80. Goodman B. A.: The effect of lattice substitutions on the derivation of quantitative site populations from the Mössbauer spectra of 2:1 layer lattice silicates. J. Phys. (France) 12, Suppl. 819–823 (1976)

81. Goodman B. A.: On the interpretation of the Mössbauer spectra of biotites. Am. Mineral. 61, 169–175 (1976)

82. Goodman B. A.: The Mössbauer spectra of a ferrian muscovite and its implication in the assignment of site in dioctahedral micas. Mineral. Mag. 40, 513–517 (1976)

83. Gorelikova N. V., Perfiliev Yu. D., Babeshkin A. M.: Iron distribution in the tourmaline structure estimated by Mössbuaer spectroscopy data. Zhapiski Vses. Mineral. obsch. 105, 418–427 (1976)

84. Gosselin I. R., Townsend M. G., Tremblay R. I., Webster A. H.: Mössbauer effect in single-crystal Fe_{1-x} S. Solid State Chem. 17, 43–48 (1976)

85. Govaert A., Dauwe C., Plinke P., De Grave E., De Sitter J.: A classification of goethite minerals based on the Mössbauer behavior. J. Phys. (France) 12, Suppl. 825–827 (1976)

86. Greaves C., Burns R. G., Bancroft G. M.: Resolution of actinolite Mössbauer spectra into three ferrous doublets. Nature Phys. Sci. 229, 60–61 (1971)

87. Greenwood N. N., Whitfield H. J.: Mössbauer effect studies on cubanite ($CuFe_2S_3$) and related iron sulfides. J. Chem. Soc. A, N 7, 1697–1698 (1968)

88. Greenwood N. N., Gibb T. C.: Mössbauer Spectroscopy. London: Chapman and Hall, 1971

89. Grubb P.L.C., Hannaford P.: Magnetism in cassiterite. Its source and paragenetic significance as exemplified by a prominent Malayan tin deposit. Miner Deposita 2, 148–171 (1966)

90. Hafner S. S., Kalvius M.: The Mössbauer resonance of Fe^{57} in troilite (FeS) and pyrrhotite ($Fe_{0.88}S$). Z. Kristallogr. 123, 443–458 (1966)

91. Hafner S. S., Virgo D.: Temperature-dependent cation distribution in lunar and terrestrial pyroxenes. Proc. Apollo 11 Lunar Sci. Conf. 3 (1970)

92. Hafner S. S., Ghose S.: Iron and magnesium distribution in cummingtonites $(Fe, Mg)_7 Si_8 O_{22}(OH)_2$. Z. Kristallogr. *133*, 301–326 (1971)

93. Hafner S. S., Janin B., Virgo D.: State and location of iron in Apollo 11 samples. In: Mössbauer Effect Methodology, Gruverman, I. J. (ed.) Vol. *6*, 193, 1971

94. Hafner S. S., Huckenholz H. G.: Mössbauer spectrum of synthetic ferri-diopside. Nature Phys. Sci. *223*, N 36, 9–11 (1971)

95. Hafner S. S., Virgo D., Warburton D.: Oxidation state of iron in plagioclase from lunar basalts. Earth Planet Sci. Lett. *12*, 159–166 (1971)

96. Hafner S. S., Virgo D., Wartburton D.: Cation distribution and cooling history of clinopyroxenes from oceanus Procellarium. Proc. 2nd Lunar Sci. Conf., The MGT Press (1971)

97. Hafner S. S.: Mössbauer spectroscopy in lunar geology and mineralogy. Top. Appl. Phys. *5*, 167–199 (1975)

98. Häggström L., Wäppling R., Annersten H.: Mössbauer study of iron-rich biotites. Chem. Phys. Lett. *4*, 107 (1969)

99. Hang N. O.: Mössbauer studies of natural apatite. Phys. Rev. *185*, 477–482 (1969)

100. Hartmann-Boutron F., Imbert P.: Mössbauer study of the electronic and magnetic properties of Fe^{2+} ions in some spinel-type compounds. J. Appl. Phys. *39*, Part I, 775–785 (1968)

101. Hargrove R. S., Kundig W.: Mössbauer measurements of magnetite below the Verwey transition. Solid State Commuu. *8*, 303 (1970)

102. Hawthorne F. C., Grundy H. D.: Resolution of the Mössbauer spectrum of oxykaersutite. Can. Mineral. *13*, 91–92 (1975)

103. Helsen J., Schmidt K., Chakupurakal Th., Conssement K., Langouche G.: Determination par effet Mössbauer du coefficient de selfdiffusion dans une vermiculite. Bull. Groupe Fr. Argiles, *24*, N 2 (1972)

104. Herr W., Sherra B.: Mössbauer spectroscopy applied to the classification of stone meteorites. In: Meteorite Research, Millman, P. M. (ed.) Holland: Derdrecht, 1969

105. Herzenberg C. L., Toms D.: Mössbauer absorption measurements in iron-containing minerals J. Geophys. Res. *71*, 2661–2677 (1966)

106. Herzenberg C. L., Riley D. L., Lamoreaux R.: Mössbuaer absorption in zinnwaldite mica. Nature (London) *219*, N5152, 364–365 (1968)

107. Herzenberg C. L., Lamoreaux R. D., Riley D. L.: Mössbafier resonant absorption in ferberite and wolframite. Z. Kristallogr. *128*, 414–417 (1969)

108. Herzenberg C. L., Riley D. L.: Analysis of first returned lunar samples by Mössbauer spectrometry. Proc. Apollo-II Lunar Sci. Conf. *3*, 2221–2241 (1970)

109. Hogarth D. D., Brown F. F., Pritchard A. M.: Biabsorption, Mössbauer spectra, and chemical investigation of five phlogopite samples from Quebec. Can. Mineral. *10*, 710–722 (1970)

110. Hoggs C. S., Meads R. E.: The Mössbauer spectra of several micas and related minerals. Mineral. Mag. *37*, 606–614 (1970)

111. Housley R. M., Blander M., Aldel-Gawad M., Grant R. W., Muir A. H.: Mössbauer spectroscopy of Apollo-11 samples. Proc. Apollo-11 Lunar Sci. Conf. *3*, 2251–2268 (1970)

112. Hrynkiewicz A. Z., Kubisz J., Kulgawczuk D. S.: Quadrupole splitting of the 14.4 KeV gamma line of ^{57}Fe in iron sulphates of the jarosite group. J. Inorg. Nucl. Chem. *27*, 2513–2517 (1965)

113. Hrynkiewicz A. Z., Kulgawczuk D. S., Tomala K.: Antiferromagnetism of α-FeOOH investigated with the Mössbauer effect. Phys. Lett. *17*, 93–95 (1965)

114. Hrynkiewicz A. Z., Kubisz J., Kulgawczuk D.: Zastosowanie efektu Mössbauera w problematyce mineralogicznej. Pr. Mineral. PAN. Krakovie *6*, 7–51 (1966)

115. Hrynkiewicz A. Z., Kulgawczuk D. S., Mazanek E. S., Wlasak J., Wyderko M. E.: Mössbauer absorption in calcium-iron olivines. Phys. Status Solidi (a) *16*, 135–240 (1973)

116. Huggins R. M.: Mössbauer studies of iron minerals under pressures of up to 200 kilobars. In: The Physics and Chemistry of Minerals and Rocks. Strens, R. G. J. (ed.), London: J. Wiley 1976

117. Imbert P., Gerard A., Wintenberger M.: Etude des sulfure, arseniosulfure et arseniure de fer naturels par effet Mössbauer. C.R. Acad. Sci. *256*, N 21 4391–4393 (1963)

118. Imbert P., Winterberger M.: Etudes des proprietes magnetiques des spectres d'absorption par effet Mössbauer de la cubanite et de la sternbergite. Bull. Soc. Fr. Mineral. Cristallogr. *90*, N 3 (1967)

119. Ingalls R.: Electric-field gradient tensor in ferrous compounds. Phys. Rev. *133*, A 787–795 (1964)

120. Ivanitsky V. P., Matiash I. V., Rakovich F. I.: Influence of radioactive radiations on the Mössbauer spectra of biotites. Geochimia *6*, 850–857 (1975)

121. Ioshioka T., Cohichi J., Kohno H.: Quantitative determination of ferrous and ferric ions using Mössbauer effect. Anal. Chem. *40*, N 3 (1968)

122. Jagnik C. M., Mathur H. B.: Mössbauer and X-ray diffraction study of the cation distribution in $FeAl_2O_4$. J. Phys. (Proc. Phys. Soc.) *C1*, N 2 (1968)

123. Jagnic C. M., Mathur H. B.: Electric field gradient in normal spinels. Mol. Phys. *16*, N 6 (1969)

124. Jamamoto N., Shinjo T., Kiyama M.: Mössbauer effect study of alpha-FeOOH and ß-FeOOH; making use of oriented particles. J. Phys. Soc. Jpn. *25*, 1267–1271 (1968)

125. Janot C., Chabanel M., Herzog E.: Etude d'une limonite par effet Mössbauer. Bull. Soc. Fr. Mineral. Cristallogr. *91*, N 2 (1968)

126. Janot C., Gibert H.: Les constituants du fer dans certaines bauxites naturelles etudiees par effet Mössbauer. Bull. Soc. Fr. Mineral. Cristallogr. *93*, 213–223 (1970)

127. Janot C.: L'effet Mössbauer et ses applications à la physique du solide et à la metallurgie physique. Paris Masson, (1972)

128. Jensen S. D., Shive P. N.: Cation distribution in sintered titanomagnetites. J. Geophys. Res. *78*, N 35 (1973)

129. Johnson C. E., Clasby G. P.: Mössbauer effect determination of particle size in microcrystalline iron-manganese nodules. Nature (London) *222*, N 5191, 376–377 (1969)

130. Kalinichenko A. M., Litovchenko A. S., Matiash I. V., Polshin E. V., Ivanitskii V. P.: Features of the Crystal-Chemistry of Sheet Silicates Based on the Radiospectroscopy [and NGR] data. Kiev: Naukova Dumka, 1973

131. Khisina N. R., Belokoneva E. L., Simonov M. A., Ivanov V. I., Makarov E. S.: Fe^{2+} − Mg^{2+} ordering in orthopyroxenes of "Luna 20" and their thermal history. Geochimia *11*, 1612–1623 (1976)

132. Khristoforov K. K., Nikitina L. P., Krizhanskii L. M., Ekimov S. P., Litvin A. P.: Mössbauer study of iron distribution in calcium amphiboles. Dokl. Akad. Nauk *210*, 931–934 (1973)

133. Khristoforov K. K., Nikitina L. P., Krizhanskii L. M., Ekimov S. P.: Kinetics of Fe^{2+} disordering in orthorhombic pyroxenes. Dokl. Akad. Nauk *214*, 909–912 (1974)

134. Kirov G. N., Tomov T. T., Ruskov T. H., Georgiev S. A.: Etude par spectrometrie Mössbauer de la chalcopyrite et des produits de son metamorphisme thermique. Dokl. Bolg. Akad. Nauk *25*, 365–368 (1972)

135. Korovushkin V. V., Korneev E. V., Belov A. F., Belov V. F., Khimich T. A., Kolesnikov I. I.: Concerning the nature of ferrites-spinels ageing. Zh. Phys. Chem. *49*, 1683–1688 (1975)

136. Kostiner E.: A Mössbauer effect study of triplite and related minerals. Am. Mineral. *57*, 1109–1114 (1972)

137. Krizhanskii L. M., Nikitina L. P., Khristoforov K. K., Ekimov S. P.: Fe^{2+} distribution and geometry of cation-oxygen polyhedra in structures of orthopyroxenes at different temperatures (according to Mössbauer spectroscopy data). Geochimia *1*, 69–79 (1974)

138. Krupianskii Yu. F., Suzdalev I. P.: Size effects in fine particles of Fe_3O_4. Zh. Exp. Techn. Phys. 736–740 (1975)

139. Kundig W., Bömmel H., Constabaris G., Lindquist R. H.: Some properties of supported small α-Fe_2O_3 particles determined with the Mössbauer effect. Phys. Rev. *142*, 327–333 (1966)

140. Kurash V. V., Goldanskii V. I., Malysheva T. V., Urusov V. S.: Low temperature Mössbauer study of wustite. Neorg. Mater. *7*, 1574–1580 (1972)

141. Kuzmin R. N., Gendler T. S.: Study of structural-magnetic phase transformations in siderite. Kristallorgaphia *15*, 736–741 (1970)

142. Law A. D.: Critical evaluation of statistical best fits to Mössbauer spectra. Am. Mineral. *58*, 128–131 (1971)

143. Leider H. R., Pipkorn D. N.: Mössbauer effect in $MgO:Fe^{+2}$; low-temperature quadrupole splitting. Phys. Rev. *165*, 494–500 (1968)

144. Lerman A., Stiller M., Hermon E.: Mössbauer quantitative analysis of Fe^{+3}/Fe^{+2} ratios in some phosphate and oxide mixtures: possibilities and limitations. Earth Planet Sci. Lett. *3*, 409–416 (1968)

145. Litvin A. L.: Crystal Chemistry and Structure Typomorphism of Amphiboles. Kiev: Naukova Dumka, 1977

146. Loseva G. V., Murashko N. V., Petukhov E. P., Povitskii V. A., Tepliakova N. V.: Transofmration of δ-FeOOH in goethite and hematite. Physica zhemli *8*, 114–120 (1975)

147. Lyubutin J. S. Magnetism and Crystal Chemistry of Garnets Studied by Mossbauer spectroscopy. Proc. Conf. Appl. Mössbauer Effect Budapest. 1971, p. 467–489.

148. Lyubutin J. S., Belyaev L. M., Grzhikhova R., Lipka I.: Mössbauer study of some aspects of crystalchemistry of garnets. Kristallographia *17*, 146–148 (1972)

149. MacKenzie K. J. D.: A Mössbauer study of the role of iron impurities in the high temperature reactions of kaolinite minerals. Clay Mineral. *8*, 151–160 (1969)

150. Maddock A. G.: Mössbauer Spectroscopy in Mineralogy. Mössbauer Spectroscopy and Applications, Vienna, 329–345. Discuss., p. 345–347.

151. Makarov E. F., Marfunin A. S., Mkrtchian A. R., Povitskii V. A., Stukan R. A.: NGR study of magnetic ordering of $CuFe_2S_3$. Physica tverd. tela *10*, 913–915 (1968)

152. Makarov E. F., Marfunin A. S., Mkrtchian A. R., Nadzharian G. N. Povitskii V.A., Stukan R. A.: NGR study of magnetic properties of Fe_3S_4. Physica tverd. tela *11*, 495–497 (1969)

153. Malysheva T. V., Kurash V. V., Ermakov A. N.: Mössbauer study of Mg-Fe^{2+} substitution in olivines. Geochimia *11*, 1405–1408 (1969)

154. Malysheva T. V., Yermakov A. N., Alexandrov S. M., Kurash V. V.: Mössbauer Study of Isomorphism in Borates of Ludvigite and Vonsenite Series. Proc. Conf. Appl. Mössbauer Effect, 745–750, Budapest, 1971

155. Malysheva T. V., Romanchev B. P., Shvagerov V. D.: Temperature of olivine formation as determined by Mössbauer spectroscopy and thermometry of inclusions. Geochimia *4*, 496–497 (1972)

156. Malysheva T. V., Lavrukhina A. K., Stakheeva S. A., Satarova L. M.: Cation Fe^{2+} – Mg ordeling in pyroxenes from olivine-hypersthene chondrites based on the Mössbauer data. Geochimia *9*, (1973)

157. Malysheva T. V.: Mössbauer spectroscopy of lunar regolith returned by the automatic station Luna 16. Proc. 3rd Lunar Sci. Conf. *1*, (1973)

158. Malysheva T. V.: Main Differences Between Lunar Regolith of Mare and Highland Origin According to Data of Mössbauer Spectroscopy. Space Res. XIV, Berlin: Akademie-Verlag, 1974

159. Malysheva T. V.: Mössbauer Effect in Geochemistry and Cosmochemistry. Moscow: Nauka, 1975

160. Manning P. G., Tricker M. J.: Optical-absorption and Mössbauer spectral studies of iron and titanium site-populations in vesuvianites. Can. Mineral. *13*, 159–265 (1975)
161. Marfunin A. S., Mkrtchian A. R.: Mössbauer Fe^{57} spectra in sulfide minerals. Geochimia *10*, 1094–1103 (1967)
162. Marfunin A. S., Mkrtchian A. R.: Mössbauer Sn^{119} effect in stannine. Geochimia *4*, 498–500 (1968)
163. Marfunin A. S., Mineeva R. M., Mkrtchian A. R., Nussik Ya. M., Fedorov V. E.: Optical and Mössbauer spectroscopy of iron in rock-forming silicates. Izv. Akad. Nauk. Ser. Geol. *10*, 86–102 (1067)
164. Marfunin A. S., Mkrtchian A. R., Nadzharian G. N., Nussik Ya. M. Platonov A. N.: Optical and Mössbauer spectra of iron in tourmalines. Izv. Akad. Nauk, Ser. Geol. *2*, 146–150 (1970)
165. Marfunin A. S., Mkrtchian A. R., Nadzharian G. N., Nussik Ya. M., Platonov A. N.: Optical and Mössbauer spectra of iron is some sheet silicates. Izv. Akad. Nauk, Ser. Geol. *7*, (1971)
166. Matsui Y., Syono Y., Maeda Y.: Mössbauer spectra of synthetic and natural calcium-rich clinopyroxenes. Mineral. J. *7*, 88–107 (1971)
167. McNab T. K., Fox R. A., Boyle A. J. E.: Some magnetic properties of magnetite (Fe$_3$O$_4$) microcrystals. J. Appl. Phys. *39*, 5703–5711 (1968)
168. Mineeva R. M.: Calculation of potentials with overlapping and their relation to cation distribution in borates of ludwigite-vonsenite series. Geochimia *2*, (1974)
169. Mizoguchi T., Tanaka M.: The nuclear quadrupole interaction of Fe^{57} in spinel type oxides. J. Phys. Soc. Jpn *18*, 1301–1306 (1963)
170. Mkrtchian V. M., Mozgova N. N., Mkrtchian A. R.: Tin in ilvaite according to Mössbauer data. Geochimia *11*, 1400–1402 (1972)
171. Mössbauer R. L.: Recoilless absorption of gamma rays and studies of nuclear hyperfine interactions in solids. In: Hyperfine Interactions. 497–551. New York - London. Academic Press 1967
172. Muir A. H. Jr., Wiedersich H.: An investigation of CuFeO$_2$ by the Mössbauer effect. J. Phys. Chem. Solids *28*, 65–71 (1967)
173. Nakamura T., Shinjo T., Endoh Y., Yamamoto N., Shiga M., Nakamura Y.: Fe^{57} Mössbauer effect in ultra fine particles of α-Fe$_2$O$_3$. Phys. Lett. *12*, 178–179 (1964)
174. Novikov G. V., Egorov V. K., Popov V. I., Bezmen N. I.: Investigation of magnetic transformation in high-temperature hexagonal iron sulfide by nuclear gamma-resonance method. Geochimia *7*, 1107–1111 (1975)
175. Novikov G. V., Egorov V. K., Popov V. I., Sipavina L. V.: α-transformation in Iron Sulfide Fe$_{1-x}$S. Proc. Int. Conf. Mössbauer Spectroscopy. Poland-Cracow, 1975, Vol. I, p.391.
176. Novikov G. V., Egorov V. K., Popov V. I., Sipavina L. V.: Kinetics and medhanism of transformations in iron-rich pyrrhotites and troilite-pyrrhotite metastable assemblages. Phys. Chem. Mineral. *1*, 1–14 (1977)
177. Ok H. N.: Mössbauer studies of natural apatite. Phys. Rev. 185 (1969)
178. Ohashi H., Hariya Y.: Order-disorder of ferric iron and aluminum in Ca-rich clinopyroxene. Proc. Jpn. Acad. *46*, 684–687 (1970)
179. Ovsyannikov E. A., Polocin A. V., Spaer A. G.: Concerning the application of nuclear gamma resonance method to phase analysis of iron ores. Proc. Vses. Inst. of Nucl. Geochem. Geophy. 19 (1974)
180. Parkin K. M., Loeffler B. M., Burns R. G.: Mössbauer spectra of kyanite, aquamarine and cordierite showing intervalence charge transfer. Phys. Chem. Minerals. *1*, 301–312 (1977)
181. Platonov A. N., Polshin E. V., Tarashchan A. N., Vorobjev I. B.: Mössbauer and optical spectroscopy of iron in some natural phosphates. Mineral. Shornik Lvov Univ. *26*, N 3, 257–268 (1975)
182. Pollack S. S.: Disordered orthopyroxene in meteorites. Am. Mineral. *51*, 1722–1726 (1966)

183. Polshin E. V., Matiash I. V.: Fe^{2+}-Mg ions distribution in octahedral sheet of biotites according to NGR data. In: Constitution and Properties of Minerals, N7, 1973
184. Prandl W., Wagner F. Die Orientierung des electrischen Feldgradienten und das innere Magnetfeld beim Almandin. Z. Kristallogr. *134*, 344–349 (1971)
185. Price D. C., Vance E. R., Smith G., Edgar A., Dickson B. L.: Mössbauer effect studies of beryl. J. Phys. Tr. *12*, Suppl., 811–817 (1976)
186. Regnard J. R.: Mössbauer study of natural crystals of staurolite. J. Phys. Fr. *12*, Suppl. 797–800 (1976)
187. Rice C. M., Williams J. M.: A Mössbauer study of biotite weathering. Mineral. Mag. *37*, 210–215 (1969)
188. Roggwiller P., Kundig W.: Mössbauer spectra of superparamagnetic Fe_3O_4. Solid State Commun. *12*, N 9 (1973)
189. Romanov V. P., Checherskii V. D., Eremenko V. V.: Peculiarities of low temperature spectra of NGR in magnetite with different content of structural vacancies. Phys. Status Solidi (a) *9*, N 2 (1972)
190. Romanov V. P., Valter A. A., Zverev N. D., Eremenko G. K.: Mössbauer effect in $(Mn_{1-x-y}Fe\ Zn_y)_4$ $(BeSiO_4)_3S$ solid solution. Phys. Status Solidi *41*, 167–171 (1970)
191. Rosencwaig A.: Double exchange and the metal-nonmetal transition in magnetite. Phys. Rev. *181*, 946 (1969)
192. Rossiter M. J.: The Mössbauer spectra of some spinel oxides containing iron. J. Phys. Chem. Solids, *26*, 775–779 (1965)
193. Rossiter M. J., Hodgson A. E. M.: A Mössbauer study of ferric oxyhydroxide. J. Inorg. Nucl. Chem. *27*, 63–71 (1965)
194. Rubinstein M., Forester D. W.: Investigation of the insulating phase of magnetite by NMR and the Mössbauer effect. Solid State Commun. *9*, 1675 (1971)
195. Sawatzky G. A., van der Woude F., Morrish A. H.: Recoilless fraction ratios for Fe^{57} in octahedral and tetrahedral sites of a spinel and a garnet. Phys. Rev. *183*, 383–386 (1969)
196. Sawatzky G. A., van der Woude F., Morrish A. H.: Mössbauer study of several ferrimagnetic spinels. Phys. Rev. *187*, N 2 (1969)
197. Schürmann K., Hafner S. S.: On the amount of ferric iron in plagioclases from lunar igneous rocks. Proc. 3rd Lunar Sci. Conf. 1 (1972)
198. Scott S. D.: Mössbauer spectra of synthetic ironbearing sphalerite. Can. Mineral. *10*, N 5 (1971)
199. Seifert F.: Compositional dependence of the hyperfine interaction of Fe^{57} in anthophyllite. Phys. Chem. Minerals. *1*, 43–52 (1977)
200. Seregin P. P., Savin E. P.: Mössbauer effect on impurity atomes I^{129} in crystals of $A_2^{III}Te_3$, $A^{II}Te$ and $A_2^{V}Te_3$ types. Physica tverd. tela *14*, 1798–1800 (1972)
201. Shimada M., Miyamoto H., Kanamaru F., Koizumi M.: Mossbauer effect and its application to mineralogy. J. Mineral. Soc. Jpn. *10*, 186–214 (1971)
202. Siegwarth J. D.: Mössbauer effect of divalent Fe^{57} in NiO and MnO. Phys. Rev. *155*, 285–296 (1967)
203. Singh S. K., Bonardi M.: Mössbauer resonance of arfvedsonite and aegirine-augite from the Joan Lake agpaitic complex, Labrador. Lithos *5*, 217–225 (1972)
204. Smith D. L., Zuckerman J. J.: ^{119}Sn Mössbauer spectra of tin-containing minerals. Nucl. Chem. *29*, N 5 (1967)
205. Sorokin V. J., Novikov G. V., Egorov V. K., Popov V. I., Sipavina L. V.: Mössbauer study of iron-containing sphalerites. Geochimia *9*, 1329–1336 (1975)
206. Sprenkel-Segel E. L., Hanna S. S.: Mössbauer analysis of iron in stone meteorites. Geochim. Cosmochim. Acta *28*, 1913–1931 (1964)
207. Sprenkel-Segel E. L.: Recoilless resonance spectroscopy of meteoritic iron oxides. J. Geophys. Res. *75*, 6618–6630 (1970)
208. Stevens J. G.: Mössbauer studies of antimony minerals. J. Phys. (Fr.) *12*, Suppl., 877 (1976)
209. Strangway D. W., McMahon B. E., Honea R. M., Larson E. E.: Superparamagnetism in hematite. Earth Planet. Sci. Lett. *2*, 367–371 (1967)

210. Stampfl P. P., Travis J. .C, Bielefeld M. J.: Mössbauer spectroscopic studies iron-doped rutile. Phys. Status Solidi (a) *15*, 181–189 (1973)

211. Stuart R. A., Donohoe A. J., Boyle A. J. F.: Determination of tin in cassiterite ores by an application of the Mossbauer effect. Proc. Aust. Inst. Min. Metallurgy *230*, 69–72 (1969)

212. Taylor G. L., Ruotsala A. P., Keeling R. O., Jr.: Analysis of iron in layer silicates by Mössbauer spectroscopy. Clays Clay Mineral. *16*, 381–391 (1968)

213. Temperley A. A., Lefevre H. W.: The Mössbauer effect in marcasite structure iron compounds. J. Phys. Chem. Solids *27*, 85–92 (1966)

214. Trooster J. M.: Mössbauer investigation of ferro-electric boracites. Phys. Stauts Solidi *32*, 179–185 (1969)

215. Tsang T., Thorpe A. N., Donnay G., Senttle F. E.: Magnetic susceptibility and tri-angle exchange coupling in the tourmaline mineral group. J. Phys. Chem. Solids *32*, N 7 (1971)

216. Tsay Fun-Dow, Manatt St. L., Chan S. I.: Magnetic phases in lunar fines: metallic Fe or ferric oxides. Geochim. Cosmochim. Acta *37*, 120–11211 (1973)

217. Valter A. A., Gorogotskaya L. I., Zverev N. D., Romanov V. P.: Two types of iron distribution in structure of pyroxenes of hedenbergite type (according to Mössbauer spectroscopy data). Dokl. Akad. Nauk *192*, 629–632 (1970)

218. Vangham D. L., Ridout M. S.: Mössbauer studies of some sulfide minerals. J. Inorg. Nucl. Chem. *33*, N 3 (1971)

219. Vereshchak M. F., Zhetbaev A. K., Kaipov D. K., Satpaev K. K.: Mössbauer effect on impurity Fe^{57} atoms in quartz single-crystals. Physica tverd tela *14*, 3082–3083 (1972)

220. Virgo D., Hafner St.: Re-evaluation of the cation distribution in orthopyroxenes by the Mössbauer effect. Earth Planet. Sci Lett. *4*, 265–269 (1968)

221. Virgo D., Hafner St. S.: Fe^{2+}, Mg order-disorder in heated orthopyroxenes. Mineral. Soc. Am. Spec. Pap. 67–81 (1969)

222. Virgo D., Hafner S. S.: Temperature-dependent Mg, Fe distribution in a lunar olivine. Earth Planet Sci. Lett. *14*, N 3 (1972)

223. Vlasov A. Ya., Gornuskina N. A.: Temperature transformation of lepidocrocite into hematite. Zhapisri Vses. Mineral. obsch. *101*, 313–317 (1972)

224. Vosnyuk P. O., Dubinin V. N.: Magnetic structure of ultrafine antiferromagnetic particles of β-FeOOH. Physica tverd. tela *15*, 1897–1899 (1973)

225. Warner B. N., Shive P. N., Allen J. L., Terry C.: A study of the hematite-ilmenite series by the Mössbauer effect. J. Geomagn. Geoelectre. *24*, 353–367 (1972)

226. Weaver C. E., Wampler J. M., Pecuil T. E.: Mössbauer analysis of iron in clay minerals. Science *156*, 504–508 (1967)

227. Wegener H.: Der Mössbauer Effect und seine Anwendung in der Physik und Chemie. Mannheim: 1965

228. Wertheim G. K.: Mössbauer Effect. London-New York: Academic Press 1964

229. Wertheim G. K., Remeika J. P.: Mössbauer effect hyperfine structure of trivalent Fe^{57} in corundum. Phys. Lett. *10*, 14–15 (1964)

230. Whitfield H. J., Feeman A. G.: Mössbauer study of amphiboles. J. Inorg. Nucl. Chem. *29*, 903–914 (1967)

231. Williams P. G. L., Bancroft G. M., Bown M. G., Turnock A. C.: Anomalous Mössbauer spectra of C2/c clinopyroxenes. Nature Phys. Sci. *230*, 149–151 (1971)

232. Wintenberger M.: Etude electrique et magnetique de composes sulfures et arsenies d'éléments de transition. Bull. Soc. Mineral. Cristallogr. *85*, 107–119 (1962)

233. Yakovlev V. V., Bashkirov Sh. Sh., Manapov R. A., Penkov I. N.: Insertion forms of iron in caustobiolites. Izv. Akad. Nauk, Ser. Geol. *1*, 131–134 (1977)

234. Zaporozhec V. M., Polosin A. V., Tkacheva T. V., Pasova F. G.: Nuclear gamma resonance study of iron oxide and hydrooxide minerals in bauxites. In: New Data in Geology of Bauxites, 1975, Vol II, 68–73.

235. Zverev N. D., Valter A. A., Romanov V. P., Gorogotskaya L. L.: Character of Fe^{2+} ion distribution in pyroxenes from eulysite. Lithos 4 (1971)

X-Ray and Electron Spectroscopy

236. Abelard P., Gabis V.: Principe et applications de la methode ESCA. L'Actualité Chimique; Octobre (1975)
237. Adams I., Thomas J. M., Bancroft G. M.: An ESCA study of silicate minerals. Earth Planet. Sci. Lett. *16*, 429–432 (1972)
238. Albee A. A., Chodos A.: Semiquantitative electron microprobe determination of Fe^{2+}/Fe^{3+} and Mn^{2+}/Mn^{3+} in oxides and silicates and its application to petrologic problems. Am. Mineral. *55*, 491–501 (1970)
239. Atkins A. J., Misell D. L.: Electron energy loss spectra for members of the mica group and related sheet silicates. J. Phys. C. Solid State Phys. *5*, 3153–3160 (1972)
240. Barinskii R. L., Nefedov V. I.: X-Ray Spectroscopy Determination of Atom Charges in Molecules. Moscow: Nauka, 1966
241. Barinskii R. L., Kulikova I. M.: Metamict transformations in some niobates and zircons according to X-ray absorption spectra. Phys. Chem. Mineral. *1*, (1977)
242. Birks L. S.: X-Ray Spectrochemical Analysis. New York: Interscience, 1969
243. Brundle C. R., Neumann D., Price W. C., Evam D., Potts A. W., Streets D. G.: Electronic structure of NO_2 studied by photoelectron and vacuum-UV spectroscopy and Gaussian orbital calculations. J. Chem. Phys. *53*, 705–715 (1970)
244. Burhop E. H. S.: The Auger Effect and Other Radiationless Transitions. Cambridge: The University Press, 1952
245. Cherkashenko V. M., Kurmaev E. Z., Fotiev A. A., Volnov V. L.: X-ray K-emission spectra of vanadium in its oxides. Physica tverd tela *17*, 280–285 (1975)
246. Compton A., Allison S.: X-Rays in Theory and Experiment. New York: Van Nostrand, 1943
247. Collins G. A. D., Cruickshanu D. W. J., Breeze A.: Ab initio calculations on the silicate ion, orthosilicic acid and their $L_{2,3}$ X-ray spectra. J. Chem. Soc., Faraday Trans. II 68, 1189–1195 (1972)
248. Coulson C. A., Zauli C.: The K_α transition in compounds of sulphur. Mol. Phys. *6*, 525–533 (1963)
249. Delgass W. N., Hughes T. R., Fadley C. S.: X-ray photoelectron spectroscopy: a tool for research in catalysis. Catalysis Rev. *4*, 179–219 (1970)
250. Dikov Yu. P., Debolsky E. I., Romashenko Yu. N., Dolin S. P., Levin A. A.: Molecular orbitals of $Si_2O_7^{6-}$, $Si_3O_{10}^{8-}$ etc. and mixed (B, Al, P, Si)$_m$ applied to clusters and X-ray spectroscopy data of silicates. Phys. Chem. Mineral. *1*, 27–41 (1977)
251. Dikov Yu. P., Nemoshkalenko V. V., Aleshin V. G., Ivanov A. V., Bogatikov O. A.: Reduced titanium in lunar regolith. Dokl. Akad. Nauk *234*, 176–179 (1977)
252. Dodd Ch. G., Glen G. L.: A survey of chemical bonding is silicate minerals by X-ray emission spectroscopy. Am. Mineral. *54*, 1299–1311 (1969)
253. Dodd Ch. G., Glen G. L.: Studies of chemical bonding in glasses by X-ray emission spectroscopy. J. Am. Ceram. Soc. *53*, 322–325 (1970)
254. Electron Probe Microanalysis: Tousmis, A. J., Marton. L. (eds.) Adv. in Electronics and Electron. Physics, Suppl. *6*, 1969
255. Electron Spectroscopy: Shirley, D. A. (ed.) Amsterdam - London: North-Holland, 1972
256. Eljashevich M. A.: Atomic and Molecular Spectroscopy. Moscow: Phys-Math. isdat, 1962
257. Ershov O. A., Goganov D. A., Lukirsky A. P.: Studies of the X-ray spectra of silica in cristalline and glassy quartz and in the litiumsilicate glass. Phys. tverd. tela 7, 2355–2361 (1965)

258. Fichter M : Über die Bindungsabhändigkeit des K-Röntgenemissionsspektrums von Phosphor. Inaugural-Dissertation. Dresden, 1966

259. Fischer D. W.: Effect of chemical combination on the X-ray K emission spectra of oxygen and fluorine. J. Chem. Phys. *42*, 3814–3821 (1965)

260. Fischer D. W.: Molecular-orbital interpretation of the soft X-ray $L_{II,III}$ emission and absorption spectra from titanium and vanadium compounds. J. Appl. Phys. *41*, 3561–3569 (1970)

261. Fischer D. W.: Soft X-ray band spectra and molecular orbital structure of Cr_2O_3, CrO_3, CrO_4^{2-} and $Cr_2O_7^{2-}$. Phys. Chem. Solids. *32*, 2455–2480 (1971)

262. Fischer D. W.: Use of soft X-ray band spectra for determining molecular orbital structure. Vanadium octahedral and tetrahedral sites. Appl. Spectrosc. *25*, 263–276 (1971)

263. Fomichev V. A.: Ultrasoft X-ray spectroscopy study of energy band structure of Al_2O_3 and AlN. Phys. tverd. tela, *10*, 763–768 (1968)

264. Freund F.: The X-ray K emission band of magnesium from $Mg(OH)_2$ single crystals. Phys. Status Solidi *66*, 271–278 (1974)

265. Glen G. L., Dodd Ch. G.: Use of molecular orbital theory of interpret X-ray K-absorption spectral data. J. Appl. Phys. *39*, 5372–5377 (1968)

266. Gold T., Bilson E., Baron R. L.: Auger analysis of the lunar soil: Study of processes which change the surface chemistry and albedo. Proc. 6th Lunar Sci. Conf. *3*, 3285–3304 (1975)

267. Handbook of X-Rays for Diffraction, Emission, Absorption and Microscopy: Kaelbe, E. F. (ed.) New York, McGraw Hill 1967

268. Hermes O. D., Ragland P. C.: Quantitative chemical analysis of minerals in thin-section with the X-ray macroprobe. Am. Mineral. *52*, 493–508 (1967)

269. Housley R. M., Grant R. W.: ESCA studies of lunar surface chemistry. Proc. 6th Lunar Sci. Conf. *3*, 3269–3276 (1975)

270. Ivanov A. V., Timakov L. D., Kuprianov V. N.: X-ray $L_{2,3}$ quantum yield spectra of sulfur photoemission in sulfides and sulfates. Opt. Spectrosc. *31*, 317–318 (1971)

271. Jacob L.: The Solid State: X-Ray Spectroscopy. London: Butterworths, 1974

272. Kolobova K. M., Nemnonov S. A., Agapova E. V.: X-ray FeK_β spectra of iron oxides. Phys. tverd. tela *10*, 729–732 (1968)

273. Kulikova I. M., Barinskii R. L., Alexandrov V. B., Proschenko E. G.: X-ray absorption spectra of niobium in some metamict minerals. Dokl. Akad. Nauk *210*, 1423–1426 (1973)

274. Laputina I. P.: Study of K_α aluminium line shift in minerals. In: X-Ray Spectra and Electron Structure of Matter 2, 281–293, Kiev, 1969

275. Läuger K.: Über den Einfluß der Bindungsart und der Kristallstruktur auf das K-Röntgenemissionsspektrum von Aluminium und Silizium. Inaug. Diss. München, 1968

276. Leroux J.: Method for finding mass-absorption coefficients by empirical equations and graphs. Adv. X-Ray Anal *5*, 153–160 (1962)

277. Long J. V. P.: Electron Probe Microanalysis. In: Physical Methods in Determinative Mineralogy, Zussman, J. (ed.) London - New York; Academic Press, 1967

278. Malissa H., Grasserbauer M.: Untersuchungen an Röntgenemissionsprofilen von leichten Elementen mit der "Makrosonde". Monatsschr. Chem. B. *102*, 1545–1557 (1971)

279. Mattson R. A., Ehlert R. C.: The application of a soft X-ray spectrometer to study the oxygen and fluorine emission lines from oxides and fluorides. Adv. X-Ray Anal. *9*, 471–486 (1966)

280. Menshikov A. Z., Brytov I. A., Kurmaev E. Z.: Crystal-field splitting of levels and X-ray spectra of transition metal monoxides. Phys. Status Solidi *35*, 89–93 (1969)

281. Miller A.: Determination of the valence state of copper in cubic $CuMn_2O_4$ spinel by X-ray absorption edge measurements. Phys. Chem. Solids *29*, 633–639 (1968)

282. Moore A. C.: A method for determining mineral compositions by measurement of the mass absorption coefficient. Am. Mineral. *54*, 1180–1189 (1969)
283. Narbut K. I.: X-ray $K_{\alpha 1,2}$ and K_β spectra of sulfur atoms in minerals and some compounds. Izv. Akad. Nauk, Ser. Phys. *38*, 548–561 (1974)
284. Nemoshkalenko V. V., Aleshin V. G.: Theoretical Basis of X-Ray Emission Spectroscopy. Kiev: Naukova dumka, 1974
285. Nemoshkalenko V. V., Aleshin V. G.: Electron Spectroscopy of Crystals. Kiev: Naukova dumka, 1976
286. Nefedov V. I.: Bestimmung von Atomladungen in Molekülen mit Hilfe der Röntgenemissionspektren. Phys. Status Solidi *2*, 904–922 (1962)
287. Nefedov V. I., Narbutt K. I.: Electronic structure of K_2PdCl_6 and K_2PdCl_4 from X-ray spectroscopy data. Zhurn. Strukt. Khim. *12*, 1019–1025 (1971)
288. Nefedov V. I., Urusov V. S., Kakhana M. M.: X-ray electron study of the chemical bond in Na, Mg, Si minerals. Geochimia *1*, 11–19 (1972)
289. Nefedov V. I.: Electronic structure of free molecules and isolated groups in crystals from X-ray spectroscopy data. Zhurn. Strukt. Khim. *13*, 352–372 (1972)
290. Nefedov V. I.: Applications of X-Ray Electron Spectroscopy in Chemistry. Moscow: Vinity, 1973
291. Norrish K., Chappell B. W.: X-ray fluorescence spectrography. In: Phys. Methods in Determinat. Mineralogy. London - New York: Academic Press, 1967
292. O'Nions R. K., Smith D. F. W.: Investigation of the $L_{II,III}$ X-ray emission spectra of Fe by electron microprobe. Am. Mineral. *56*, 1452–1463 (1971)
293. Pantelides S. T., Harrison W. A.: Electronic structure, spectra, and properties of 4:2 coordinated materials. I. Crystalline and amorphous SiO_2 and GeO_2. Phys. Rev. B*13*, 2667–2691 (1976)
294. Parilis E. S.: Auger Effect. Tashkent: FAN, 1969
295. Raymond M., Virgo D.: X-ray photoelectron spectroscopy study of sillimanite (Al_2SiO_5). Carnegie Inst. Annu. Rept. Dir. Geophys. Lab., 1971–1972. Washington, 504–506, 1972
296. Regler F.: Einführung in die Physik der Röntgen und Gammastrahlen. München: Thienig, 257, 1967
297. Reynolds R. C.: Matrix corrections in trace elements analysis by X-ray fluorescence: estimation of the mass absorption coefficient by Compton scattering. Am. Mineral. *48*, 1133–1143 (1963)
298. Petrovich E. V., Smirnov Yu. P., Zykov V. S., Grushko A. I., Sumbaev O. I., Band I. M., Trzhavskovskaya M. B.: Chemical shifts of X-ray $K_{\alpha 1,2}$, $K_{\beta 1,3}$ and $K_{\beta 2,4}$ lines in heavy elements connected with s, p, d, or f valent electrons. Zhurn. Eksp. Teor. Phys. *61*, 1756–1768 (1971)
299. Petrovich R., Berner R. A., Goldhaber M. B.: Rate control in dissolution of alkali feldspars. I. Study of residual feldspar grains by X-ray photoelectron spectroscopy. Geochim. Cosmochim. Acta *40*, 537–548 (1976)
300. Ribble T. J.: L_{II} and L_{III} emission spectra of copper compounds. Phys. Status Solidi (a) *6*, 473–478 (1971)
301. Riviere J. C.: Auger spectroscopy and its applications to some technical problems. Bull. Soc. Tr. Mineral. Cristallogr. *94*, 187–194 (1971)
302. Röntgenspektren und Chemische Bindung: Leipzig, 1966
303. Seka W., Hanson H. P.: Molecular orbital interpretation of X-ray absorption edges. J. Chem. Phys. *50*, 344–350 (1969)
304. Senemand C., Costalima M. T., Roger J. A., Cachard A.: X-ray K absorption spectra of silicon in Si, SiO and SiO_2. Chem. Phys. Lett. *26*, 431–433 (1971)
305. Sevier K. D.: Low Energy Electron Spectroscopy. New York: Wiley Interscience, 1972
306. Shuvaev A. T.: Determination of ion charge in compounds of third row elements from X-ray emission spectra. Izv Akad. Nauk, Ser. Phys. *28*, 758–764 (1964)

307. Siegbahn M.: Spektroskopie der Röntgenstrahlen. Berlin, 1931
308. Siegbahn K., Nordling C., Fahlman A., Nordberg E., Hamrin K., Hedman J., Johansson G., Bergmark T., Karlsson S.-E., Lindgren J., Lindberg B.: ESCA. Atomic, Molecular and Solid State Structure Studies by Means of Electron Spectroscopy. Nova Acta Regiae Societatis Scientiarum Upsaliensis, Ser. IV, v. 20, 1967
309. Siegbahn K., Nordling C., Johansson G., Hedman J., Heden P. F., Hamrin K., Gelius U., Bergmark T., Werme L. O., Manne R., Baer Y.: ESCA Applied to Free Molecules. Amsterdam - London: North-Holland, 1971
310. Siivola J.: The aluminium K_β-band structure of andalusite, sillimanite and kyanite. Bull. Geol. Soc. Finl. *43*, 1–6 (1971)
311. Siivola J.: The specific features of the TiK$_\beta$ spectrum of some titanium bearing compounds. Phys. Jenn. *9*, 111–119 (1974)
312. Smith D. G. W., O'Nions R. K.: Investigations of bonding by oxygen K_α-emission spectroscopy: further evidence concerning the true character of the oxygen K emission band. Chem. Geol. *9*, 145–146 (1972)
313. Sommer Sh.: X-ray photoelectron spectra of C_{1s} and O_{1s} in carbonate minerals. Am. Mineral. *60*, 483–484 (1975)
314. Sumbaev O. I., Smirnov Yu. P., Petrovich E. V., Zykov V. S., Egorov A. I., Grushko A. I.: Chemical shifts of X-ray K_α lines in oxidating of rare-earth metals: role of f electrons. In: X-Ray Spectra and Electronic Structure of the Matter 2, 172–178. Kiev, 1969
315. Sumbaev O. I., Petrovich E. V., Smirnov Yu. P., Egorov A. I., Zykov V. S., Grushko A. I.: Chemical shifts of K_α lines and valent structure of transition metals of 5th and 6th rows. Zh. Exp. Teor. Phys. *53*, 1545–1552 (1967)
316. Thomassin J.-H., Baillif P., Caplapkulu F., Gabis V., Touray J.-C.: Application de l'analyse par spectrometrie d'electron (ESCA) à l'etude des echanges entre mineraux et solutions. C. R. Acad. Sci. Paris D*281*, 1067–1070 (1975)
317. Thomassin J.-H., Goni J., Baillif P., Touray J. C., Jaurand M. C.: An XPS study of the dissolution kinetics of chrysolite in 0.1 N oxalic acid at different temperatures. Phys. Chem. Mineral (1977)
318. Thomas J. M., Evans E. L., Barber M., Swift P.: Determination of the occupancy of valence bands in graphite, diamond and less-ordered carbons by X-ray photoelectron spectroscopy. Trans. Faraday Soc. *67*, 1875–1886 (1971)
319. Thomas S.: Electron-irradiation effect in the Auger analysis of SiO_2. J. App. Phys. *45*, 161–166 (1974)
320. Tossell J. A.: Interpretation of K X-ray emission spectra and chemical bonding in oxides of Mg, Al and Si using quantitative molecular orbital theory. Geochim. Cosmochim. Acta *37*, 583–594 (1973)
321. Tossell J. A.: Molecular orbital interpretation of X-ray emission and ESCA spectral shifts in silicates. Phys. Chem. Solids *34*, 307–319 (1973)
322. Tossell J. A., Vaughan D. J., Johnson K. H.: X-ray photoelectron, X-ray emission and UV spectra of SiO_2 calculated by the SCF X$_\alpha$ scattered wave method. Chem. Phys. Lett. *20*, 329–334 (1973)
323. Urusov V. S.: Relation of X-ray lines shift to atom ionization. Third row elements and their compounds. Dokl. Akad. Nauk *166*, 660–663 (1966)
324. Vainstein E. E.: Methods of Quantitative X-Ray Spectral Analysis. Moscow: Akad. Nauk, 1956
325. Vinogradov A. S., Zimkina T. M., Fomichev V. A.: Special features of X-ray spectra of sulfur and fluorine in SF_6. Zh. Strukt. Khim. *12*, 899–904 (1971)
326. White E. W., Gibbs G. V.: Structural and chemical effects on the SiK$_\beta$ X-ray line for silicates. Am. Mineral. *52*, 985–993 (1967)
327. X-ray Wavelengths and X-ray Atomic Energy Levels. Washington, 1967
328. Yin L., Ghose S., Adler I.: Core binding energy differences between bridging and nonbridging oxygen atoms in a silicate chain. Science *173*, 633–635 (1971)

329. Yin L., Tsany T., Adler I.: ESCA studies on solarwind reduction mechanisms. Proc. 6th Lunar Sci. Conf. *3*, 3277–3284 (1975)
330. Zhmudsky A. Z.: Origin of X-Ray Satellites and Structures of Diagram Lines. Kiev: Naukova dumka, 1966
331. Zimkina T. M., Fomichev V. A.: Ultrasoft X-ray spectroscopy. Leningrad: University edition, 1971

Electron Paramagnetic Resonance (EPR of Electron-Hole Centers See pp. 338–345)

332. Abdulsabirov P. Yu., Vinokurov V. M., Zaripov M. M., Stepanov V. G.: EPR of Fe^{3+} ions in natrolite. Phys. tverd. tela *9*, 689–690 (1967)
333. Abragam A., Bleaney B.: Electron Paramagnetic Resonance of Transition Ions. Oxford: Clarendon Press 1970
334. Alger R. S.: Electron spin resonance in chemistry. Techniques and applications. New York: Interscience Publ, 1968
335. Altshuler S. A., Kozyrew B. M.: Paramagnetische Elektronenresonanz. Leipzig: Teubner, 1963
336. Anufrienko V. F., Yandralova L. G., Tarasova D. V.: EPR spectra and state of Fe^{3+} ions in SnO_2. Phys. tverd. tela *13*, 2353–2356 (1971)
337. Ayscough P. B.: Electron Spin Resonance in Chemistry. London: Methuen, 1971
338. Azarkin V. A., Lushnikov V. G., Sorokina L. P.: Electron paramagnetic resonance of trivalent gadolinium and iron ions in synthetic calcite. Phys. tverd. tela *7*, 2367–2369 (1965)
339. Baker J. M., Bleaney B., Hayes W.: Paramagnetic resonance of S-state ions in calcium fluoride. Proc. R. Soc. A*247*, 141–151 (1958)
340. Ball D.: Paramagnetic resonance of Er^{3+} and Dy^{3+} in natural single cristals of zircon. Phys. Status Solidi (b) *46*, 635–641 (1971)
341. Barry W. R., Troup G. J.: EPR of Fe^{3+} ions in chrysoberyl. Phys. Status Solidi *38*, 229–234 (1970)
342. Barry W. R., Holuj F.: ESR spectrum of Fe^{3+} in topaz. III. ENDOR of ^{19}F. Can. J. Phys. *51*, 95–101 (1973)
343. Benedek G. B.: Magnetic resonance at high pressure. New York, 1963
344. Bershov L. V., Marfunin A. S.: Estimation of the chemical bond state from Mn^{2+} hyperfine structure of EPR spectra. Dok. Akad. Nauk *166*, 632–635 (1964)
345. Bershov L. V., Marfunin A. S.: Vanadile ion in minerals Izv. Akad. Nauk, Ser. Geol. *9*, 42–52 (1965)
346. Bershov L. V., Marfunin A. S., Mineeva R. M.: Electron paramagnetic resonance of tetrahedral complex $/MnF_4/^{2-}$ in sheelite. Zh. Exp. Teor. Phys. *49*, 743–746 (1965)
347. Bershov L. V., Marfunin A. S., Mineeva R. M.: EPR of Mn^{2+} in apophyllite. Dokl. Adad. Nauk *164*, 1141–1142 (1965)
348. Bershov L. V., Marfunin A. S., Mineeva R. M.: EPR of Mn^{2+} in tremolite. Geochimia *4*, 464–466 (1966)
349. Bershov L. V., Vinokurov V. M., Zaripov M. M., Stepanov V. G., Kropotov V. S.: EPR of Mn^{2+} in datolite. Geochimia *1*, 122–123 (1966)
350. Bershov L. V., Mineeva R. M., Nussik Ya. M.: Forms of copper insertion in some minerals. Geochimia *12*, 1398–1400 (1968)
351. Bershov L. V., Mineeva R. M., Tarashchan A. N.: EPR and luminescence of d^9 ions in single crystals of calcite. Exp. Teor. Chim *6*, 395–397 (1969)
352. Bersohn M., Baird J. C.: An Introduction to Electron Paramagnetic Resonance. New York - Amsterdam: Benjamin 1966
353. Boesman E., Schoemaker D.: Resonance paramagnetique de l'ion Fe^{3+} dans la kaolinite. C. R. Acad. Sci., Paris, 1931–1933 (1961)
354. Brun E., Hafner S., Loeliger H., Waldner F.: Zur paramagnetischen Resonanz von Cr^{3+} in Spinell $(MgAl_2O_4)$, Helv. Phys. Acta *33*, 966–968 (1960)

355. Burley S. P., Troup G. J.: Paramagnetic resonance of Fe^{3+} in benitoite. Br. J. Appl. Phys. *16*, 315–318 (1965)

356. Burns G.: Concentration-dependent electron spin resonance. Phys. Rev. *135*A, 479–481 (1964)

357. Carrington A., McLachlan A. D.: Introduction to Magnetic Resonance. New York: Harper and Row, 1967

358. Carter D., Okaya A.: EPR of Fe^{3+} in TiO_2 (rutile). Phys. Rev. *118*, N 4 (1960)

359. Chatelain A., Weeks R. A.: Electron paramagnetic resonance study of ordered Mn^{2+} in Mg_2SiO_4. J. Chem. Phys. *52*, 5682–5687 (1970)

360. Chen I., Kikuchi C., Watanabe H.: Superhyperfine structures in ESR and ENDOR of cubic $CdTe: Mn^{2+}$. J. Chem. Phys. *42*, 189–190 (1965)

361. Danilov A. G., Manoogian A.: Electron spin resonance of Mn^{2+} impurities in monticellite. Can. J. Phys. *47*, 839–846 (1969)

362. Dmitrieva L. V., Zonn Z. N., Ioffe V. A.: NMR Li^7 and EPR Fe^{3+} spectra in synthetic eucryptite single crystal. Neorg. Mater. *5*, 1269–1272 (1969)

363. Denning R. M., Poindexter E. H.: Crystallographic implications of EPR in neutron-irradiated diamond. Am. Mineral. *49*, 277–285 (1964)

364. Depireux I., Duchesne I., Kaa I. M. van der: Resonance electronique paramagnetique dans les fossiles vegetaux. J. Chim. Phys. Phys.-Chim. Biol. *56*, 810–811 (1959)

365. Diza J., Farach H. A., Poole C. P., Jr.: An electron spin resonance and optical study of turquoise. Am. Mineral. *56*(773–781 (1971)

366. Dickinson A. C., Moore W. J.: Paramagnetic resonance of metal ions and defect centers in topaz. J. Phys. Chem. *71*, 231–240 (1967)

367. Donner J. C., Ranon U., Stamires D. N.: Hyperfine, superhyperfine and quadrupole interactions of Gd^{3+} in YPO_4. Phys. Rev. B*3*, 2141–2149 (1971)

368. Dvir M., Low W.: Paramagnetic resonance and optical spectrum of iron in beryl. Phys. Rev. *119*, 1507–1591 (1960)

369. Dzionara M., Kahle H. G., Schedewie F.: Paramagnetic resonance of Er^{3+} in YPO_4. Phys. Status Solidi (b) *47*, 135–136 (1960)

370. From W. H. Electron paramagnetic resonance of Cr^{3+} in SnO_2. Phys. Rev. *131*, 961–964 (1963)

371. Gainon D., Lacroix R.: Electron paramagnetic resonance of Fe^{3+} ion in anatase. Proc. Phys. Soc. *79*, 658–659 (1962)

372. Gaite J. M., Michoulier J.: Application de la résonance paramagnétique électronique de l'ion Fe^{3+} a l'étude de la structure des feldspaths. Bull. Soc. Franc. Mineral. Cristallogr. *93*, 341–356 (1970)

373. Gaite J. M.: Etude de propriétés locales dans les cristaux de basse symmetrie à l'aide de la résonance paramagnétique électronique d'ions à l'état S. Thèse, Université d'Orleans, 1973

374. Gavrilov I. A., Litovkina L. P., Meilman M. L.: EPR of impuirty iron ions in diaspore (α-AlOOH) single crystals. Phys. tverd. tela *10*, 2765–2770 (1968)

375. Geusic J. E., Peter M., Schulz-DuBois E. O.: Paramagnetic resonance spectrum of Cr^{3+} in emerald. Bell System Techn. J. *38*, 291–296 (1959)

376. Ghose S., Schindler P.: Determination of the distribution of trace amounts of Mn^{2+} in diopsides by electron paramagnetic resonance. Mineral. Soc. Am. Spec. Pap. 51–58 (1969)

377. Gilinskaya L. G., Scherbakova M. Ya: Isomorphous substitutions and structure defects in apatite according to electron paramagnetic resonance data. In: Physics of Apatite (Spectroscopy Studies of Apatite), 7–63. Novosibirsk: Nauka, 1975

378. Golding R. M., Newman R. H., Rae A. D., Tennant W. C.: Single crystal ESR study of Mn^{2+} in natural tremolite. J. Chem. Phys. *57*, 1912–1918 (1971)

379. Goodman B. A., Raynor J. B.: Electron spin resonance of transition metal complexes. Adv. Inorg. Chem. Radiochem. *13*, 135–362 (1970)

380. Grunin V. S.: EPR of V^{4+} ion in crystobalite. Phys. tverd. tela *12*, 2243–2238 (1970)

381. Grunin V. S., Ioffe V. A., Zonn Z. N.: EPR study in the SiO_2 system. J. Non-Cryst. Solids *11*, 341–349 (1973)

382. Grunin C. S., Davtian G. D , Ioffe V. A., Patrina I. B.: EPR of Cr^{3+} in anatase. Phys. tverd. tela *17*, 2174–2176 (1975)

383. Grunin V. S., Davtian G. D., Ioffe V. A., Patrina I. B.: EPR of Cu^{2+} and radiation centers in anatase (TiO_2). Phys. Status. Solich (b) *77*, 85–92 (1976)

384. Henning J. C. M., Liebertz J., Van Stapele R. P.: Evidence for Cr^{3+} in four-coordination: ESR- and optical investigation of Cr-doped $AlPO_4$-crystals. J. Phys. Chem. Solids, *28*, 1109–1114 (1967)

385. Hillmer W.: Paramagnetic resonance of Yb^{3+} and Nd^{3+} in YPO_4. Phys. Status Solidi (b) 133–134 (1971)

386. Holuj F., Thyer J. R., Hedgecock N. E.: ESR spectra of Fe^{+3} in single crystals of andalusite. Can. J. Phys. *44*, 509–523 (1966)

387. Holuj F.: EPR of Mn^{++} in spodumene. I. Natural crystals. Can. J. Phys. *46*, 287–302 (1968)

388. Holuj F., Manoogian A.: EPR of Mn^{++} in spodumene. II. Heated crystals. Can. J. Phys. *46*, 303–306 (1968)

389. Horn M., Schwerdtfeger C. F.: EPR of substitutional and charge compensated Fe^{3+} in anatase (TiO_2). J. Phys. Chem. Solids *32*, 2529–2538 (1971)

390. Howling D. H.: Correlations of electron spin parameters with Fe^{3+} sites in topaz. Phys. Status. Solidi. *40*, 667–676 (1970)

391. Hubin R.: La thorite métamicte: application de la résonance paramagnétique électronique (R.P.E.) à l'étude de la recristallisation. Bull. Soc. Fr. Mineral. Cristallogr. *97*, 417–421 (1974)

392. Hutton D. R.: Paramagnetic resonance of Fe^{3+} in amethyst and citrine quartz. Phys. Lett. *12*, 310 (1964)

393. Hutton D. R., Troup G. J.: Paramagnetic resonance of Gd^{3+} in zircon. Br. J. Appl. Phys. *15*, 405–406 (1964)

394. Hutton D. R., Troup G. J.: Paramagnetic resonance of Cr^{3+} in kyanite. Br. J. Appl. Phys. *15*, 275–280 (1964)

395. Hutton D. R.; Paramagnetic resonance of VO^{2+}, Cr^{3+} and Fe^{3+} in zoisite. J. Phys.; C: Solid State Phys. *4*, 1251–1257 (1971)

396. Hutton D. R. Le Marshall J., Troup G. J.: Paramagnetic resonance of Fe^{3+} in adamite. Phys. Status Solidi (a), *14*, K147–K148 (1972)

397. Ingram D. J. E.: Spectroscopy at Radio and Microwave Frequencies. London: Butterworths, 1967

398. Ibers J. A., Swalen J. D.: Paramagnetic resonance line chapes and magnetic parameters of polycrystalline substances. Phys. Rev. *127*, 1914–1917 (1962)

399. Ja Y. H.: Electron paramagnetic resonance of Fe^{3+} and Mn^{2+} in natural single crystals of petalite, $LiAlSi_4$. Aust J. Phys. *23*, 299–310 (1970)

400. Kedzie R. W., Lyons D. H., Kestigian M.: Paramagnetic resonance of the Fe^{3+} ion in $CaWO_4$ (strong tetragonal crystal field). Phys. Rev *138*, 918–924 (1965)

401. Kemp R. C.: Electron spin resonance of Fe^{3+} in phlogopite .J. Phys. C:Solid State Phys. *5*, 3566–3572 (1972)

402. Kemp R. C.: Electron spin resonance of Fe^{3+} in muscovite. Phys. Status Solidi (b) *57*, K79–K81 (1973)

403. Kiggins B., Manoogian A.: Electron spin resonance of Mn^{2+} impurities in blodites Can. J. Phys. *49*, 3174–3179 (1971)

404. Kneubühl F. K.: Line shapes of electron paramagnetic resonance signals produced by powders, glasses and viscous liquids. J. Chem. Phys. *33*, 1074 (1960)

405. Kokoszka G. F., Gordon G.: Electron paramagnetic resonance .Techn Inorg. Chem. *7*, 151–272 (1968)

406. Kurkin I. N.: EPR of trivalent rare earth ions in homological series of crystals, with $CaWO_4$ structure. In: Paramagnetic Resonance 5, 31–73. Kazan: University edition, 1969

407. Kuska H. Z., Rogers M. T.: Electron spin resonance of first row transition metal complex ions. In: Radical Ions, Kaiser, E. T. Kevan, L. (eds.) 579–745. New York: J. Wiley, 1968

408. Lebedev Ya. S., Chernikov D. M., Tikhomirova N. P.: Atlas of EPR Spectra: Calculated Multicomponent Symmetric Spectra Moscow: Nauka, 1962

409. Lebedev Ya. S : EPR applications in chemical kinetics. Usp. Chim. 37, 934–967 (1968)

410. Low W.: Paramagnetic resonance in solids. Solid State Phys. Suppl. 2 (1960)

411. Low W.: Electron spin resonance. A tool in mineralogy and geology. Adv. Electronics Electron Phys. 24, 51–108, (1968)

412. Low W., Zeira S.: ESP spectra of Mn^{2+} in heat-treated aragonite. Am .Mineral. 57, 1115–1124 (1972)

413. Manoogian A.: The electron spin resonance of Mn^{2+} in tremolite. Can. J. Phys. 46, 129–133 (1968)

414. Manoogian A., Hsu Y.: Electron spin resonance of Mn^{2+} impurities in morinite Can. J. Phys. 47, 1869–1875 (1969)

415. Manoogian A., Kiggins B.: Electron spin resonance of Mn^{2+} impurities in newberyite. Am. Mineral. 57, 52–61 (1972)

416. Marfunin A. S., Bershov L. V.: Application of Electron Paramagnetic Resonance in Mineralogy. Moscow: Viniti, 1964

417. Marfunin A. S.: Radiospectroscopy of minerals. Geol. J. (Liverpool) 4, 361–372 (1965)

418. Marfunin A. S., Bershov L. V., Mineeva R. M.: La résonance paramagnétique électronique de l'ion VO^{2+} dans le sphene et l'apophyllite et de l'ion Mn^{2+} dans la tremolite, l'apophyllite et la scheelite. Bull. Soc. Fr. Mineral. Crist. 89, 177–183 (1966)

419. Marfunin A. S., Bershov L. V., Meilman M. L., Michoulier J.: Paramagnetic resonance of Fe^{3+} in some feldspars. Schweiz. Mineral. Petrogr. Mitt. 47, 13–20 (1967)

420. Marshall S. A.. Serway R. A.: Electron-spin-resonance absorption spectrum of trivalent gadolinium in single crystal calcite. Phys. Rev. 171, 345–349 (1968)

421. Masykin V. V., Matiash I. V., Polshin E. V.: Structure position of Fe^{3+} ions in kaolinites by the electron paramagnetic resonance data. Dokl. Ukr. Adad. Nauk 12, 1066–1070 (1976)

422. Matiash I. V., Bagmut N. N., Respalko N. A., Brik A. B,. Lazarenko N. E., Fedotov Yu. U.: Isomorphous ion-radical NH_3^+ in feldspars. In: Problems of Regional and Genetic Mineralogy, 120–122. Kiev: Naukova dumka, 1977

423. McBride M. B., Mortland M. M., Pinnavaia T. J.: Exchang ion positions in smectite: effects on electron spin resonance of structural iron. Clays Clay Mineral. 23, 162–164 (1975)

424. Meads R. E., Malden P. J.: Electron spin resonance in natural kaolinites containing Fe^{3+} and other transition metal ions. Clay Mineral. 10, 313–345 (1975)

425. Meilman M. L., Samoilovich M. I.: Introduction into EPR Spectroscopy of Activated Single Crystals. Moscow: Atomisdat, 1977

426. Memory J. D.: Quantum Theory of Magnetic Resonance Parameters. New York. McGraw Hill, 1968

427. Michoulier J., Gaite J -M., Maffeo B.: Résonance paramagnétique de l'ion Mn^{2+} dans un monocristal de forsterite. C. R. Acad. Sci. Paris 269, 535–538 (1969)

428. Michoulier J.: Contribution à l'étude des propriétés cristallines des mineraux de basse symmétrie à l'aide de la résonance paramagnétique électronique de l'ion Fe^{3+}: application aux feldspaths. Dissertation; Grenoble 1970

429. Michoulier J., Gaite J. M.: Site assignment of Fe^{3+} in low symmetry crystals. Application to $NaAlSi_3O_8$. J Chem. Phys. 56, 5205–5209 (1972)

430. Mitsuk B. M., Bagmut N .N., Matiash I. V., Fedotov Ya. V.: On the nature of EPR spectra of the black opals from Volyn In: Constitution and Properties of Minerals 8, 105–107, Kiev, 1974

431. Nepsha V· I., Sherstkov Yu. A., Legkikh N. V., Meilman M. L.: ESR spectra of Gd^{3+}-doped $SrMoO_4$ and $CaWO_4$ in external electric field. Phys. Status Solidi. *35*, 627–633 (1969)

432. Neibuhr H. H., Zeira S., Hafner S. S.: Ferric iron in plagioclase crystals from anorthosite 15415 .Proc. 4th Lunar Sci. Conf. *1*, 971–982 (1973)

433. Niebuhr H. H.: Elektronenspin-Resonanz von Dreiwertigem Eisen in Forsterit. Marburg/Lahn: University edution, 1975

434. Nisamutdinov N. M., Bulka G. R., Vinokurov V. M.: Classification of point paramagnetic defect centers in crystals base on the space symmetry groups analysis.In: Composition, Structure and Properties of Minerals. Kazan: University edition, 1973

435. Nisamutdinov N. M., Bulka G. R., Vinokurov V. M.: Structure information from superhyperfine interactions in crystals. In: Composition, Structure and Properties of Minerals. Kazan: University edition, 1973

436 Novozhilov A. I., Samoilovich M. I., Anikin I. N., Matveev S. I.: Electron paramagnetic resonance and optical absorption spectra of fluor-phlogopite with titanium and vanadium impurities. Neorg. Mater. *8*, 1465–1469 (1972)

437. Orton J. W.: Electron Paramagnetic Resonqnce. An Introduction to Transition Group Ions in Crystals. London: Iliffe Books, 1968

438. Pieczonka W. A., Petch H. E., McLay A. B.: An electron spin resonance study of Mn impurity in brucite. Can. J. Phys. *39*, 145–157 (1961)

439. Pifer J. H.: Magnetic resonance of Mn^{2+} in PbS, PbSe and PbTe, Phys. Rev. *157*, 272–276 (1967)

440. Piper W. W., Prener J. S : Electron-paramagnetic-resonance study of Mn^{2+} in calcium chlorophosphate. Phys. Rev. *6*, 2547–1554 (1972)

441. Poole Ch. P.: Electron spin resonance. A Comprehensive Treatise on Experimental Technique. New York: Interscience Publishes, 1967

442. Räuber A., Schneider J.: Paramagnetische Resonanz von Fe^{3+}-Ionen in synthetischen kubischen ZnS-Kristallen. Z. Naturforsch. *17*a, N 3, 266–270 (1962)

443. Retcofsky H. L., Sharkey A. G., Jr., Friedel R. A.: Electron paramagnetic resonance of coals during electron irradiation. Fuel *46*, 109–114, 155, 157 (1967)

444. Rhein W., Rosinski Ch.: ENDOR investigations on the superhyperfine interaction of Fe^{3+} in SnO_2. Phys. Status. Solidi. (b) *49*, 667–672 (1972)

445. Richardson R. J., Sock Lee, Menne T. J.: Electron spin resonance of Mn^{2+} in BaF_2, CdF_2 and CaF_2. Phys. Rev. *6*, 1065–1066 (1972)

446. Rostworowski J. A., Horn M., Schwerdtfeger C. F.: EPR of substitutional Fe^{3+} in TiO_2 (brookite). J. Phys. Chem. Solids *34*, 231–234 (1973)

447. Samoilovich M. I., Meilman M. L., Novozhilov A. I.: EPR of Fe^{3+} in topaz. Proc. Vses. Inst. Miner. Materials (VIMS) *11*, 121–124 (1970)

448. Scala C. M., Hutton D. R.: Site assignment of Fe^{3+} in quartz. Phys. Status Solidi. (b) *73*, K115–K117 (1976)

449. Schindler P., Ghose S.: Electron paramagnetic resonance of Mn^{2+} in dolomite and magnesite, and Mn^{2+} distribution in dolomites. Am. Mineral. *55*, 1889–1896 (1970)

450. Schneider J., Sircar S. R.: Paramagnetische Resonanz von Mn^{2+} Ionen in synthetischen und natürlichen ZnO - Kristallen. Z. Naturforsch. *17*a, 570–577, 651–654 (1962)

451. Schneider J., Sircar S. R., Räuber A.: Electronen-Spin-Resonanz von Mn^{2+} - Ionen im kubischen und trigonalen Kristallfeld des ZnS. Z. Naturforsch. *18*a, 8/9, 980–993 (1963)

452. Shcherbakova M. Ya., Istomin V. E.: Calculation of EPR spectra of Fe^{3+} with high zero-field splitting in polycrystalline materials. Phys. Status. Solidi. *67*, 461–469 (1975)

453. Shuskus A. J.: Electron spin resonance of Fe^{3+} and Mn^{2+} in single crystals of CaO. Phys. Rev. *127*, N 5 (1962)

454. Shuskus A. J.: Electron spin resonance of Gd^{3+} and Eu^{2+} in single crystals of CaO. Phys. Rev. *127*, N 6 (1962)

455. Simanek E., Müller K. A.: Covalency and hyperfine structure constant A of iron group impurities in crystals. J. Phys. Chem. Solids *31*, 1027 (1970)
456. Solntsev V. P., Shcherbakova M. Ya., Sotnikov V. I.: EPR of Gd^{3+} in natural sheelite. Geol. Geoph. (Novosibirsk) *2*, 125–127 (1969)
457. Slichter C. P.: Principles of Magnetic Resonance. London - New York: Harper and Row, 1963
458. Standley K. J., Vaughan R. A.: Electron Spin Relaxation Phenomena in Solids. London: Hilger, 1969
459. Tikhomirova N. N., Dobryanov S. N., Nikolaeva I. V.: The calculation of ESR spectrum of Mn^{2+} ions in polycrystalline samples. Phys. Status Solidi. *10*, 593–603 (1972)
460. Thyer J. R., Quick S. M., Holuj F.: ESR spectrum of Fe^{+3} in topaz. Can. J. Phys. *45*, 3597–3610 (1967)
461. Title R. S., Sorokin P. P., Stevenson M. J., Pettit G. D., Scardefield J. E., Lanhard J. R.: Optical spectra and paramagnetic resonance of U^{4+} ions in alkaline earth fluoride lattices. Phys. Rev. *128*, 1 (1962)
462. Title R. S.: Paramagnetic resonance measurements on rare earth doped II-VI compounds. Proc. Int. Conf. Luminescence (Budapest, 1966), *2*, 1576–1587 (1968)
463. Troup G. I., Rutton D. R.: Paramagnetic resonance of Fe^{3+} in kyanite. Br. J. Appl. Phys. *15*, 1493–1499 (1964)
464. Valishev R. M., Vinokurov V. M., Zaripov M. M., Stepanov V. G.: EPR of Er^{3+} ions in zircon crystals. Geochimia *10*, 1265–1266 (1965)
465. Vinokurov V. M., Zaripov M. M., Stepanov V. G., Polsky Yu. E., Chirkin G. K., Shekun L. Ya.: EPR in natural chrysoberyl. Phys. tverd. tela *3*, 2475–2480 (1961)
466. Vinokurov V. M., Zaripov M. M., Polsky Yu. E., Stepanov V. G., Chirkin G. K.: Paramagnetic resonance of Cr^{3+} in andalusite. Phys. tverd. tela *4*, 646–649 (1962)
467. Vinokurov V. M., Zaripov M. M., Polsky Yu. E., Stepanov V. G., Chirkin G. K., Shekun L. Ya.: EPR study of isomorphous insertion of Fe^{3+} ions in andalusite. Kristallographia *7*, 318–320 (1962)
468. Vinokurov V. M., Zaripov M. M., Polsky Yu. E., Stepanov V. G., Chirkin G. K., Shekun L. Ya.: Radiospectroscopy detection of small quantities of Eu^{2+}, Gd^{3+}, Nb^{4+} and their isomorphism in fluorite and zircon. Geochimia *11*, 1002–1007 (1963)
469. Vinokurov V. M., Zaripov M. M., Kropotov V. S., Stepanov V. G.: EPR study of Mn^{2+} ion isomorphism in beryl. Geochimia *1*, 104–109 (1965)
470. Vinokurov V. M., Zaripov M. M., Kropotov V. S., Stepanov V. G.: EPR of Mn^{2+} ions in cordierite. Geochimia *12*, 1486–1490 (1965)
471. Vinokurov V. M., Gainullina N. M., Nisamutdinov N. M., Krasnobaev A. A.: Distribution of Fe^{3+} impurity ions in zircon single crystals from kymberlite pipe "Mir". Geochimia *11*, 1402–1405 (1972)
472. Wauchope R. D., Haque R.: ESR in clay minerals. Nature. Phys. Sci. *233*, 141–142 (1971)
473. Weeks R. A.: Paramagnetic resonance spectra of Ti^{3+}, Fe^{3+}, and Mn^{2+} in lunar plagioclases. J. Geophys. Res. *78*, 2393–2401 (1973)
474. Wieringen J. S. van.: Paramagnetic resonance of divalent manganese incorporated in various lattices. Discuss. Faraday. Soc. *19*, 118 (1955)
475. Wildeman Th. R.: The distribution of Mn^{2+} in some carbonates by electron paramagnetic resonance. Chem. Geol. *5*, 167–177 (1970)
476. Wilmshurst T. H.: Electron Spin Resonance Spectrometers. London: Hilger, 1967
477. Zaripov M. M., Shekun L. Ya.: Electron paramagnetic resonance in crystals. In: Paramagnetic Resonance, 5–41. Kazan: University edition, 1964
478. Zaripov M. M., Kaibianen V. K., Meikliar V. P., Falin M. L.: Electron-nuclear double resonance of Mn^{2+} in CaF_2. Phys. tverd. tela, *17*, 1691–1695 (1975)
479. Zeira S., Hafner S. S.: The location of Fe^{3+} ions in forsterite (Mg_2SiO_4). Earth Planet Sci. Lett. *21*, 201 (1974)

480. Zhitomirsky A. N.: Relation between ionicity and EPR hyperfine structure parameter of impurity manganese in binary crystals. Zh. Struckt. Chim. *9*, 612–615 (1968)

Nuclear Magnetic and Nuclear Quadrupole Resonance

481. Abragam A.: The Principles of Nuclear Magnetism. Oxford: Clarendon Press, 1961
482. Alexandrov N. M., Skripov F. I.: Structure studies in crystals by nuclear magnetic resonance method. Usp. Phys. Nauk *75*, 585–628 (1961)
483. Andrew R.: Nuclear Magnetic Resonance. Cambridge: University Press, 1956
484. Andrew E. R., Eades R. G.: Possibilities for high resolution nuclear magnetic resonance spectra of crystals. Discuss. Faraday Soc. *34*, 38 (1962)
485. Averbuch P., Ducros P., Pare X.: Resonance magnetic nucleaire de protons dans la chabasite. C. R. Acad. Sci. Paris *250*, 322 (1960)
486. Balogh B., Wilson D. M., Burlingame A. L.: ^{13}C nuclear magnetic resonance in organic geochemistry. Adv. Org. Geochem. 163–171 (1972)
487. Belitsky I. A., Gabuda S. P., Lundin A. G.: Proton magnetic resonance of the hydrogen atoms in natural eddingtonite. Dokl. Akad. Nauk *172*, 1318–1320 (1967)
488. Belitsky I. A., Gabuda S. P.: Classification of water types in zeolites based on nuclear magnetic resonance data. Geol. Geoph. *6*, 3–14 (1968)
489. Blinc R., Maricic S., Pinter M.: A proton magnetic resonance and infrared study of colemanite and inyoite. In: Hochfrequenzspetrroskopie. Berlin: Akademic Verlag, 1961
490. Bondar A. M., Revokatov O. P.: NMR study of some borates. In: Experimental Studies of Mineral Formation. Moscow: Nauka, 1971
491. Bray P. J., O'Keefe J. G., Barnes R. G.: As^{75} pure quadrupole resonance in synthetic claudetite. J. Chem. Phys. *25*, 742–744 (1956)
492. Bray P. J., Edwards J. O., O'Keefe J. G., Ross V. F., Tatsuzaki I.: Nuclear magnetic resonance study of B^{11} in crystalline borates. J. Chem. Phys. *35*, N 2 (1961)
493. Brinkmann D., Staehli J. L., Ghose S.: Nuclear magnetic resonance of ^{27}Al and ^{1}H in zoisite, $Ca_2Al_3Si_3O_{12}$ (OH). J. Chem. Phys. *51*, 5128–5133 (1969)
494. Brinkmann D., Ghose S., Laves F.: Nuclear magnetic resonance of ^{23}Na, ^{27}Al, and ^{29}Si and cation disorder in nepheline. Z. Kristallogr. *135*, 208–218 (1972)
495. Brun E., Hafner St., Hartmann P., Laves F., Staub H.: Magnetische Kernresonanz zur Beobachtung des Al, Si-Ordnung - Unordnungsgrades in einigen Feldspäten. Z. Kristallogr. *113*, 65–74 (1959)
496. Brun E., Hafner St., Hartmann P.: Electrische Quadrupolwechselwirkungen in einigen Feldspäten. Helv. Phys. Acta *33*, 495 (1960)
497. Brun E., Hartmann P., Laves F., Schwarzenbach D.: Elektrische Quadrupolwechselwirkung von Al^{27} in $AlPO_4$. Helv. Phys. Acta *34*, 388–391 (1961)
498. Brun E., Hafner St.: Die elektrische Quadrupolaufspaltung von Al^{27} in Spinell $MgAl_2O_4$ und Korund Al_2O_3. Z. Kristallogr. *117*, 37–62 (1962)
499. Cuthbert J. D., Petch H. E.: NMR studies of hydrated sodium tetraborate minerals. I. Boron-oxygen polyion in borax and tincalconite. J. Chem. Phys. *38*, 1912–1919 (1963)
500. Cuthbert J. D., MacFarlane W. T., Petch H. E.: ^{11}B quadrupole coupling tensors in the hydrated borates. I. The tunellite polyanion. J. Chem. Phys. *43*, 173–177 (1965)
501. Dmitrieva L. V., Zonn Z. N., Shakhdinarov G. M.: NMR and EPR spectra of synthetic crystals of eucryptite and spodumene. Phys. tverd. tela *12*, 42–45 (1970)
502. Das T. P., Hahn E. L.: Nuclear quadrupole resonance spectroscopy. Solid State Phys. Suppl. 1 (1958)
503. Day S. M., Grimes G. G., Jr., Weatherford W.: Nuclear resonance of solids doped with paramagnetic impurities. Phys. Rev. *139*, 515–518 (1965)

504. Ducros P.: Etude de la mobilite de l'eau et des cations dans quelques zeolites par relaxation dielectrique et resonance magnetique nucleaire. Bull. Soc. Fr. Mineral. Cristallogr *83*, 85 (1960)

505. Ducros P. et Dupont M.: Etude par resonance magnetique nucleaire des protons dans les argiles. Bull. Groupe Fr. Argiles *13*, N 8, 59 (1962)

506. Eades R. G.: An investigation of the nuclear magnetic resonance absorption spectrum of Al^{27} in a single crystal of euclase. Can. J. Phys. *33*, 286 (1955)

507. Elleman D. D., Williams D.: Proton positions in brucite crystals. J. Chem. Phys. *25*, 742 (1956)

508. Emsley J. W., Phillips L.: Fluorine Chemical shifts. Progress in nuclear magnetic resonance spectroscopy, 7 (1971)

509. Farrar Th. C., Becker E. D.: Pulse and Fourier Transform NMR. Introduction to Theory and Methods. New York - London: Academic Press, 1971

510. Fedin E. I., Semin G. K.: Applications of nuclear quadrupole resonance in crystallochemistry. Zh. Strukt. Chim. 1 (1960)

511. Gabathuler Ch., Hundt E. E., Brun E.: Dynamic nuclear polarization of Ti^{47}, Ti^{49} and O^{17} in rutile TiO_2:Cr^{3+}. In: Magnetic Resonance and Related Phenomena. 499–501. Amsterdam - London. North-Holland 1973

512. Gabuda S. P., Lundin A. G., Mikhailov G. M., Alexandrov K. S.: Hydrogen atom position in natrolite. Kristallographia *8*, 388–392 (1963)

513. Gabuda S. P., Mikhailov G. M.: Proton magnetic resonance of the water in zeolites at low temperatures. Zh. Struckt. Chim. *4*, 446–449 (1963)

514. Gabuda S. P., Mikhailov G. M., Alexandrov K. S.: Zeolite water behavior and symmetry of harmotome. Dokl. Akad. Nauk *156*, 129–132 (1963)

515. Genser E. E.: NMR studies of ^7Li, ^{23}Na, and ^{27}Al in Y-faujasite. J. Chem. Phys. *54*, 4612–4616 (1971)

516. Graham J., Walker G. F., West G. W.: Nuclear magnetic resonance study of interlayer water in hydrated layer silicates. J. Chem. Phys. *40*, 540 (1964)

517. Hafner St., Hartmann P., Laves F.: Magnetische Kernresonanz von Al^{27} in Adular. Zur Deutung der Adularstruktur. Schweiz. Mineral. Petrogr. Mitt. *42*, 277 (1962)

518. Hafner St., Hartmann P.: Measurements and calculations of electric field gradients in feldspars. In: Magnetic and Electronic Resonance and Relaxation Amsterdam: North-Holland 674–677, 1963

519. Hafner St., Hartmann P.: Elektrische Feldgradienten und Sauerstoff-Polarisierbarkeit in Alkali-Feldspäten (Na Al Si_3O_8 und K Al Si_3O_8). Helv. Phys. Acta *3$*, 348–360 (1964)

520. Hafner St., Raymond M.: The nuclear quadrupole coupling tensors of Al^{27} in kyanite. Am. Mineral. *52*, 1632–1642 (1967)

521. Hafner S. S., Raymond M., Ghose S.: Nuclear quadrupole coupling tensors of ^{27}Al in andalusite (Al_2SiO_5). J. Chem. Phys. *52*, 6037–6041 (1970)

522. Hatton J., Rollin B. V., Seymour E. F. W.: Nuclear magnetic resonance measurements on Be^9, Al^{27} and Si^{29} in beryl. Phys. Rev. *83*, 672 (1951)

523. Henderson D. M., Gutowsky H. S.: A nuclear magnetic resonance determination of the hydrogen positions in $Ca(OH)_2$. Am. Mineral. *47*, 1231–1251 (1962)

524. Hockenberry J. H., Brown L. C., Williams D.: Nuclear resonance spectrum of Al^{27} in chrysoberyl. J. Chem. Phys. *28*, 367 (1958)

525. Holuj F., Petch H. E.: A nuclear magnetic resonance study of colemanite. Can. J. Phys. *38*, 515 (1960)

526. Hutton G., Pedersen B.: Proton and deuteron magnetic resonance in partly deuterated crystals. III. Gypsum. J. Phys. Chem. Solids *30*, 235–242 (1969)

527. Ivlev V. F., Zavarzina N. I., Gabuda S. P.: Proton position and water molecule diffusion in apophyllite $KFCa_4Si_8O_{20}$ $8H_2O$. Kristallographia *13*, 815–820 (1968)

528. Jeffrey G. A., Sakurai T.: Applications of nuclear quadrupole resonance. Progress in Solid State Chem. *1*, 380–416 (1964)

529. Kalinichenko A. M., Pavlishin V. I., Matiash I. V.: Iron ions distribution in litium-iron micas according to proton magnetic resonance data. In: Constitution and Properties of Minerals 9, 48–52 (1975)

530. Kalinichenko A. M., Banzaraksheev N. Yu., Matiash I. V., Litvin A. L., Polshin E. V.: Refinement of structural features of hornblendes according to proton magnetic resonance data. Kristallographia 22, 397–401 (1977)

531. Kvlividze V. I., Krasnushkin A. V.: Water mobility at the surface of clay minerals from O^{17} NMR data. Dokl. Akad. Nauk 222, 388–391 (1975)

532. Lal K. C., Petch H. E.: Quadrupole coupling tensor for ^{11}B in danburite. J. Chem. Phys. 40, 2741–2742 (1964)

533. Lal K. C., Petch H. E.: ^{11}B quadrupole coupling tensors in the hydrated borates. II. The pentaborate polyanion. Chem. Phys. 43, 178–184 (1965)

534. Lösche A.: Kerninduction. Berlin: Deutscher Verlag der Wissenschaft, 1957

535. Lösche A.: Kristallwasseruntersuchungen an Seignettesalz mit Hilfe der magnetischen Kernresonanzabsorption. Exp. Techn. Phys. 3, 18 (1955)

536. Lugt van der W., Poulis N. J.: Proton magnetic resonance in azurite. Physica 25, 1313 (1959)

537. Lugt W. van der, Poulis N. J.: The splitting of the nuclear magnetic resonance lines in vivianite. Physica 27, 733 (1961)

538. Lugt W. van der, Knottnerus D. I. M., Perdok W. G.: Nuclear magnetic resonance investigation of fluoride ions in hydroxyapatite. Acta Crystallogr. 27, 1509–1516 (1971)

539. Marfunin A. S.: Nuclear Magnetic and Nuclear Quadrupole Resonance in Minerals. Moscow: VINITI, 1966

540. Marfunin A. S., Mkrtchian A. R.: Nuclear magnetic resonance of Li^7, Al^{27}, P^{31}, H^1, F^{19} in amblygonite. Izv. Akad. Nauk, Ser. Geol. 2, 42–54 (1966)

541. Matiash I. V., Piontkovskaya M. A., Tarasenko L. M.: Proton magnetic resonance of water in zeolites. Zh. Strukt. Chim. 3, 214–215 (1962)

542. Matiash I. V.: Water in Condensed Matter. Kiev: Naukova dumka, 1971

543. Matiash I. V., Litovchenko A. S., Proshko V. Ya., Bagmut N. N.: Water in inclusions in feldspars according to proton magnetic resonance data. Dokl. Ukr. Akad. Nauk 1B, 18–20 (1977)

544. Meerssche van M., Dereppe J. M., Lobo P. W.: Resonance protonique de $MgSO_4$ $7H_2O$ et $ZnSO_4$ $7H_2O$. Bull. Soc. Chim. Fr. 5, 1035 (1962)

545. Mikhailov G. M., Lundin A. G., Gabuda S. P.: State of water in opal. Proc. Sibir. Techol. Inst. 36, 29–33 (1963)

546. Moiseev B. M., Fedorov L. I., Petropavlov M. V.: Proton magnetic resonance in erythrite $Co_3(AsO_4)_2$ $8H_2O$. Neorg. Mater. 3, 1442–1446 (1967)

547. Moiseev B. M., Portnov A. M., Fedorov L. I.: F^{19} nuclear magnetic resonance in metamict minerals. Geochimia 6, 749–751 (1971)

548. Moiseev B. M., Milovidova N. D., Fedorov L. I.: Proton magnetic resonance in chrysotile-asbest. Neorg. Mater. 3, 1514–1515 (1967)

549. Nisamutdinov N. M., Vedenin S. V., Zakharchenko T. A., Vinokurov V. M.: NMR and EPR in herderite. Geochimia 3, 361–365 (1971)

550. Pake G. E.: Paramagnetic Resonance. New York: Benjamin, 1962

551. Pare X., Ducros P.: Etude par resonance magnetique nucleaire de l'eau dans le beryl. Bull. Soc. Fr. Mineral. Cristallogr. 87, 429–433 (1964)

552. Penkov I. N., Safin I. A.: Nuclear quadrupole resonance in realgar. Dokl. Akad. Nauk 153, 692–694 (1963)

553. Penkov I. N., Safin I. A.: Nuclear quadrupole resonance in orpiment. Dokl. Akad. Nauk, 146, 158–160 (1964)

554. Penkov I. N., Safin I. A.: Nuclear quadrupole resonance in proustite and pyrargirite. Phys. tverd. tela 6, 2467–2470 (1964)

555. Penkov I. N., Safin I. A., Mozgova N. N.: NQR study of impurity insertion in antimonite. Izv. Akad. Nauk, Ser. Geol. 12, 69–75 (1969)

556. Penkov I. N., Safin I. A.: NQR study of franckeite $Sn_3Pb_5Sb_2S_{14}$. Geochimia *1*, 118–120 (1971)
557. Penkov I. N.: Application of nuclear quadrupole resonance in mineral investigations. Izv. Akad. Nauk SSSR, Ser. Geol. *12*, 41–52 (1966)
558. Pennington K. S., Petch H. E.: Nuclear magnetic resonance spectrum of B^{11} in inderite. J. Chem. Phys. *33*, 329 (1960)
559. Pennington K. S., Petch H. E.: Nuclear magnetic resonance spectum of B^{11} in lesserite. J. Chem. Phys. *36*, 2151 (1962)
560. Petch H. E., Cranna N. G., Volkoff G. M.: Second order nuclear quadrupole effects in single crystals. II. Experimental results for spodumene. Can. J. Phys. *31*, 837 (1955)
561. Petch H. E., Pennington K. S.: Nuclear quadrupole coupling tensors for Na^{23} and Al^{27} in natrolit, a fibrous zeolite. J. Chem. Phys. *36*, 1216 (1962)
562. Petch H. E., Pennington K. S., Cuthbert J. D.: On Christ's postulated boron - oxygen polyions in some hydrated borates of unknown crystal structures. Am. Mineral. *47*, N 3–4 (1962)
563. Pfeifer H., Przyborowski F., Schirmer W., Stach H.: Kernspinresonanzmessungen an Zeolithen. Z. Phys. Chem. *236*, 345–361 (1967)
564. Pople, J. A., Schneider W. G., Bernstein H. J.: High-Resolution Nuclear Magnetic Resonance. New York: McGraw-Hill, 1959
565. Poulis N. J., Lugt W. van der: The splitting of the NMR lines in vivianite $Fe_3(PO_4)_2$ $8H_2O$. J. Phys. Soc. Jpn. *17*, Suppl. B. *1*, 505–506 (1962)
566. Progress in Nuclear Magnetic Resonance Spectroscopy. Emsley J. W., Feeney J., Sutcliffe L. H., (eds) 1966–1971, v. 1–7. New York: Pergamon Press, 1966–1971
567. Rao L. D. V., Rao D. V., Narasimha G. L.: Quadrupole interactions in chrysoberyl. Phys. Rev. *160*, 274–280 (1967)
568. Raymond M., Hafner S. S.: Nuclear quadrupole coupling tensors of ^{27}Al in sillimanite (Al_2SiO_5). J. Chem. Phys. *53*, 4110–4111 (1970)
569. Reaves H. L., Gilmer Th. E., Jr.: Quadrupole splitting of the magnetic resonance spectrum of ^9Be in chrysoberyl. J. Chem. Phys. *42*, 4138–4140 (1965)
570. Reeves L. W.: The study of water in hydrate crystals by nuclear magnetic resonance. Progress in Nuclear Magnetic Resonance Spectroscopy, *4*, (1970)
571. Retcofsky H. L., Friedel R. A.: Carbon-13 magnetic resonance in diamonds, coals, and graphite. J. Phys. Chem. *77*, 68–71 (1973)
572. Ryskin N. I., Stavitskaya G. P., Mitropolsky N. A.: Infrared and proton magnetic resonance spectra of xonotlite. Neorg. Mater. *5*, 577–581 (1969)
573. Ross V., Edwards J. O.: Tetrahedral boron in teepleite and bendylite. Am. Mineral. *44*, 875 (1959)
574. Schnabel B., Jungnickel B., Taplick T., Heide K.: Bestimmung der Lage der Wassermolekule im Kieserit $MgSO_4$ H_2O mit Hilfe der magnetischen Protonenresonanz. Krist. Techn. *6*, 193–202 (1971)
575. Spence R. D., Muller J. H.: Proton resonance in dioptase ($CuSiO_3$ H_2O). J. Chem. Phys. *29*, 961 (1958)
576. Turov E. A., Petrov M. P.: Nuclear Magnetic Resonance in Ferro- and Antiferromagnetics. Moscow: Nauka, 1969
577. Tsang T., Ghose S.: Nuclear magnetic resonance of ^{27}Al in topaz, $Al_2SiO_4(F, OH)_2$. J. Chem. Phys. *56*, 261–262 (1972)
578. Tsang T., Ghose S.: Nuclear magnetic resonance of ^1H, ^7Li, ^{11}B, ^{23}Na and ^{27}Al in tourmaline (elbaite). Am. Mineral. *58*, 224–229 (1973)
579. Volkoff G. M., Petch H. E., Smellic D. W.: Nuclear electric quadrupole interaction in single crysials. Can. J. Phys. *30*, 270 (1952)
580. Watton A., Petch H. E., Pintar M. M.: Proton dynamics in ferroelectric colemanite. Can. J. Phys. *48*, 1081–1085 (1970)
581. Wu Th.: A nuclear magnetic resonance study of water in clay. J. Geophys. Res. *69*, 1083 (1964)

582. Waugh J. C., Gibby M. G., Kaplan S., Pines A.: Proton-enhanced NMR of dilute spins in solids. In: Magnetic Resonance and Related Phenomena. Amsterdam - London: North-Holland, 1973

Luminescence and Thermoluminescence

583. Aitken M. J.: Thermoluminescent dating in archaeology: introductory review. In: Thermoluminescence of Geoloical Material. McDougall, D. J. (ed.) London - New York: Academic Press, 1968
584. Aitken M. J., Fleming S. J., Doell R. R., Tanguy J. C.: Thermoluminescent study of lavas from MtEtna and other historic flows: preliminary results. In: Thermoluminescence of Geological Material. 359–366. London - New York: Academic Press, 1968
585. Alexandrov V. I., Voronko Yu. K., Maximova G. V., Osiko V. V.: $CaWO_4$:Nd^{3+} single crystals. Neorg. Mater., *3*, 368–371 (1967)
586. Angino E. E.: The effects of non-hydrostatic pressures on radiation-damage thermoluminescence. Geochim. Cosmochim. Acta *28*, 381–388 (1964)
587. Angino E. E.: Some effects of pressure on the thermoluminescence of amblygonite, pectolite, orthoclase, scapolite and wollastonite. Am. Mineral. *49*, 387–394 (1964)
588. Anderson W. W., Razi S., Walsh D. J.: Luminescence of rare-earth-activated zinc sulfide. J. Chem. Phys. *43*, 1153–1162 (1965)
589. Anderson W. W.: Luminescence of rare-earth-activated cadmium sulfide. J. Chem. Phys. 3283, 3288 (1966)
590. Antonov A. V., Kulevsky L. V., Melamed Sh. G.: The use of luminescence of phosphor crystals in quantitative determination of microimpurities of REE. Zavod. Lab. 37, 518–521 (1971)
591. Antonov-Romanovsky V. V.: Kinetics of Photoluminescence of Phosphor Crystals. Moscow: Nauka, 1966
592. Argunov K. P., Bartoshinsky Z. V.: Luminescence properties of diamonds from kymberlites. Mieral Sb. Lvov Univ. *24*, N 2 ,185–190 (1970)
593. Arkhangelskaya V. A.: Trapping centers in single crystals of Ca, Sr, Ba fluorides activated by rare earths. Izv. Akad. Nauk, Ser. Phys. *29*, 454–459 (1965)
594. Arkhangelskaya V. A., Maxanov B. I., Feofilov P. P.: Additively reduced divalent rare earths in fluorite crystals. Phys. tverd. tela *7*, 2260–2262 (1965)
595. Arkangelskaya V. A., Kiseleva M. N.: Radiochemical reduction of MeF_2:RE^{3+} crystals. Phys. tverd. tela *9*, 3523–3526 (1967)
596. Arkhangelskaya V., Feofilov P. P.: Thermo- and phototransfer of charge between activator centres in fluorite-type crystals. Proc. Int. Conf. Luminescence, Budapest, *2*, 1682–1688 (1968)
597. Arkhangelskaya V. A., Kiseleva M. N., Shreiber V. M.: Thermic ionization potentials of divalent rare earths in fluorite type crystals. Phys. tverd. tela *1*, 869–876 (1969)
598. Artamonova M. V., Briskina Ch. M., Zalin V. F.: Excitation energy transfer in ion chain $UO_2^{2+} \rightarrow Nd^{3+} \rightarrow Yb^{3+} \rightarrow Er^{3+} \rightarrow Tu^{3+}$ in glass. Zh. Prikl. Spectros *6*, 112–114 (1967)
599. Arpiarian N.: The centenary of the discovery of luminescent zinc sulphide. Proc. Int. Conf. Luminescence, Budapest *1*, 903–906 (1968)
600. Babko A. K., Dubovenko L. I., Lukovskaya N. M.: Chemiluminescence Analysis. Kiev: T ekhnika, 1966
601. Bächtiger K.: Die Thermolumineszenz einiger skandinavischer und nordamerikanischer Plagioklase. Schweiz. Mineral. Petrogr. Mitt. *47*, 365–384 (1967)
602. Barabanov V. F., Goncharov G. N.: Concerning the relation of luminescence spectra of fluorite to its geologic origin. Dokl. Akad. Nauk *173*, 1408–1410 (1967)
603. Barnes D. F.: Infrared luminescence of minerals. U. S. Geol. Surv. Bull., p. 1052 (1958)

604. Barsanov G. P., Sheveleva V. A.: Contributions to study of luminescence of minerals. Proc. Mineral. Museum *4*, 3–35; *5*, 56–89; *6*, 29–48; 7, 3–1. (1952–1955)
605. Batygov S. Kh.: Electron-hole processes in CaF_2 crystals activated by rare earths. In: Spectroscopy of Laser Crystals with Ionic Structure. Moscow: Nauka, 1972
606. Belyaev L. M., Martyshev Yu. N.: On the triboluminescence spectrum of LiF crystals. Zhurn. Prikl. Spectros *6*, 114–117 (1967)
607. Bettinali C., Ferraresso G.: Adsorption thermoluminescence of $CaCO_3$. In: Thermoluminescence of Geological Materials. 143–152. London - New York: Academic Press, 1968
608. Bhalla R. J. R. S. B., White E. W.: Polarized cathodoluminescence emission from willemite/Zn_2SiO_4(Mn)/single crystals. J. Appl. Phys. *41*, 2267–2268 (1970)
609. Billington C.: Phosphorescence mechanisms. Phys. Rev. *120*, 697–715 (1960)
610. Birman J. L.: Theory of luminescent centers and processes in ZnS type II-VI compounds. Proc. Int. Conf. Luminescence, Budapest, 1966, *1*, 919–961 (1968)
611. Blair I.M., Edgington J. A.: Luminescent properties of rocks, metorites and natural glasses under proton bombardment. Nature (London) *217*, 157–160 (1968)
612. Blanchard F. N.: Thermoluminescence of fluorite and age of deposition. Am. Mineral *51*, 474–485 (1966)
613. Blanchard F. N.: Thermoluminescence of synthetic fluorite. Am. Mineral. *52*, 371–379 (1967)
615. Booth A. H.: Calculation of electron trap depths from thermoluminescence maxima. Can. J. Chem. *32*, 214 (1954)
616. Borchardt H. J.: Rare-earth tungstates and 1:1 oxytungstates. J. Chem. Phys. *39*, 504–511 (1963)
617. Borsi S., Fornaca-Rinaldi G.: Relationship between crystallization temperature and some properties of thermoluminescence in natural calcites. In: Thermoluminescence of Geological Materials 225–231. London - New York: Academic Press, 1968
618. Botden Th. P. J.: Transfer and transport of energy by resonance processes in luminescent solids. Philips Res. Rep. *6*, 425–473 (1951)
619. Bowen E. J.: Luminescence in Chemistry. London, 1968
620. Bozhevolnov E. A.: Luminescence Analysis of Inorganic Materials. Moscow: Chimia, 1966
621. Bramanti D., Mancini M., Ranfagni A.: Molecular orbital model for KCl:Tl. Phys. Rev. *133*, 3670–3676 (1971)
622. Bràunlich P.: Thermoluminescence and thermally stimulated current-tools for the determination of trapping parameters. In: Thermoluminescence of Geological Materials 61–88. London - New York: Academic Press, 1968
623. Bryant F. J., Cox A. F. J.: Energy level structure for the infra-red luminescence of cadmium sulphide and zinc sulphide. Br. J. Appl. Phys. *16*, 463–469 (1965)
624. Burchell D., Fremlin J. H.: Relative efficiency of different radiations in storing thermoluminescent energy. In: Thermoluminescence of Geological Materials 407–412. London - New York: Academic Press, 1968
625. Burns G., Geiss E. A., Jenkins B. A., Nathan M. I.: Cr^{3+} fluorescence in garnets and other crystals. Phys. Rev. *139*A, 1687–1693 (1965)
626. Carnall W. T., Fields P. R.: Lanthanide and actinide absorption spectra in solution. In: Lanthanide/Actinide chemistry. (Adv. in Chem. Series), 86–101, 1967
627. Charlet J. M.: La thermoluminescence des roches quartzofeldspathiques. Application à l'etude des series sedimentaires detritiques; interet dans la datation des granites. Bull. Bur. Rech. Geol. Minier. Sec. 2, 51–97 (1969)
628. Chesnokov B. V.: Luminescence of pyrochlor from Vishnevy mountains, Ural. Zap. Vses. Mineral. obshch. *89*, 96–99 (1960)
629. Christodoulides C., Ettinger K. V.: Some useful data from the Randall-Wilkins equation. Mod. Geol. *2*, 235–238. (1971)

630. Christodoulides C., Ettinger K. V., Fremlin J. H.: The use of TL glow peaks at equilibrium in the examination of the thermal and radiation history of materials. Mod. Geol. *2*, 275–280 (1971)
631. Claffy E. W.: Composition, tenebrescence and liminescence of spodumen minerals. Am. Mineral. *38*, 919–931 (1953)
632. Coy-Yll R.: Quelques aspects de la cathodoluminescence des mineraux. Chem. Geol. *5*, 243–254 (1970)
633. Curie D.: Luminescence et champ cristalline. Paris: Dunod, 1960
634. Dake H. C.: The Uranium and Fluorescent Minerals, a Handbook of Uranium Minerals and a Field Guide for Uranium Prospecting. Portland, Oregon: Mineralogist Publ. Co., 1953
635. Daniels F.: Kinetics and thermoluminescence in geochemistry. Geochim. Cosmochim. Acta. London, 1961, *22*, 65–74 (1961)
636. Dean P. J.: Lattices of the diamond type. In: Luminescence of Inorganic Solids, 120–205. New York - London: Academic Press, 1966
637. De Ment J.: Handbook of Fluorescent Gems and Minerals. Oregon: Mineralogist Publ. Co., 1949
638. Dicke G. H., Duncan A. B. F.: Spectroscopic Properties of Uranium Compounds. New York: McGraw Hill, 1949
639. Dicke G. H.: Spectra and Energy Levels of Rare Earth Ions in Crystals. New York: Interscience Publishers, 1968
640. Diehl H., Grasser R., Scharmann A.: Discussion of a simplified model for the thermoluminescence of inorganic photoconducting phosphors. In: Thermoluminescence of Geological Materials. 39–50. London - New York: Academic Press, 1968
641. Dieleman J.: On the utilization of electron spin resonance results: illustration with the aid of the (3d) transition elements in some II-VI compounds. Proc. Int. Conf. Luminescence, Budapest, *2*, 2131–2142 (1968)
642. Dielman J.: Electron spin resonance in luminescent solids. In: Luminescence of Inorganic Solids. 338–384. New York - London: Academic Press, 1966
643. Dobrolyubskaya T. S.: Luminescence Methods of Uranium Determination. Moscow: Nauka, 1968
644. Douglas G., Morency M., McDougall D. J.: Detection of strain in rocks using an intrinsic "semi-insulator" characteristic of some minerals. Mod. Geol. *1*, 211–217. (1970)
645. Dudkin O. B.: Rare earths minerals absorption in visible region. In: Materials on Mineralogy of the Kola Peninsula, 218–224. Moscow-Leningrad: Nauka, 1965
646. Durrani S. A., Christodoulides C., Ettinger K. V.: Thermoluminescence in tektites. J. Geophys. Res. *75*, 983–995 (1970)
647. Durrani S. A.: Thermoluminescence in meteorites and tektites. Mod. Geol. *2*, 247–262 (1971)
648. Dyatkina M. E., Markov V. P., Capkina I. V., Mikhailov Yu. N.: Electronic structure of UO_2 group in uranyl compounds. Zh. Neorg. Chim. *6*, 575–580 (1961)
649. Dyatkina M. E., Mikhailov Yu. N.: Uranyl and its analogues electronic structure. Zh. Strukt. Chim. *3*, 724–727 (1962)
650. Edelstein N., Easley W., McLaugklin R.: Optical and electron paramagnetic resonance spectroscopy of actinide ions in single crystals. In: Lanthanide - Actinide Chemistry. (Adv. in Chem. Series) 203–210, 1967
651. Eisenbrand J.: Entwicklungslinien der Luminescenzanalyse in den letzten 30 Jahren. Z. Analyt. Chem. *192*, 83–91 (1963)
652. Electroluminescence and related effects: Ivey, H. F. (ed.) Adv. in Electronics and Electron Physics, Suppl. 1, 1963
653. Elyashevich M. A.: Spectra of Rare Earths. Moscow: GITTL. 1953
654. Eremenko G. K.: New electron-vibronic series in luminescence spectrum of natural diamond. Zh. Priul. Spectros. *5*, 82–84 (1971)

655. Fair H. D., Ewing R. D., Williams F. E.: Electron paramagnetic resonance of photo-excited donor-acceptor pairs in zinc-sulfide systems. Phys. Rev. Lett *15*, 355–356 (1965)

656. Fair H. D., Ewing R. D., Williams F. E.: Photo-induced paramagnetic resonance of chromium in zinc sulfide crystals. Phys. Rev. *144*, 298–301 (1966)

657. Feofilov P. P.: Polarized Luminescence of Atoms, Molecules and Crystals. Moscow: Phys.-Math. isdat, 1959

658. Feofilov P. P.: The luminescence of rare earths. Proc. Int. Conf. Luminescence, Budapest, *2*, 1727–1736 (1968)

659. Fieschi R., Scaramelli P.: Photostimulated thermoluminescence in alkali halide crystals. In: Thermoluminescence of Geological Materials. 291–308. London - New York: Adademic Press, 1968

660. Fischer A. G.: Electroluminescence in II-VI compounds. In: Luminescence of Inorganic Solids. 541–602. New York - London: Academic Press, 1966

661. Flanigen E. M., Breck D. W., Mumbach N. R., Taylor A. M.: Characteristics of synthetic emeralds. Am. Mineral. *52*, 744 (1967)

662. Fleming S. J., Thompson J.: Quartz as a heat-resistant dosimeter. Health Phys. *18*, 567–568 (1970)

663. Fluorescence and phosphorescence analysis. Principles and applications: Hercules, D. M. (ed.) New York, 1966

664. Fock M. V.: Introduction to Kinetics of Luminescence of Crystallophosphors. Moscow: Nauka, 1964

665. Fong F. K., Cape J. A., Wong E. Y.: Monovalent samarium in potassium chloride. Phys. Rev. *151*, 299–303 (1966)

666. Fornaca-Rinaldi G.: Some effects of heating on the radiation sensitivity of natural crystals. In: Thermoluminescence of Geological Materials. 103–110. London - New York: Academic Press, 1968

667. Fred M.: Electronic structure of the actinide elements. In: Lanthanide - Actinide Chemistry. (Adv. in Chem. Series), 180–202, 1967

668. Galanin M. D.: Resonance transfer of excitation energy im luminescent solutions. Proc. Lebedev Phys. Inst. *12*, 3–53 (1960)

669. Galkin L. N., Korolev N. V.: Photoluminescence of PbS in infrared region. Dokl. Akad. Nauk *92*, 529–530 (1963)

670. Garlick G. F. S.: Luminescence in solids. Sci. Progress *52*, 3–25 (1964)

671. Garlick G. F. J.: Cathodo- and radioluminescence. In: Luminescence of Inorganic Solids. 685–832, New York - London: Academic Press, 1966

672. Gilfanov F. Z., Lux R. K., Stolov A. L.: Excitation of centers with non-local charge compensation in $MeF_2:TR^{3+}$ system. Phys. tverd. tela *11*, 2502–2505 (1969)

673. Gobrecht H., Hoffmann D.: Spectroscopy of traps by fractional glow technique. J. Phys. Chem. Sci. *27*, 509–522 (see also: Physik Cond. Materie *5*, 39–47) (1966)

674. Gomon G. O.: Diamonds. Moscow: Mashinostroenie, 1966

675. Goni J., Remond G.: Localization and distribution of impurities in blende by cathodoluminescence. Mineral Mag. *37*, 153–155 (1969)

676. Gool W. van: Fluorescence centres in ZnS. Philips Res. Repts. Suppl. N *3*, 1–119 (1961)

677. Görlich P., Karras H., Lehmann R.: Über die optischen Eigenschaften der Erdalkalihalogenide vom Flusspat-Typ. Phys. Status. Solid. *1*, 389–440, 525–553 (1961)

678. Görlich P., Karras H., Kotitz G., Lehmann R.: Spectroscopic properties of activated laser crystals. Phys. Status. Solid. *5*, N *3*; *6*, N 2, 8 N 2 (1965)

679. Gorobets B. S., Novozhilov A. I., Samoilovich M. I., Shamovsky L. M.: Monovalent europium in NaCl:Eu and KCl:Eu. Dokl. Akad. Nauk *180*, 1351–1353 (1968)

680. Gorobets B. S., Portnov A. M.: Green luminescence centers in silex, chalcedones and allophanes from Paleozoic rocks of Russian platform. Zap. Vses. Mineral. obshch. *102*, 357–360 (1973)

681. Gorobets B. S., Sidorenko G. A.: Luminescence of secondary uranium minerals at low temperatures. Atomic Energy *36*, 6–13 (1974)

682. Gorobets B. S.: Reabsorption of luminescence dans les mineraux related to neodym impurities. Zap. Vses. Mineral. abshch. *104*, 357–359 (1975)

683. Gorobets B. S., Kudenko M. A.: Tipomorphous features of sheelite by rare earths photoluminescence spectra. In: Constitution and Properties of Minerals *10*, 82–88 (1976)

684. Greet R. T., Weber J. N.: Correlation of mineral luminescent phenomena and its selenological implications. Icarus *11*, 55–65 (1969)

685. Gumlich H. E., Schulz H. J.: Optical transitions in ZnS type crystals containing cobalt. J. Phys. Chem. Solids *27*, 187–195 (1966)

686. Gurov E. P., Gurova E. P.: Some features of thermoluminescence spectra of fluorites. In: Constitution and Properties of Minerals *4*, 78–81 (1970)

687. Gurvich A. M.: Evolution of understanding of the chemical nature of luminescent centers in zinc sulfide luminophors. Usp. Chim. *35*, 1495–1526 (1966)

688. Gurvich A. M.: Introduction to Physical Chemistry of Crystallophosphors. Moscow: Vyssh. Shcola, 1971

690. Hall M. R., Ribbe P. H.: An electron microprobe study of luminescence centers in cassiterite. Am. Mineral *56*, 31–45 (1971)

691. Hargreaves W. A.: Energy levels of uranium ions in calcium fluoride crystals. Phys. Rev. B. *2*, 2273–2284 (1970)

692. Harvey E. N.: A History of Luminescence from the Earliest Times Until 1900. Philadelphia: Memoirs Am. Philosoph. Soc. Vol. 44, 1957

693. Hautermans F., Ziener A.: Thermoluminescence of meteorites. J. Geophys. Res. *71*, 3387–3396 (1966)

694. Henisch H. K.: Electroluminescence. Oxford: Pergamon Press, 1962

695. Holloway W. W., Prohofsky E. W., Kestigian M.: Magnetic ordering and the fluorescence of concentrated Mn systems. Phys. Rev. *139A*, 954–961 (1965)

696. Holton W. C., Schneider J., Estle T. L.: Electron paramagnetic resonance of photosensitive iron transition group impurities in ZnS and ZnO. Phys. Rev. **133A**, 1638–1641 (1964)

697. Huber-Schausberger I., Schroll E.: UV-Lumineszens und Seltenerdgehalte in Flusspaten. Geochim. Cosmochim. Acta *31*, 1333–1341 (1967)

698. Hutchison Ch. S.: The calibration of thermoluminescence measurements. New Zeal. J. Sci. *8*, 431–445 (1965)

699. Hutchison Ch. S.: The dating by thermoluminescence of tectonic and magmatic events in orogenesis. In: Thermoluminescence of Geological Materials 341–358. London - New York: Academic Press, 1968

700. Hwang F. S. W., Göksü H. Y. A further investigation on the thermoluminescence dating technique. Mod. Geol. *2*, 225–230 (1971)

701. Hwang F. S. W.: Thermoluminescence dating applied to volcanic lavas. Mod. Geol. *2*, 231–243 (1971)

702. Ilyin V. E., Yuryeva O. P., Sobolev E. V.: Excitation and absorption spectra of S_2 center of natural diamonds. Zh. Prikl. Spectrosk. *14*, 158–160 (1971)

703. Ioffe V. A., Yanchevskaya I. S.: Study of thermoluminescence and EPR in irradiated aluminosilicates. Opt. Spectrosk. *23*, 494–496 (1967)

704. Ivey H. F.: Sensitized luminescence and its application in laser materials. Proc. Int. Conf. Luminesce, Budapest *2*, 2027–2049 (1968)

705. Johnson L. F., Van Uitert L. G., Rubin J. J., Thomas R. A.: Energy transfer from Er^{3+} to Tm^{3+} and Ho^{3+} ions in crystals. Phys. Rev. *133*A, 494–501 (1964)

706. Johnson L. F.: Optical maser characteristic of rare earth ions in crystals. J. Appl. Phys. *34*, 897–909 (1963)

707. Johnson N. M.: Radiation dosimetry from natural thermoluminescence. In: Thermoluminescence of Geological Materials 451–451. London - New York: Academic Press, 1968

708. Kaminsky A. A., Osiko V. V.: Neorganic laser materials with ionic structure. Neorg. Mater. *1*, 2049–2087; *3*, 417–463; *6*, 629–696 (1965, 1967, 1970)

709. Kang Ch. S., Beverley P. P., Bube R. H.: Photoelectronic processes in ZnS single crystals. Phys. Rev. *156*, 998–1009 (1967)
710. Karapetian G. O.: Luminescence of glasses with rare earth activators. Izvestia Akad. Nauk. Ser. Phys. *27*, 799–802 (1963)
711. Karapetian G. O., Reichakhrit A. L.: Luminescent glasses as materials for optic quantum generators. Neorg. Mater., *3*, 217–259 (1967)
712. Kasai P. H., Otomo Y.: Electron paramagnetic resonance studies of the ZnS-A and -B centres. J. Chem. Phys. *37*, 1263–1275 (1962)
713. Kats M.L.: Luminescence and Electron-Hole Processes in Photochemically Colored Alkali Halide Compounds. Saratov Univ. edition, 1960
714. Kaufhold J., Herr W.: Factors influencing dating CaF_2 - minerals by thermoluminescence. In: Thermoluminescence of Geological Materials 153–167. London - New York: Academic Press, 1968
715. Kaul I. K., Bhattacharya P. K., Tolpadi S.: Age determination by study of the thermoluminescence of smoky quartz. J. Geophys. Res. *71*, 1275–1282 (1966)
716. Kaul I. K., Bhattacharya P. K., Tolpadi S.: Factors in age determination by thermoluminescence of smoky quartz. In: Thermoluminescence of Geological. Materials 327–340. London - New York: Academic Press, 1968
717. Kayser H.: Handbuch der Spectroscopie, t. IV, Ch. V (p. 579–839), Ch. VI (p. 839–1214), 1908
718. Kazakov V. P., Korobeinikova V. N., Kobets L. I.: Electroluminescence of CaF_2: Tb^{3+} crystals. Opt. Spectrosk. *26*, 319–321.(1969)
719. Kelly P., Bräunlich P.: Phenomenological theory of thermoluminescence. Phys. Rev. *1*, 1587–1595 (1970)
720. Kelly P., Laubitz G., Bräunlich P.: Exact solutions of the kinetic equations governing thermally stimulated luminescence and conductivity. Phys. Rev. *4*, 1960–1968 (1971)
721. Kingsley J. D., Prener J. S., Segall B.: Spectroscopy of MnO_4^{3-} in calcium halophosphates. Phys. Rev. *137*A, 189–203 (1965)
722. Kirk R. D.: The luminescence and tenebrescence of natural and synthetic sodalite. Am. Mineral. 40, 22–31 (1955)
723. Kirton J., McLaughlan S. D.: Correlation of electron paramagnetic resonance and optical-absorption spectra of CaF_2:Yb^{3+}. Phys. Rev. *155*, 279–284 (1967)
724. Kiss Z. J., Yocom P. N.: Stable divalent rare-earth-alkaline-earth halide systems. J. Chem. Phys. *41*, 1511–1512 (1964)
725. Klick C. C., Schulman J. H.: Luminescence in solids. Solid State Phys. *5*, 97–172 (1957)
726. Kluev Yu. A., Nepsha V. I., Dudenkov Yu. A., Zvonkov S. D., Zubkov V. M.: Absorption spectra of different type diamonds. Dokl. Akad. Nauk *203*, N 5 (1972)
727. Kluev Yu. A., Nepsha V. I., Dudenkov Yu. A.: On Physical Classification of Diamond. Proc. VNIIAlmas 3, 1974
728. Köhler A., Leitmeier H.: Die natürliche Thermolumineszenz bei Mineralien und Gesteinen. Z. Kristallogr. *87*, 146–180 (1934)
729. Komov I. L., Khetchikov L. N.: Thermoluminescence of natural quartz. In: Physical Studies of Quartz. Moscow: Nedra, 1975
730. Komovsky G. F., Lozhnikova O. N.: Luminescence Analyses. In: Ores and Minerals Studies: Moscow: Gosgeoltekhisdat, 1954
731. Komovsky G. F.: Thermoluminescence of chondritic meteorites. Meteoritica *21*, 64–70 (1961)
732. Komiak A. I., Sevchenko A. N., Sidorenko M. M.: Luminescence and absorption spectra of uranyl chloride crystals in polarized light. Izv. Akad. Nauk, Ser. Phys. *34*, 576–581 (1970)
733. Korsunsky M. I., Osipov A. N.: Dependence of the electron lifetime in deep trap on temperature. Izv. Kazakh. Akad. Nauk, Ser. Phys.-Math. 2, 67–70 (1969)
734. Krasnobaev A. A.: On thermoluminescence of zircons. Zap. Vses. Mineral. obshch. *93*, 713–720 (1964)

735. Kuznetsov G. V., Tarashchan A. N.: Luminescence centers in natural apatites. In: Constitution and Properties of Minerals 9, 120–124 (1975)

736. Labeyrie J., Lalou C., Nordemann D.: High sensitivity apparatus to detect thermoluminescence induced by very weak irradiations. In: Thermoluminescence of Geological Materials 175–181. London - New York: Academic Press, 1968

737. Larach S.: Group II-VI phosphors with rare earth activators. Proc. Int. Conf. Luminescence, Budapest, 2, 1549–1569 (1968)

738. Laud K. R., Gibbons E. F., Tien T. Y., Stadler H. L.: Cathodoluminescence of Ce^{3+}- and Eu^{2+}-activated alkaline earth feldspars. J. Electrochem. Soc. 118, 918–923 (1971)

739. Leverenz H. W.: An Introduction to Luminescence of Solids. New York: Wiley, 1950

740. Levshin V. L.: The processes of energy transfer in the physics of inorganic crystal phosphors. Proc. Int. Conf. Luminescence, Budapest 1, 23–25 (1966)

741. Levy P. W.: A brief survey of radiation effects applicable to geology problems. In: Thermoluminescence of Geological Materials 25–38. London - New York: Academic Press, 1968

742. Levy P. W., Mattern P. L., Lengweiler K.: Three dimensional thermoluminescent analysis of minerals. Mod. Geol. 2, 295–297 (1971)

743. Lewis D. R.: Exoelectron-emission phenomena and geological applications. Bull. Geol. Soc. 77, 761–769 (1966)

744. Lewis D. R.: Effect of grinding on thermoluminescence of dolomite, calcite and halite. In: Thermoluminescence of Geological Materials. 125–132. London - New York: Adademic Press, 1968

745. Lewis D. R.: Special considerations in the design of equipment for use in geological studies-thermoluminescence and exoelectron apparatus. In: Thermoluminescence of Geological Materials. 183–190. London - New York: Academic Press, 1968

746. Livanova L. D., Saitkulov I. G., Stolov A. L.: Energy transfer from Gd^{3+} to Pr^{3+} in fluorite crystal. Phys. tverd. tela 11, 857–864 (1969)

747. Livanova L. D., Saitkulov I. G., Stolov A. L.: Quantum summation in CaF$_2$ and SrF$_2$ crystals activated by Tb^{3+} and Yb^{3+} ions. Phys. tverd. tela 11, 918–923 (1969)

748. Loh E.: Lowest 4f $\rightarrow$ 5d transition of trivalent rare earth ions in CaF$_2$ crystals. Phys. Rev. 147, 332–335 (1966)

749. Long J. V. P.: Electron probe microanalysis. In: Physical Methods in Determinative Mineralogy. 215–260. London - New York: Academic Press, 1967

750. Lozykowski H., Holuj F.: Luminescence in phenacite. J. Chem. Phys. 51, 2315–2321 (1969)

751. Luminescence of Inorganic Solids: Goldberg, P (ed.). London - New York: Adademic Press, 1966

752. Lushchik Ch. B.: Studies of Trapping Centers in Alkali Halide Crystals. Tartu: University edition, 1955

753. Lushchik Ch. B.: Electron excitations and electron processes in luminescent ionic crystals. Proc. Inst. Phys. Ast. Akad. Nauk Est. S1K, 31, 19–83 (1966)

754. Lüty F.: Elektronenubergänge an Farbzentren. In: Halbleiterprobleme, VI, 1961

755. Lysakov V. S., Sakhu I. E., Serebryannikov A. I., Solntsev V. P.: On the nature of trapping and emission centers in quartz and nepheline. Dokl. Akad. Nauk 186, 177–180 (1969)

756. MacDiarmid R. A.: Natural thermoluminescence as a prospecting tool for hydrothermal ores in carbonate host rocks. In: Thermoluminescence of geological Material 547–556. London - New York: Academic Press, 1968

757. Markovsky L. Ya., Pekerman F. M., Petushina K. N.: Luminophors. Moscow: Chimia, 1966

758. Maxwell J. A.: The laser as a tool in mineral identification. Can. Mineral. 7, 727–737 (1965)

759. McClure D. S., Kiss Z.: Survey of the spectra of the divalent rare-earth ions in cubic crystals. J. Chem. Phys. 39, 3251–3257 (1963)

760. Medlin W. L.: Thermoluminescence in dolomite. J. Phys. Chem. *34*, N 2 (1961)
761. Medlin W. L.: Thermoluminescence in anhydrite. J. Phys. Chem. Solids. **18**, 238–252 (1961)
762. Medlin W. L.: Thermoluminescence in quartz. J. Chem. Phys. *38*, 1132–1143 (1963)
763. Medlin W. L.: Trapping centers in thermoluminescent calcite. Phys. Rev. *135*, 1770–1779 (1964)
764. Medlin W. L.: Thermoluminescence growth curves in calcite. In: Thermoluminescence of Geological Material 91–101. London - New York: Academic Press, 1968
765. Medlin W. L.: The nature of traps and emission centers in thermoluminescent rock materials. In: Thermoluminescence of Geological Materials 193–223. London - New York: Academic Press, 1968
766. Melton Ch. E., Giardini A. A.: Experimental evidence that oxygen is the principal impurity in natural diamonds. Nature (London) *263*, 309–310 (1976)
767. Merz J. L., Pershan P. S.: Charge conversion of irradiates rare-earth ions in CaF_2. Phys. Rev. *162*, 217–235, 235–247 (1967)
768. Millson H. E., Millson H. E., Jr.: Duration of phosphorescence. J. Opt. Soc. Am. *54*, 638–640 (1964)
769. Morency M.: An investigation of the effects of pressure on thermoluminescence. In: Thermoluminescence of Geological Material 233–239. London - New York: Academic Press, 1968
770. Moskvin A. V.: Catodoluminescence. I General Properties of the Phenomenon. II. Catodoluminophors and Screens. Moscow: Gostekhisdat, 1948, 1949
771. Muer D. de: Development of a universal method for calculating the thermoluminescence parameters. Physics *48*, 1–12 (1970)
772. Murata K. J., Smith R. L.: Manganese and lead as coactivators of red fluorescence in halite. Am. Mineral *31*, 527–538 (1946)
773. Nash D. B.: Proton-excited luminescence of silicates: experimental results and lunar implications. J. Geophys. Res. *71*, 2517–2534 (1966)
774. Nauchitel M. A.: Photoluminescence of minerals of sheelite-povellite series. In: Relationships of Impurity Center Distribution in Ionic Crystals 3, 71–79, 1974
775. Nicholas J. V.: Origin of the luminescence in natural zircon. Nature (London) *215*, 1476 (1967)
776. Nicholas J. V.: Luminescence of hexavalent uranium in CaF_2 and SrF_2 powders. Phys. Rev. *155*, 151–156 (1967)
777. Osiko V. V.: Optical centers statistics in fluorite, corundum, garnet and sheelite types crystals. Neorg. Mater. *5*, 433–440 (1969)
778. Ovcharenko V. K., Eremenko G. K.: Luminescence of zircon from Octyabrsky alkali massiv. In: Constitution and Properties of Minerals 4, 58–62 (1970)
779. Ovchinnikov L. N., Maxenkov V. G.: Relation of limestone thermoluminescence with endogenic mineralization. Geol. Rud. Mestorozhd. *6*, 3–22 (1965)
780. Ovchinnikov L. N., Maxenkov V. G.: Thermoluminescence of some chondritic meteorites. Meteoritica *27*, 58–62 (1966)
781. Ovsyankin V. V., Feofilov P. P.: Cooperative luminescence of condense materials. Zh. Prikl. Spectrosk *7*, 498–506 (1967)
782. Parker C.: Photoluminescence of Solutions. Amsterdam. Elsevies, 1968
783. Passwater R. A.: Guide to Fluorescence Literature. New York: Plenum Press, Vol 1 1967; vol 2, 1970; vol 3, 1974
784. Peterson G. E., Bridenbaugh P. M.: Study of relaxation processes in Nd using pulsed excitation. J. Opt. Soc. Am. *54*, 644–650 (1964)
785. Platonov A. N., Tarashchan A. N.: On the red photoluminescence of natural fluorites. Dokl. Akad. Nauk *177*, 415–417 (1967)
786. Platonov A. N., Tarashchan A. N.: Concerning the form of thallium insertion in collomorph sphalerites. Mineral. Sb. Lvov Univ. *23*, N 1, 74–77 (1969)
787. Platonov A. N., Belichenko V. P.: Color and thermoluminescence of Volyn topaz. Mineral. Sb. Lvov Univ. *18*, N 4, 412–421 (1964)

788. Platonov A. N., Marfunin A. S.: Isomorphous substitutions in sphalerites from spectroscopy data. In: Problem of Isomorphous Substitutions of Atoms in Crystals, 268–281. Moscow: Nauka, 1971
789. Porter G. B., Schläfer H. L.: Lumineszenz von Übergangsmetallverbindungen. Ber. Bunsenges. Phys. Chem. 68, 316 (1964)
790. Poluektov N. S., Kononenko L. I.: Spectrophotometric Methods of Determination of Individual Rare Earths. Kiev: Naukova dumka, 1968
791. Prizbram K.: Irradiation colours and luminescence. A Contribution to Mineral Physics. London: Pergamon Press, 1956
792. Rabbiner N.: Fluorescence of Dy^{3+} in CaF_2. Phys. Rev. 132, 224–228 (1963)
793. Rabinowitch E., Belford R. L.: Spectroscopy and Photochemistry of Uranyl Compounds. Oxford: Pergamon Press; New York: Macmillan, 1964
794. Radiation Dosimetry.: New York: Academic Press, 1968
795. Ralph J. E., Townsend M. G.: Near-infrared fluorescence and absorption spectra of Co^{2+} and Ni^{2+} in MgO. J. Chem. Phys. 48, 149–154 (1968)
796. Räuber A., Schneider J.: Electron spin resonance of a luminescent center in aluminium-activated cubic ZnS single crystals. Phys. Lett. 3, 230–231 (1963)
797. Rebane K.: Luminescence. Tartu: University edition, Vol. I, II, III, 1965, 1966, 1969
798. Rebane K.: Secondary glow of impurity center in crystal. Tartu: University edition, 1970
799. Recker K., Neuhaus A., Leckebusch R.: Vergleichende Untersuchungen der Farb- und Lumineszenzeigenschaften natürlicher und gezuchteter, definiert dotierter fluorite. Papers and Proc. 5th Gen. Meet. Int. Mineral. Assoc., Cambridge, 145–152, 1968
800. Richman I., Kisliuk P., Wong E. Y.: Absorption spectrum of U^{4+} in zircon ($ZrSiO_4$). Phys. Rev. 155, 262–267 (1967)
801. Rinck B.: Übergangswahrscheinlichkeiten für strahlende und strahlungslose Prozesse im kristallinen $Eu_2(SO_4)_3$ $8H_2O$. Z. f. Naturforsch. 3a, 406–412 (1948)
802. Rolfe J.: Low-temperature emission spectrum of O_2^- in alkali halides. J. Chem. Phys. 40, 1664–1670 (1964)
803. Ronca L. B.: Thermoluminescence as a paleoclimatological tool. Am. J. Sci. 262, 767–781 (1964)
804. Scharmann A.: Exo-electron emission. In: Thermoluminescence of Geological Materials 281–290 London - New York: Academic Press, 1968
805. Scharmann A., Schwarz G.: Luminescence decay and delayed spectra of $CaWO_4$ single crystals. Phys. Status. Solid 42, 781–785 (1970)
806. Schneider J., Holton W. C., Estle T. L.: Electron spin resonance in self-activated ZnS crystals. Phys. Lett. 5, 312–315 (1963)
807. Schneider J., Räuber A., Dischler B., Estle T. L., Holton W. C.: Direct confirmation of the A - center model in ZnS by observation of hyperfine structure in EPR spectra. J. Chem. Phys. 42, 1839–1841 (1965)
808. Schulman J. H., Evans L. W., Ginther R. J., Murata K. J.: The sensitized luminescence of manganese-activated calcite. J. Appl. Phys. 18, 732–739 (1947)
809. Schulman J. H., Kirk R. D.: Luminescent centers in alkali halides. Solid State Commun 1, 105 (1964)
810. Schulman J. H.: Survey of luminescence dosimetry. In: Luminescence Dosimetry. U. S. Atomic Energy. Comiss., 1967
811. Sevchenko A. N., Volodko L. V., Umreiko D. S., Komyak A. I.: Nature of luminescence spectra of uranyl compounds. Vestn. Beloruss. Univ. 1, 23–28 (1969)
812. Shamovsky L. M.: Phosphor crystals on the base of anomalous solid solutions. In: Relationships of Impurity Centers Distribution in Ionic Crystals 3, 7–15, 1974
813. Shionoya Sh.: Luminescence of lattices of the ZnS type. In: Luminescence of Inorganic Solids, 206–286. New York - London: Academic Press, 1966
814. Shionoya Sh.: Review of luminescence in II-VI compounds. J. Luminescence 1–2, 17–38 (1970)

815. Schwartz K. K., Grant Z. A., Megis T. K., Grube M. M.: Thermoluminescent Dosimetry. Riga: Zinatne, 1968
816. Siegel F. R., Vaz J. E., Ronca L. B.: Thermoluminescence of clay minerals. In: Thermoluminescence of Geological Materials 635–641. London - New York: Academic Press, 1968
817. Sild O. I.: Theory of Luminescence Center in Crystal. Tartu: University edition, 1968
818. Sippel R. F.: Geologic applications of cathodoluminescence. Proc. Int. Conf. Luminescence, Budapest, 2, 2079–2084 (1968)
819. Sobolev E. V., Ilyin V. E., Lenskaya S. V., Yurieva O. P.: On platelets defects effects in absorption and luminescence excitation spectra of natural diamonds. Zh. Prikl. Spectrosk 9, 654–657 (1968)
820. Sobolev E. V., Ilyin V. E., Yurieva O. P.: Electronphonon interactions in some electron-vibrational series of diamond luminescence spectra. Phys. tverd. tela 11, 1152–1158 (1969)
821. Sokolov V. A.: Candoluminescence. Tomsk Univ. edition, 1967
822. Sorokin P. P.: Transitions of RE^{+2} ions in alkaline earth halide lattices. In: Electronique Quantique; C. R. 3 Conf. Int. 985–997. Paris - New York: Columbia Univ. Press, 1964
823. Steinmetz H., Brüll E.: Über die Thermolumineszenz des Fluorits. Neues Jahrb. Mineral. Monats 11, 333–346 (1967)
824. Stepanov B. I., Gribnovsky V. P.: Introduction to Theory of Luminescence. Minck: Akad. Nauk, 1963
825. Stolyarov K. P., Grigoriev N. N.: Introduction to Luminescent Analysis of Inorganic Compounds. Leningrad: Chimia, 1967
826. Sugibuchi K.: Photosensitive electron spin resonance of Sn^{3+} in zinc sulfide. Phys. Rev. 153, 404–406 (1967)
827. Sunta C. M.: Thermoluminescence spectrum of gamma irradiated natural calcium fluoride. J. Phys.; C: Solid State Phys. 3, 1978–1983 (1970)
828. Tarashchan A. N., Marfunin A. S.: On the luminescence of apatites. Izv. Akad. Nauk, Ser. Geol. 3, 102–108 (1969)
829. Tarashchan A. N., Platonov A. N.: On the linear luminescence spectra of natural calcites. In: Constitution and Properties of Minerals 5, 77–82, 1971
830. Tarashchan A. N., Serebrennikov A. I., Platonov A. N.: Features of lead ions luminescence in amazonite. In: Constitution and Properties of Minerals 7, 106–111, 1973
831. Tarashchan A. N., Platonov A. N.: Isomorphism and valence transoformations of rare earths in natural anhydrites. Geochimia 8, 1203 (1973)
832. Tarashchan A. N., Platonov A. N., Bershov L. V., Belichenko V. P.: On the luminescence of molecule centers in sodalite. In: Constitution and Properties of Minerals 4, 63–65, 1970
833. Tarashchan A. N., Krasilshchikova O. A., Platonov A. N., Povarennych A. S.: Luminescence of uranyl minerals. In: Constitution and Properties of Minerals 8, 78–85, 1974
834. Tarashchan A. N.: Luminescence of Minerals. Kiev: Naukova dumka, 1978
835. Teegarden K.: Halide lattices. In: Luminescence of Inorganic Solids, 53–119. New York: Academic Press, 1966
836. Tite M. S.: Some complicating factors in thermoluminescent dating and their implications. In: Thermoluminescence Geological Materials 389–405. London - New York: Academic Press, 1968
837. Townsend P. D., Clark C. D., Levy P. W.: Thermoluminescence in lithium fluoride. Phys. Rev. 155, 908–917 (1967)
838. Townsend P. D.: Measurement of charge trapping centers in geological specimens. In: Thermoluminescence of Geological Materials 51–60. London - New York: Academic Press, 1968

839. Trofimov A. K.: Origin of the linear luminescence spectrum of zircons. Geochimia *11*, 972–981 (1962)
840. Trukhin A. N., Silin A. P., Landa L. M.: Study of quartz luminescence. Izv. Akad. Nauk, Ser. Phys. **33**, 911–914 (1969)
841. Turin V. I.: Correlation of carbonates deposits by gamma-thermoluminescence method. Proc. Kuibyshev Inst. Neft. Prom. *26*, 105–117 (1964)
842. Uitert L. G. Van: Mechanisms of energy transfer involving trivalent Sm, Eu, Tb, Dy and Yb. Proc. Int. Conf. Luminescence, Budapest, *2*, 1588–1603 (1968)
843. Vasilkova N. N., Solomkina S. G.: Typomorph features of Fluorite and Quartz. Moscow: Nedra (1965)
844. Vaz J., Eduardo, Senftle F. E.: Thermoluminescence study of the natural radiation damage in zircon. J. Geophys. Res. *76*, 2038–2050 (1971)
845. Vaz J. E., Senftle F. E.: Geologic age dating of zircon using thermoluminescence. Mod. Geol. *2*, 239–245 (1971)
846. Vilutis E. S., Penzina E. E.: Excitation and emission spectra of diamonds from Yakutia. Opt. Spectrosk. *18*, 446–449 (1965)
847. Volodko L. V., Komyak A. I., Sleptsov L. E.: Absorption spectra and tentative classification of electronic states of uranyl compounds. Opt. Spectrosk. *23*, 730–736 (1967)
848. Voronko Yu. K., Denker B. I., Osiko V. V.: X-ray luminescence of CaF_2:TR^{3+} crystals. Phys. tverd. tela *13*, 2193–2197 (1971)
849. Voronko Yu. K., Dmitruk M. V., Maximova G. V., Osiko V. V., Timoshechkin M. I., Shcherbakova I. A.: Reduced absorption of Nd^{3+} ion in different matrices. Zh. Exp. Teor. Phys. *57*, 117–124 (1969)
850. Whote Ch. E., Arganer R. J.: Fluorescence analysis. A Practical Approach. New York: Dekker, 1970
851. Wight D. R., Dean P. J.: Extrinsic recombination radiation from natural diamond: exciton luminescence associated with the N 9 center. Phys. Rev. *154*, 689–696 (1967)
852. Williams F.: Theoretical basis for solid-state luminescence. In: Luminescence of Inorganic Solids, 1–52. London - New York: Academic Press, 1966
853. Williams F.: Theoretical aspects of point and associated luminescent centers. In: Proc. Int. Conf. Luminescence, Budapest *1*, 113–123 (1968)
854. Zeller E. J., Wray J. L., Daniels F.: Factors in age determination of carbonate sediments by thermoluminescence. Bull. Am. Assoc. Petrol. Geol. *41*, 1212–1219 (1957)

Electron-Hole Centers in Minerals

855. Alig R. C.: Theory of photochromic centers in CaF_2. Phys. Rev. *3*, 536–545 (1971)
856. Anderson J. H., Weil J. A.: Paramagnetic resonance of color centers in quartz. J. Chem. Phys. *31*, 427 (1959)
857. Anderson C. H., Sabisky E. S.: EPR studies of photochromic CaF_2. Phys. Rev. *3*, 527–536 (1971)
858. Andersson L. O.: EPR of hydrogen atoms in beryl. Proc. 18th Congress AMPERE *18*, 129–130 (1974)
859. Atkins P. W., Symons M. C. R.: The structure of inorganic radicals. An Application of Electron Spin Resonance to the Study of Molecular Structure. Amsterdam: Elsevier, 1967
860. Archangelskaya V. A., Erofeichev V. G., Kiseleva M. N.: Autolocalised hole centers in fluorite type crystals activated with rare earths. Phys. tverd. tela *11*, 2008–2010 (1969)
861. Bakhtin A. I., Khasanov R. A., Vinokurov V. M.: EPR and optical absorption spectra of several defect centers in barytes and celestines. In: Composition, Structure, and Properties of Minerals, 84–90. Kazan: University edition, 1973

862. Balitsky V. S., Samoilovich M. I.: EPR of Ga⁴⁺ in irradiated synthetic quartz. In: Physical Studies of Quartz, 27–31. Moscow: Nedra, 1975

863. Ballentyne D. W. G., Bye K. L.: The nature of photochromism in chlorosodalites from optical data. J. Phys. D. Appl. Phys. *3*, 1438–1443 (1970)

864. Barry T. I., McNamara P., Moore W. J.: Paramagnetic resonance and optical properties of amethyst. J. Chem. Phys. *42*, 2599–2606 (1965)

865. Bambauer H. U.: Spurenelementengehalte und γ-Farbzentren in Quarzen aus Zerrklüften der Schweizer Alpen. Schweiz. Mineral Petrogr. Mitt. *41*, 337–369 (1961)

866. Bershov L. V., Marfunin A. S.: Paramagnetic resonance of electron-hole centers in minerals. Dokl. Akad. Nauk *173*, 410–412 (see also: Papers and Proc. 5th General IMA Meeting, Cambridge, 1966) (1967)

867. Bershov L. V., Tarashchan A. N., Samoilovich M. I., Lushnikov V. G.: Electron-hole centers in natural calcites with phosphors impurities. Zh. Strukt. Chim. *5*, 309–311 (1968)

868. Bershov L. V., Mineeva R. M.: Hyperfine interaction of Pb³⁺ in calcite. Phys. tverd. tela *11*, 803–804 (1969)

869. Bershov L. V., Martirosian V. O., Platonov A. N., Tarashchan A. N.: Stable inorganic radicals study in sodalite single crystals. Neorg. Mater. *5*, 1780–1784 (1969)

870. Bershov L. V.: Isomorphism of titanium in natural minerals. Izv. Akad. Nauk *12*, 47–54 (1970)

871. Bershov L. V.: Methans and atomic hydrogen in some natural minerals. Geochimia *7*, 853–856 (1970)

872. Bershov L. V.: On the isomorphism of Tb⁴⁺, Tu²⁺ and Y³⁺ in zircons. Geochimia *1*, 48–53 (1971)

873. Bershov L. V., Martirosyan V. O., Zavadovskaya E. K., Starodubtsev V. A.: Point defects and radiation damage in alumino-phosphate glasses. Neorg. Mater *7*, 476–480 (1971)

874. Bershov L. V., Martirosyan V. O., Marfunin A. S., Speranskii A. V.: The yttrium-stabilised electron-hole centers in anhydrite. Phys. Status. Solid (b) *44*, 505–512 (1971)

975. Bershov L. V., Martirosyan V. O., Marfunin A. S., Speranskii A. V.: EPR and structure models for radical ions in anhydrite crystals. Fortschr. Mineral. *52*, 591–604 (1975)

876. Biederbick R., Born G., Hofstaetter A., Scharman A.: EPR investigation on the hole centers in CaWO₄ at T = 4.2 K. Phys. Status. Solidi. (b) *69*, 55–62 (1975)

877. Bill H.: Investigation on colour centrs in alcaline earth fluorides. Helv. Phys. Acta *42*, 771–797 (1969)

878. Born G., Hofstaetter, Scharman A., Vitt B.: Anisotropic hyperfine interaction of Pb³⁺ ions in S₁/₂ - state EPR. Phys. Status. Solidi. (b) *66*, 305–308 (1974)

879. Bower H.J., Symons M. C. R., Tinling D. J. A.: The structure of inorganic radicals. In: Kaiser, E. T., Kevan, L. (eds): Radical Ions. New York: Interscience Publishers, 1969

880. Castner T. G., Jr.: Spin-lattice relaxation of the O₂⁻ molecule - ion in the potassium halides. Phys. Kondens. Mater. *12*, 104–130 (1970)

881. Carrington A.: Microwave Spectroscopy of Free Radicals. London - New York: Academic Press, 1974

882. Chakraborty D., Lehmann G.: On the structures and orientations of hydrogen defects in natural and synthetic quartz crystals. Phys. Status. Solidi. (a) *34*, 467 (1976)

883. Chantry G. W., Horsfield A., Morton J. R., Rowlands J. R., Whiffen D. H.: The optical and electron resonance spectra of SO₃⁻. Mol. Phys. *5*, 233–240 (1962)

884. Chen Y., Sibley W. A.: Study of Ionization-Induced radiation domage in MgO. Phys. Rev. *154*, 842–850 (1967)

885. Clark H. C., Horfield A., Symons M. C. R.: Unstable intermediates. P. XII. The radical-ions SO₂⁻ and NO₂²⁻. J. Chem. Soc. 7–11 (1961)

886. Cohen A. J., Hassan F.: Ferrous and ferric ions in synthetic α:quartz and natural amethyst. Am. Mineral. *59*, 719–729 (1974)

887. Collongues R.: La Non-Stoechiometrie. Paris: Masson, 1971

888. Compton W. D., Rabin H.: F-aggregate centers in alkali halide crystals. Solid State Phys. 16 (1964)

889. De Boer J. H.: Über die Natur der Farbzentren in Alkalihalogenid-Kristallen. Rec. Trav. Chim. *56*, 301–309 (1937)

890. Delbecq Ch. J., Kolopus J. L., Yasaitis E. L., Yuster Ph. H.: Correlation of the optical and electron-spin resonance absorptions of the H center in KCl. Phys. Rev. *154*, 866–872 (1967)

891. De Lisle J. M., Golding R. M.: ESR study of X-irradiated sodium thiosulfate single crystals. J. Chem. Phys. *43*, 3298–3303 (1965)

892. Dennen W. H.: Stoichiometric substitution in natural quartz. Geochim. Cosmochim. Acta *30*, N 12 (1966)

893. Dines G. J., Hatcher R. D., Smoluchowski R.: Structure and stability of H-centers. Phys. Rev. *157*, 692–700 (1967)

894. Dreybrodt W., Silber D.: EPR measurements on diatomic halogen centers in alkali halides. Phys. Status. Solidi. *16*, 215–223 (1966)

895. Deigen M. F., Ruban M. A., Gromovoi Yu. S.: Electron nuclear double resonance of F centers in KCl at room temperature. Phys. tverd. tela *8*, 826–831 (1966)

896. Ebert I., Henning H. P.: Electron spin resonance of mechanically activated quartz. Phys. Chem. *225*, 812–814 (1974)

897. Edgar A., Vance E. R.: Electron paramagnetic resonance, optical absorption, and magnetic circular dichroism of the CO_3^- molecular-ion in irradiated natural beryl. Phys. Chem. Mineral. *1*, 165–177 (1977)

898. Evgrafova L. A., Gainullina N. M., Nisamutdinov N. M., Vinokurov V. M.: On the nature of the electron and hole centers in phenacite single crystal. Phys. Mineral. (Kazan) *3*, 14–22(1971)

899. Feigl F. J., Anderson J. H.: Defects in crystalline quartz: electron paramagnetic resonance of E vacancy centers associated with germanium impurities. J. Phys. Chem. Solids *31*, 575–596 (1970)

900. Fong F. K.: Lattice defects, ionic conductivity, and valence change of rare-earth impurities in alkaline earth halides. Progr. Solid. State. Chem. *3*, 135–212 (1967)

901. Free radicals in inorganic chemistry: Papers presented at the symposium on inorganic free radicals and free radicals inorganic chemistry. Adv. Chem. Series, 36 (1962)

902. Gainullina N. M., Evgrafova L. A., Nisamutdinov N. M., Vinokurov V. M.: EPR of electron-hole centers in zircon crystals. Phys. Mineral. (Kazan) *3*, 3–13 (1971)

903. Gamble F. T., Bartram R. H., Young C. G., Gilliam O. R., Levy P. W.: Electron-spin resonance in gamma-ray-irradiated aluminum oxide. Phys. Rev. *134*A, 589–595 (1964)

904. Gilinskaya L. G., Shcherbakova M. Ya., Zanin Yu. N.: Carbon in the structure of apatite according to electron paramagnetic resonance data. Sov. Phys.: Crystallography *15*, 1016–1019 (1971)

905. Gilinskaya L. G., Zanin Yu. N., Shcherbakova M. Ya.: Isomorphism in apatite of continental phosphorites according to electron paramagnetic resonance data. Litol. Polezn. Iskop. *6*, 111–120 (1973)

906. Görlich P., Karras H., Symanowski Ch., Ullmann P.: The colour centre absorption of X-ray coloured alkaline earth fluoride crystals. Phys. Status. Solidi. *25*, 93–101 (1968)

907. Gool W. van: Principles of defect chemistry of crystalline solids. L. (1966)

908. Griffiths J. H. E., Owen J., Ward J. M.: Magnetic resonance in irradiated diamond and quartz. Reports of the Bristol Conference. London: Phys. Soc (1954)

909. Gromov V. V., Karaseva L. G.: Defects in β, γ-irradiated sulfates of second row elements. In: Radiation Physics 4, 49–54. Riga (1966)

910. Grunin V. S., Ioffe V. A. The radiation centers in quartz, cristobalite and vitreous silica. J. Non-Cryst. Solids, 6, N 2, 163–169 (1971)
911. Gupta N. M., Luthra J. M., Sastry M. D.: Thermoluminescence and ESR of γ-ray and pile-irradiated natural barite. J. Luminescence 10, 305–312 (1975)
912. Hassan F., Cohen A. J.: Biaxial color centers in amethyst quartz. Am. Mineral. 59, 709–718 (1974)
913. Hayes W., Twidell J. W.: The self-trapped hole in CaF_2. Proc. Phys. Soc. 79, 1295–1296 (1962)
914. Hee Y. L.: Nonthermoluminescent smoky color center in γ-irradiated α-quartz. J. Appl. Phys. 39, 4850–4851 (1968)
915. Holton W.G., Blum H.: Paramagnetic resonance of F-centers in alkali halides. Phys. Rev. 125, 89–103 (1962)
916. Holzrichter J. F., Emmett J. L.: Transient color centers in fused quartz. J. Appl. Phys. 40, 159–163 (1969)
917. Horsfield A., Morton J. R., Whiffen D. H.: Electron spin resonance and structure of the ionic radical, PO_3^{2-}. Mol. Phys. 4, 475–480 (1961)
918. Hulton D. D.: Paramagnetic resonance Fe^{3+} in amethyst and citrine quartz. Phys. Lett. 12, 310–311 (1964)
919. Ikeya M.: Dating a stalactite by electron paramagnetic resonance. Nature (London) 255, N 5503, 48–50 (1975)
920. Ingram D. J. E.: Free Radicals as Studied by Electron Spin Resonance. London: Butterworths Scientific Publ., 1958
921. Ioffe V. A, Yanchevskaya I. S : EPR and thermoluminexcence study of irradiated $NaAlSi_3O_8$ and $LiAlSiO_4$ single crystals. Phys. tverd. tela 10, 472–477 (1968)
922. Jain V. K.: Behavior of divalent cation impurities in alkali halide crystals. Phys. Status Solidi (b) 44, 11–28 (1971)
923. Johnson P. D., Prener J. S., Kingsley J. D.: Apatite: origin of blue color. Science 141, 1179–1180 (1963)
924. Känzig W., Woodruff T. O.: The electronic structure of an H-center. J. Phys. Chem. Solids 9, 70–92 (1958)
925. Känzig W.: Trapped holes in alkali halides. Proc. Int. Confer. Semicond. Phys., Prague, 705–711 (1961)
926. Kasai P.H.: Electron spin resonance studies of γ-and X-ray - irradiated zeolites. J. Chem. Phys. 43, 3322–3327 (1965)
927. Kats A.: Hydrogen in α-quartz. Phillips Res. Rep. 17, N 1–2, 133–168 (1962)
928. Knottnerus D. I. M., Hartog H. W. den, Lanjouw A.: F-centers in calcium chlorapatite doped with NaCl. Phys. Status Solidi. 32, 287–295 (1975)
929. Kolopus J. L., Lapeyre G. J.: ESR studies of radiation damage centers in barium sulfide. Phys. Rev. 176, 1025–1029 (1968)
930. Kolopus J. L., Finch C. B., Abraham M. M.: ESR of Pb^{3+} centers in ThO_2. Phys. Rev. B2, 2040–2045 (1970)
931. Komov I. L., Samoilovich M. I., Khetchikov L. N., Tsinober L. I.: Electron paramagnetic resonance in natural quartz crystals. In: Physical Studies of Quartz, 47–51. Moscow: Nedra, 1975
932. Koryagin V. F., Grechushnikov B. N.:Electron paramagnetic resonance spectrum of atomic hydrogen in beryl. Sov. Phys.: Solid State 7, 2010–2012 (1966)
933. Koryagin V. F., Grechushnikov B. N.: Photosensitive paramagnetic centers in calcite crystals. Kristallographia 15, 985–987 (1970)
934. Kozlovsky V. Kh., Kuznetsova N. N., Snopko Ya. P.: Study of the nature of yellow color of Iceland spar crystals and mechanism of its bleaching. Proc. Vses. Inst. Synth. Miner. 10, 118–129 (1969)
935. Kreiskop V. N., Samoilovich M. I., Tsinober L. I.: EPR and optical spectra of point defects in crystalline quartz. In: Phys. Studies of Quartz, 5–11. Moscow: Nedra, 1975
936. Lehmann G.: Farbzentren des Eisens als Ursache der Farbe von Amethyst. Z. Naturforsch. 22a, 2080–2085 (1967)

937. Lehmann G.: Yellow color centers in natural and synthetic quartz. Phys. Kondens. Mater. *13*, 297–306 (1971)

938. Lehmann G., Bambauer H. U.: Quartz crystals and their colors. Angew. Chem. *85*, 281–289 (1973)

939. Lehmann G.: On the color centers of iron in amethyst and synthetic quartz: a discussion. Am. Mineral. *60*, 335–337 (1975)

940. Loubser J. H. N., Wright A. C. J.: A singly ionized N-C-N centre in diamond. J. Phys. D: Appl. Phys. *6*, 1129–1141 (1973)

941. Lüty F.: Farbzentren in Fluβspat. Z. Phys. *134*, 596–603 (1953)

942. Lushchik Ch. B., Elango M. A.: On the mechanisms of radiation coloring of ionic crystals. Proc. Inst. Phys. Astr. Est. Akad. Nauk *26*, 93–111 (1964)

943. Mackey J. H.: ESR study of impurity-related color centers in germanium doped quartz. J. Chem. Phys. *33*, 74–83 (1963)

944. Mackey J. H., Boss J. W., Wood D. E.: EPR study of substitutional-aluminum-related hole centers in synthetic α-quartz. J. Magn. Reson. *3*, 44–54 (1970)

945. Marfunin A. S., Bershov L. V.: Electron-hole centers in feldspars and their possible crystalchemical and petrological significance. Dokl. Akad. Nauk *193*, 412–414 (1970)

946. Marfunin A. S.: Les impurétés de substitution isomorphe et la formation de radicaux libres dans les minéraux (centres à electrons et centres à trous). Bull. Soc. Fr. Mineral. Crist. *97*, 194–201 (1974)

947. Markham J. J.: F-centers in alkali halides. Solid State. Phys. Suppl. 8 (1966)

948. Marshall S. A., McMillan J. A.: Electron spin resonance absorption spectrum of CO_2^- molecule ions associated with F^- ions in single-crystal calcite. J. Chem. Phys. *1*, 4887–4890 (1968)

949. Marshall S. A., McMillan J. A., Serway R. A.: Electron spin resonance absorption spectrum of Y^{3+}-stabilized CO_3^{3-}molecule-ion in single-crystal calcite. J. Chem. *48*, 5131–5137 (1968)

950. Mason D. R., Koehler H. A., Kikuchi C. Identification of defect sites in $CaWO_4$ from the correlation of ESR and thermoluminescence measurements. Phys. Rev. Lett. *20*, 451–452 (1968)

951. McLaughlan S. D., Marshall D. J.: Paramagnetic resonance of sulfur radicals in synthetic sodalites. J. Phys. Chem. *74*, 1359–1363 (1970)

952. McMillan J. A., Marshall S. A.: Motional effects in the ESR absorption spectrum of CO_2^- in single-crystal calcite. J. Chem. Phys. *48*, 1471–1476 (1968)

953. McMorris D. W.: ESR detection of fossil alpha damage in quartz. Nature (London) *226*, 146–148 (1970)

954. Meistrich M. L.: ESR and optical studies of O_2^- in NaF. J. Phys. Chem. Solids *29*, 1111–1118 (1968)

955. Mitchell E. W., Paige E. G.: The optical effects of radiation induced atomic damage in quartz. Phil. Mag. *1*, 1–10 (1956)

956. Minkoff G.J.: Frozen Free Radicals. New York: Interscience Publ., 1960

957. Morton J. R.: Identification of some sulfur containing radical trapped in single crystals. J. Phys. Chem. *71*, 89–92 (1967)

958. Nassau K., Prescott B. E.: Blue and brown topaz produced by gamma irradiation. Am. Mineral. *60*, 705–709 (1975)

959. Nassau K.: Gamma ray irradiation induced changes in the color of tourmalines. Am. Mineral. *60*, 710–713 (1975)

960. Nassau K., Prescott B. E., Wood D. L.: The deep blue Maxixe-type color center in beryl. Am. Mineral. *61*, 100–107 (1976)

961. Neeley V. I., Gruber J. B., Gray W. J.: F Centers in thorium oxide. Phys. Rev. *158*, 809–813 (1967)

962. Nelson R. L., Tench A. J., Harmsworth B. J.: Chemisorption on some alkaline earth oxides. P. I. Surface centres and fast irreversible oxygen adsorption on irradiated MgO, CaO and SrO. Trans. Faraday. Soc. *63*, P. 6, 1427–1446 (1967)

963. Novozhilov A. I., Samoilovich M. I., Mikulskaya E. K.: On the nature of blue color in cancrinite group minerals. Zap. Vses. Mineral. obshch. *95*, 736–738 (1966)

964. Novozhilov A. I., Samoilovich M. I., Sergeev-Babr A. A., Anikin I. N.: EPR of irradiated fluorphologopite. Zh. Strukt. Chim. *10*, 450–453 (1969)

965. Novozhilov A. I., Voskresenskaya I. E., Samoilovich M. I.: EPR study of tourmalines. Kristallographia *14*, 507–509 (1969)

966. O'Brien M. C. M., Pryce M. H. L. Paramagnetic resonance in irradiated diamond and quartz: interpretation. Defects in crystalline solids, L., 1955.

967. Ovenall D. W., Whiffen D. H. Electron spin resonance and structure of the CO_2^- radical ion. Mol. Phys. *4*, 135–144 (1961)

968. Fowler, W. B. (ed.): Physics of color centers, 1968

969. Pick, H.: Farbzentren-Assoziate in Alkalihalogeniden. Z. Physik *159*, 68–76, 1960

970. Piper W. W., Kravitz L. C., Swand R. K.: Axially symmetric paramagnetic color centers in fluorapatite. Phys. Rev. *138*A, 1802–1814 (1965)

971. Platonov A. N., Krasilshchikova O. A., Tarashchan A. N.: Color centers in apatite. In: Constitution and Properties of Minerals *7*, 64–75, 1973

972. Pustylnikov A. M.: On the origin of blue color of halite from Cambrium salt deposits of sibirian platform. Litol. Polezn. Iskop. *3*, 152–157 (1975)

973. Rabo J. A., Angell C. L., Kasai P. H., Schomaker V.: Studies of cations in zeolites: adsorption of carbon monoxide; formation of Ni ions and Na_4^{3+} centers. Discuss Faraday Soc. *41*, 328–349 (1966)

974. Radhakrishna S., Chowdari B. V. R.: Z centers in impurity doped alkali halides. Phys. Status Solidi (a) *14*, 11–39 (1972)

975. Royce B. S. H.: The creation of point defects in alkali halides. Progress in Solid-State Chemistry *4*, New York: Pergamon Press, 1967

976. Samoilovich M. I., Novozhilov A. I.: Electron paramagnetic resonance in irradiated topaz. Zh. Strukt. Chim. *6*, 461–463 (1965)

977. Samoilovich M. I., Novozhilov A. I.: EPR spectrum of CH_3, H_2O^+ and OH radicals and atomic hydrogen in beryl. Zh. Neorg. Chim. *15*, 84–86 (1970)

978. Samoilovich M. I., Novozhilov A. I., Radyansky V. M., Davydchenko A. G., Smirnova S. A.: On the nature of blue color of lazurite. Izv. Akad. Nauk *7*, 95–102 (1973)

979. Sayer M., Lynch G. F.: Ultra-violet excited paramagnetic centres in calcium tungstate. Phys. Status Solidi *37*, 673–681 (1970)

980. Schneider J., Dischler B., Räuber A.: Electron spin resonance of sulfur and selenium radicals in alkali halides. Phys. Status. Solidi. *13*, 141–157 (1966)

981. Schoemaker D.: $\langle III \rangle$ - Oriented FCl^-, FBr^-, and FJ^- centers in mixed alkali halides. Phys. Rev. *149*, 693–704 (1966)

982. Schulman J. H., Compton W. D.: Color Centers in Solids. New York: Pergamon Press, 1963

983. Seitz F.: Color centers in alkali halides crystals. Rev. Mod. Phys. *26*, 7 (1954)

984. Serway R. A., Marshall S. A.: Electron spin resonance absorption spectrum of orthorhombic CO_3 molecule-ions in irradiated single crystal calcite. J. Chem. Phys. *47*, 868–869 (1967)

985. Serway R. A., Chan S. S. L., Marshall S. A.: Temperature dependence of the hyperfine structure splittings of XO_3 molecule - ions in single-crystal calcite. Phys. Status Solidi (b) *57*, 269–276 (1973)

986. Shcherbakova M. Ya., Sobolev E. V., Samsonenko N. D., Axenov V. K.: Electron paramagnetic resonance of ionised nitrogen pair in diamond. Phys. tverd. tela *11*, 1364–1367 (1969)

987. Shcherbakova M. Ya., Sobolev E. V., Nadalinny W. A., Axenov V. K.: Defect in plastic deformated diamonds according to optical and EPR spectra. Dokl. Akad. Nauk *225*, 566–569 (1975)

988. Sierro J.: Paramagnetic resonance of the V_F center in CaF_2. Phys. Rev. *138*A, 648–650 (1965)

989. Silsbee R. N.: Electron spin resonance in neutron irradiated quartz. J. Appl. Phys. *32*, 1459–1462 (1961)
990. Silsbee R. H.: R center in KCl stress effects in optical absorption. Phys. Rev. *138*, A180–197 (1965)
991. Smith W. V., Sorokin P. P., Gelles I. L., Lasher G. J.: Electron-spin resonance of nitrogen donors in diamond. Phys. Rev. *115*, 6 (1959)
992. Sobolev E. V., Samsonenko N. D., Ilyin V. E., Axenov V. K., Shcherbakova M. Ya.: On most common state of nitrogen in natural diamonds. Zh. Strukt. Chim. *10*, 552–553 (1969)
993. Sobolev E. V.: On the nature of yellow color of diamonds. Geol. Geoph. *12*, 127–129 (1969)
994. Solntsev V. P., Shcherbakova M. Ya.: EPR study of structure defects in irradiated zircons. Dokl. Akad. Nauk *212*, 156–158 (1973)
995. Solntsev V. P., Shcherbakova M. Ya.: EPR study of structure defects in $CaWO_4$. Zh. Strukt. Chim. *14*, 222–229 (1973)
996. Solntsev V. P., Shcherbakova M. Ya., Dvornikov E. V.: Radicals SiO_2^-, SiO_3^- and SiO_4^{5-} in $ZrSiO_4$ structure according to EPR data. Zh. Strukt. Chim. *15*, 217–221 (1974)
997. Solntsev V. P., Lysakov V. S.: EPR and luminescence study of electron trapping centers in irradiated quartz. Zh. Prikl. Spektrosk. *22*, 450–452 (1975)
998. Solozhenkin P. M., Glembotsky V. A., Emelyanov A. F., Kopitsya N. I.: Effect of γ-irradiation on the flotation of some aluminosilicates and phosphates. Izv. Akad. Nauk Tadzh. SSR, Ser. Phys.-Math.-Geol.-Chim. *2*, 58–62 (1971)
999. Staebler D. L., Schmatterly S. E.: Optical studies of a photochromic color center in rare-earth doped CaF_2. Phys. Rev. *3*, 516–526 (1971)
1000. Stamires D. N., Turkevich J.: Paramagnetic resonance absorption of γ-irradiated synthetic zeolites. J. Am. Chem. Soc. *86*, 757–761 (1964)
1001. Sternlicht H.: Polycrystalline resonance line shape of π-electron radicals. J. Chem. Phys. *33*, 1128 (1960)
1002. Stevels J. M., Kats A.: Systematics of imperfections in silicon-oxygen networks. Phillips Res. Rep. *11*, 103–112 (1956)
1003. Stoneham A. M.: Theory of Defects in Solids. Oxford University. Press, 1975
1004. Speit B., Lehmann G.: Hole centers in the feldspar sanidine. Phys. Status. Solidi. (a) *36*, 471–481 (1976)
1005. Swank R. K.: Color centers in X-irradiated halophosphate crystals. Phys. Rev. *135* A, 280–282 (1964)
1006. Tench A. J., Nelson R. L.: Electron spin resonance of F centres in irradiated ^{43}CaO and other alkaline earth oxides. Proc. Phys. Soc. *92*, 1055–1063 (1967)
1007. Tench A. J., Lawson T., Kibblewhite J. F. J.: Oxygen species adsorbed on oxides. J. Chem. Soc. Faraday Trans. I, *68*, 1169 (1972)
1008. Tsay Fun-Dow, Chan S. I., Manatt S. L.: Electron paramagnetic resonance of radiation damage in a lunar rock. Nature. Phys. Sci., *237*, 121–122 (1972)
1009. Ursu J., Mistor S. V.: ESR studies of some radiation paramagnetic centers in alkali halides. In: Magnetic Resonance and Related Phenomena 166–176. Amsterdam-London: North - Holland, 1973
1010. Unruh W. P., Culvahouse J. W.: Electron-nuclear double resonance of F centers in MgO. Phys. Rev. *154*, 861–866 (1967)
1011. Vannotti L. E., Morton J. R.: Paramagnetic-resonance spectra of S_2^- in alkali halides. Phys. Rev. *161*, 282–284 (1967)
1012. Vehse W. E., Sibley W. A., Keller F. J., Chen Y.: Radiation damage in ZnO single crystals. Phys. Rev. *167*, 828–836 (1968)
1013. Wang K. M., Lunsford J. H.: An electron paramagnetic resonance study of Y-type zeolites. III. O_2^- on AlHY, ScY, and LaY zeolites. Phys. Chem. *75*, 1165–1168 (1971)

1014. Weeks R. A.: Paramagnetic spectra of E_2 centers in crystalline quartz. Phys. Rev. *130*, 570–576 (1963)

1015. Weeks R. A., Abraham M.: Electron spin resonance of irradiated quartz: atomic hydrogen. J. Chem. Phys. *42*, 68–71 (1965)

1016. Weeks K. A.: Some defect electron states of pure fourfold coordination oxides: expectations and realization. In: Interaction of Radiation with Solids. New York: Plerun Press, 1967

1017. Wickramasinghe N. C.: Irradiated quartz particles as interstellar grains. Nature Phys. Sci. *234*, 7–10 (1971)

1018. Wright P. M., Weil J. A., Buch T., Anderson J. H.: Titanium color centers in rose quartz. Nature (London) *197*, 246–248 (1963)

1019. Wong H.-B., Lunsford J. H.: EPR study of O_3^- on magnesium oxide. J. Chem. Phys. *56*, 2664–2666 (1972)

1020. Yatsiv S., Peled S., Rosenwaks S.: Spectra of H^- in CaF_2 containing trivalent rare earth ions. Optical properties of ions in crystals. Crosswhite, H. M., Moos, H. W. (eds.) 1967, 1967

1021. Zeldey H., Livingston R. Paramagnetic resonance study of irradiated single crystals of calcium tungstate. J. Chem. Phys. *34*, 247–252 (1961)

1022. Zeller E. J., Levy P. W., Mattern P. L.: Geologic dating by electron spin resonance. Radioactive Dating and Meth. Low-Level Count. Vienna, 531–539. Discuss. 539–540, 1967

1023. Zwingel D.: The electronic structure of trapped hole centers in SnO_2. Phys. Status Solidi (b) *77*, 171–170 (1976)

Subject Index

Absorption edge in X-ray spectra   47, 52–54
Acceptors   179–181, 185
Actinides
  absorption and luminescence spectra   211
  electron configurations   212
  EPR   85
Adamite, EPR of Fe^{3+} 117
Adularia
  absorption and luminescence of Fe^{3+} 197
  EPR of Fe^{3+} 116
  NMR of Al^{27} 135
Aegirine-augite, NGR of Fe^{57} 26
Akaganeite, NGR of Fe^{57} 32, 35
Albite
  electric field gradient   135
  EPR of Fe^{3+} 116
  mass absorption coefficient   52
  NMR and order-disorder of Na^{23} and Al^{27} 134
  O^- center   278
Alkali halide crystals
  defects and centers   242–246, 289
  luminescence   216, 298
  models of the centers   299
  thermoluminescence   231
Altaite, EPR of Mn^{2+} 118
Amblygonite
  EPR of Ti^{3+}, V^{4+}, Fe^{3+} 117
  NMR of Li^7, Al^{27}, P^{31}, H^1, F^{19} 122
  NMR and Li distribution   135, 136
  O^- center   257
Amethyst, EPR of the electron centers and color   272, 278
Analcime, proton magnetic resonance   131
Andalusite
  electric field gradient   135
  EPR of Fe^{3+}, Cr^{3+} 117
Anhydrite
  dosimeter   235
  luminescence   211
  models of the centers   284
  thermoluminescence   235, 236
Anorthite EPR of Fe^{3+} 116
  mass absorption coefficient   52
  O^- center   278

Antibonding molecular orbitals and electron centers   248
Antophyllite, NGR of Fe^{57} 24, 25
Apatite
  EPR of Mn^{2+} in three sites   113, 117
  laser material   211
  luminescence   208–211
  models of the centers and colors   258, 286
  NMR of OH–O–F configuration   135
  NGR of implanted Fe^{57} 20, 35
Apophyllite, EPR of V^{4+}, Mn^{2+}, Ti^{3+} 117
Aragonite, EPR of Mn^{2+} 117
Arfvedsonite, cation distribution and NGR 26
Arsenolite, NQR of As^{75} 139
Arsenopyrite, complex radicals and NGR 31
Assignment of X-ray transitions in molecular orbital schemes   59, 60
Astrakhanite, EPR of Mn^{2+} 117
Asymmetry parameter
  in EPR   102
  in NMR   128
  in NQR   137
Augelite
  EPR of V^{4+}, Mn^{2+}, Ti^{3+} 117
  O^- center   258
Auger spectroscopy   58
Augite
  assignment of NGR doublets   24
  NGR spectrum   23

Band theory
  in luminescence   147, 176–185
  in thermoluminescence   224, 226, 234
  in X-ray spectroscopy   54, 55, 59
Bands $X_1 K_1 Q$ in EPR   81
Barite
  models of the centers   284, 285
  radiation stability   241
  thermoluminescence   238
Bauxites, phase analysis by means of NGR spectra   35
Berlinite, EPR of Cr^{3+} 117
Berthierite
  complex radicals and NGR of Fe^{57} 29–31
  quadrupole splitting   13

Beryl
  atomic hydrogen   267, 284
  EPR of Cr^{3+}, Fe^{3+}, Mn^{2+} 117
  EPR of Ti^{3+} 117, 266
  luminescence   195
  methyl radical and CO_3^- radicals   284
  water molecules in structure channels
    (NMR, EPR)   131, 267
Biotite, cation positions and NGR of Fe^{57}
  22, 23–25
Bismite, NQR of Bi^{209} 139
Bismuthinite, NQR of Bi^{209} 139
Bohr magneton   79, 80
Boracite, NGR and ferroelectric properties
  35
Bornite, NGR of Fe^{57} 13, 28–31
Brucite, proton positions (NMR)   133

Calcite
  configuration curves   152
  energy levels and luminescence and
    excitation spectra of Mn^{2+} 150
  EPR of Mn^{2+}, Fe^{3+}, Cu^{2+}, Ag^{2+}, Gd^{3+}
    117, 284
  luminescence of Pb^{2+} 215
  sensitization of luminescence   166
  thermoluminescence   226, 238
  types of the centers and radicals   262, 265,
    266, 268, 284
  uranyl   213
Candoluminescence   189, 193
Cassiterite
  EPR of Fe^{3+}, Cr^{3+} 117
  NGR of Sn^{119} 35
  weak ferromagnetism and NGR of Fe^{57}
    37
Cathodoluminescence   189, 192
Chalcopyrite, NGR of Fe^{57} 28–31
Chemical shift
  in NGR   7–12
  in NMR   125
  in X-ray spectra   72–75
Chemiluminescence   189, 193
Chromites, NGR of Fe^{57} 34
Chrysotile-asbestos, proton magnetic
  resonance   132
Cinnabar, edge luminescence   220
Claudetite, NQR of As^{75} 139
Coesite, X-ray SiK_β spectrum   65
Colemanite
  NMR and B^{11} positions   134, 135
  proton magnetic resonance and ferro-
    electric properties   131
Configuration curves   151–153
Cordierite, EPR of Mn^{2+} 117
Corundum
  electric field gradient calculations   135

energy levels, luminescence and excitation
  spectra, laser emission   149, 195
EPR of Fe^{3+}, Cr^{3+}, Mn^{2+} 117
NGR of iron impurity   18
NMR of Al^{27} 135
X-ray emission and absorption spectra
  65
Cristobalite
  EPR of V^{4+} 117
  X-ray spectrum   65
Crystal field theory
  and EPR spectra   82, 84, 88–90, 94–98
  rare earth ions luminescence   201, 202
  transition metal ion luminescence   194
Cubanite, NGR spectra   29–31
Cummingtonite, cation distribution, and
  NGR spectra   24, 25
Cyanite
  EPR of Cr^{3+}, Fe^{3+}, Ti^{3+} 117
  luminescence and excitation spectra of
    Cr^{3+} 195
  NMR of Al^{27} and electric field gradient
    135
  O^- center   284

Danalite, NGR of Fe^{57} 28
Danburite
  EPR of the centers   283
  NMR of B^{11} and electric field gradient
    135
Datolite
  electron-hole centers   283
  EPR of Mn^{2+} 117
Decay of luminescence   159, 160, 184
Defects (point)   244
Delocalization of electron in radicals   246
Desmine, water mobility, and proton
  magnetic resonance   132
Diamond
  colors   221
  luminescence spectra   220, 221
  models of the centers   220
  NGR of implanted Fe   20, 36
  thermoluminescence   238
Diaspore
  EPR of Fe^{3+} 117
  proton magnetic resonance   133
Diopside
  cation positions   22, 23
  EPR of Mn^{2+} 117
Dioptase, proton magnetic resonance   134
Dolomite, thermoluminescence   238
Donor-acceptor pairs, luminescence in
  sphalerite   185, 219, 220
Donors   179, 180, 185
Doppler shift   7

Effective charge 72–75
Electric field gradient
  calculation 15
  in EPR 99
  in NGR 12
  in NMR 128, 134, 135
  in NQR 137
Electroluminescence 189, 192
Electron Paramagnetic Resonance (EPR)
  comparison with universal stage 251
  fine structure 90, 94, 95
  hyperfine structurs (hfs) 104–109
  identification of the centers 252
  spin sublevels 77, 78, 83
  substances which can be studied 85–87
  superhyperfine structure (shfs) 110, 111
  units of measurements 80, 81
Electron-hole centers 242–246, 254
Emission centers 224
Energy level diagrams
  EPR 77, 78, 83, 105, 107
  in luminescence
    Pb^{2+} 216
    rare earths 151, 152, 198
    uranyl 212
  NGR 8, 14
  NMR 119
  NQR 137
  of the radicals and centers 249–251, 258, 260
  in X-ray spectroscopy (molecular orbitals) 41, 61–63
Epidote, NGR of Fe^{57} 23, 24
Euclase
  EPR of Fe^{3+} 117
  NMR of Al^{27} 135
  types of the centers 283
Excitation of luminescence
  spectra of 148–150
  types of 188–194

g-Factor 77, 88, 89, 91–94
F-aggregate centers 298
Faujasite, proton magnetic resonance 131
F-center
  in alkali halide crystals 289–294
  luminescence 216
  the model 242, 243, 289
  in other compounds 294–297
Feldspars
  electron-hole centers 278–279
  EPR 117
  luminescence 197, 215
  NGR 27
  NMR 134, 135
  thermoluminescence 237, 238
  X-ray spectra 65

Fine structure 94
  intensities relation (EPR) 97
Fluorescence 153, 158, 160
Fluorite
  EPR of Mn^{2+}, Gd^{3+}, Eu^{2+}, RE^{3+}, U^{4+} 117
  laser material 210
  luminescence 206, 208–210
  models of the centers 287, 288
  shfs of EPR of Mn^{2+} 110
  thermoluminescence 233, 234
Free radicals 242, 246, 247
Frequency factor (attempt-to-escape frequency) 228–230, 240

Galena, EPR of Mn^{2+} 117
Goethite
  NGR spectra 32–34, 36
  superparamagnetic 34
Gypsum
  crystal structure and proton magnetic resonance 127
  fine structure in NMR spectra 132
  radiation stability 241
Gyromagnetic ratio 121

Hackmannite, photochromism and types of the centers 281, 282
Halite
  models of the centers 267, 289–294
  sensibilized luminescence of Pb^{2+} 166
Hamiltonian 111, 112
Harmotome, proton magnetic resonance 131
Helvine, NGR of Fe^{57} 28
Hematite
  aluminous in bauxites 35
  electric field gradient calculations 18
  NGR 18, 32, 33
  superparamagnetism 34
Hole center 245, 246, 258, 266
Holes 245
Hybridization, parameters from EPR spectra 104, 109, 248
Hydride ion H^-
  in alkali halide crystals 267
  in fluorite 288
Hydrogen, atomic, in crystals 267, 268, 270, 284, 287, 288, 299
Hydroxyl radical 284
Hyperfine structure (hfs)
  in EPR spectra 104–109
  in NGR 18
Hypersthene, cation distribution and NGR spectra 23, 24

Ilmenite, NGR of ilmenites from different
    localities 33
Ilvaite, Curie point and NGR spectra 28
Inderite, NMR of B^{11} and electric field
    gradient 134
Interactions
    dipole-dipole (in NMR) 124, 125
    electrons with nuclei in NGR and EPR 9
    hyperfine in EPR 104, 106–108
        in NMR 125
        in NQR 137
    hyperfine in EPR 104, 106–108
    with ligand ion nuclei (shfs) 110
    quadrupole in NGR 15
Isomer shift 7–12, 17
    in different sources 9

Kaolinite
    EPR of Fe^{3+} 117
    proton magnetic resonance 132
Kramers and non-Kramers ions 95, 96
Kurnakovite, NMR of B^{11} in trigonal and
    tetrahedral coordination 134

Laser materials 87, 149, 195, 196, 206, 210
Lautite, NQR of As^{75} 139
Lazurite
    luminescence of the molecule ion S_2^- 217
    models of the centers and color 281, 282
Lepidocrocite, NGR spectra of Fe^{57} 32–35
Life time
    of a level (in luminescence) 155–157, 164,
        166, 172
    of a trap center 229, 230
Loellingite, NGR spectra of Fe^{57} 29–31
Luminescence
    band schemes 176–188
    configuration curves 152, 153
    crystal field theory 194–196
    energy level diagrams 151, 152
    energy transfer 160
    excitation of luminescence 188–194
    photosensitive EPR and luminescence
        187, 188
    types of luminescent systems 194
Lunar minerals
    electron-hole centers 269
    luminescence 192
    NGR of Fe^{57} 37

Maghemite, parameters of the NGR spectra
    32, 33
Magnetic fields at nuclei 17, 124
Magnetite, cation distribution, vacancies and
    NGR spectra 32, 33
Maser materials 87

Meteorites
    NGR and classification of meteorites 36
    thermoluminescence 238
Miargyrite, NQR of Sb 139
Microcline
    crystal structure and hole centers 278,
        279
    EPR of Fe^{3+} 116
    order-disorder and NMR spectra 135
    thermoluminescence 237
Models of the centers
    in alkali halide crystals 299
    in anhydrite 284
    in apatite 258, 286
    in baryte 284
    in diamond 220
    in feldspars 278, 279
    in fluorite 287
    in quartz 270–278
    in sheelite 286, 287
    in sodalite-lazurite 281, 282
    in sphalerite 217
    in zircon 283
Molecular ions
    EPR and energy level diagrams 260, 263,
        301–303
    luminescence of S_2^-, O_2^- 216
Molecular orbitals
    EPR parameters of the centers 247, 248
    MO diagrams for AB_2 249, AB_3 250,
        AB_4 251
Muscovite, cation distribution and NGR
    spectra 25

Natrolite
    electric field gradient from NMR of Na^{23}
        135
    EPR of Fe^{3+}, Ti^{3+} 117
    proton magnetic resonance 131
Nepheline, NMR of Na^{23} and Al^{27} 135
Nitrogen centers in diamond 220–227
Nomenclature
    of Auger-electrons 58, 59
    of the transitions from and to molecular
        orbitals 60
    of X-ray emission spectra lines 43–45
Nuclear gamma-resonance spectroscopy
    (NGR) or Mössbauer spectroscopy
    energy levels splitting 8
    isomer transitions 2
    parameters of spectra 7–19
Nuclear magnetic resonance (NMR)
    high-resolution in solids 130
    resonance condition 120
    types of interactions 123

Nuclei
  in NGR 6
  in NMR 123
  in NQR 137
  superhyperfine structure (EPR) 253
Nuclear magneton 120
Nuclear quadrupole resonance (NQR)
  quadrupole levels 137
  parameters of spectra 139
  spin-echo 139
Nuclei
  in NGR 6
  in NMR 123
  in NQR 137
  superhyperfine structure (EPR) 253

Olivines
  crystal structure and NGR spectra 21–25
  EPR of Mn^{2+} (in forsterite) 117
Order-disorder
  EPR spectra 116
  in silicates and NGR spectra 20
Orpiment, nonequivalent As^{75} nuclei in
  NQR spectra 138, 139
Orthoclase
  EPR of Fe^{3+} 116
  NGR of Fe^{3+} in iron orthoclase 27
  O^- center 278
Oxyanions of third row elements
  molecular orbital diagrams 61
  X-ray and ESCA spectra 72
Ozonide-ion 261, 284

Pentlandite, NGR of Fe^{57} 28–31
Periclase
  EPR of Mn^{2+}, Fe^{3+}, Cr^{3+} 117
  F-center, g-factor shift and hyperfine
    structure 295, 296
Periodic system of electron-hole centers
  256
Phenakite, types of the centers 283
Phosphorescence 153, 160
Plagioclases
  EPR of Fe^{3+} 116
  Fe^{2+} in lunar plagioclases (NGR) 27
  Ti^{3+}, O^- centers 278
Precursors of electron-hole centers 244
Proton, magnetic moment 121
Proton magnetic resonance 126–128,
  130–134
Protonoluminescence 189, 192
Proustite, NQR of As^{75} 139
Pseudobrookite, NGR spectra 32
Pyrargyrite, NQR of Sb 139
Pyrite, low-spin iron and NGR spectra
  28–31
Pyrrhotite, NGR spectra 28–31

Quadrupole splitting
  in NGR 8, 12–16
  in NMR 119, 128–130
  in NQR 14, 137, 138
Quantum yield
  in luminescence 156–158
  in X-ray spectra 50
Quartz
  EPR of Ti^{3+}, Fe^{3+} 117, 270, 271
  luminescence 196
  models of the centers 258, 264, 265,
    270–278
  thermoluminescence 236
  X-ray emission SiK_β spectra 65
  X-ray and ESCA spectra related to
    molecular orbital diagrams 61

Radioluminescence 189, 191
Rare earths
  absorption spectra 203
  electron configurations 197, 200
  energy level diagrams 198, 201
  luminescence 161–175, 187, 196–200, 204
  relative lines intensities for similar
    concentrations 205
  terms positions 199, 200
Realgar, nonequivalent As^{75} atoms (NQR)
  139
Recombination of electrons and holes
  181–183, 226, 227
Resonance conditions
  EPR 76
  NMR 120
  NQR 137
Rhombicity-axiality 102, 103
Roentgenoluminescence 189, 191
Rutile
  EPR of Fe^{3+} 117
  NMR of high-resolution 131

Sanidine
  EPR of Fe^{3+} 116
  hole centers 278
  NMR 135
Sapphirine, Fe^{3+} in tetrahedral position
  (NGR) 27
Satellites in X-ray spectra 48
Scapolite, types of the centers 279
Senarmontite, NQR of Sb 139
Sensitization of luminescence 160–175
Sheelite
  EPR of Mn^{2+}, Fe^{3+}, Gd^{3+}, RE^{3+} 117
  luminescence 142, 146, 176
  models of the centers 286
Sheet silicates
  proton magnetic resonance 132, 133
  thermoluminescence 238

Siderite, magnetic transformations 36
Sillimanite, NMR of Al^{27} 135
Smithite, NGR of Fe^{57} 29
Sodalite
  luminescence of S_2^- 217
  models of the centers 279
Sphalerite
  EPR of Mn^{2+}, Fe^{3+}, Co^{2+}, RE^{3+} 118
  luminescence 142, 146, 177–188, 217–220
  models of the centers 219
  NGR of Fe^{57} 29–31
  photosensitive EPR 187
  S^- center 266
Spin-Hamiltonian 111, 112
  parameters of 98–100
Splitting of energy levels in crystal field of
  different symmetry 97
Spodumene
  electric field gradient calculation 129
  EPR of Mn^{2+} 117
  NMR of Li^6 and Li^7 135
  X-ray spectra 65
Stannite
  NGR of Fe^{57} 29–31
  of Sn^{119} 37
Staurolite, NGR of Fe^{2+} in tetrahedral
  position 28
Stishovite, X-ray spectra of Si in octahedral
  position 65
Superhyperfine structure (shfs) 109, 110,
  253
Superoxides ion O_2^- 216, 258, 259, 283, 284,
  299
Superparamagnetism 34

Tectites, NGR of Fe^{57} 36
Thermoluminescence
  energy band schemes 226, 234
  factors of 239
  parameters 228–230
  spectra 233
Tin, NGR spectra of Sn^{119} 2, 15, 20, 37
Topaz
  EPR of Fe^{3+}, Ti^{3+} 117
  types of the centers 283
Trap depth 228–230
Triboluminescence 189, 193

Units of measurement
  in radiospectroscopy 80–81
  in X-ray spectroscopy 49, 50

Uranyl
  absorption series 212
  luminescence and absorption spectra 213
  molecular orbitals 214, 215

Vacancies and formation of centers 243,
  244, 255, 257
V-centers 243, 301–303
Vermiculite, NGR spectra 28
Vivianite
  Fe^{3+} : Fe^{2+} ratio from NGR spectra 35
  proton magnetic resonance 134
Voncenite, NGR of Fe^{57} 35

Willemite, luminescence of Mn^{2+} 196
Wolframite, cation distribution and NGR
  spectra 35
Wurtzite, NGR of Fe^{57} 29, 30

X-ray spectroscopy
  assignment of X-ray lines in molecular
    orbital schemes 61–63
  band decomposition in components 70
  chemical shifts 72–75
  emission lines nomenclature 45
  energy level diagrams and X-ray
    transitions 40, 41
  K-, L-, M-, N-, O-, P-, Q-shells 40
  position of absorption edges and emission
    lines 47
  units of measurements 49

Zeolites
  EPR of Ti^{3+}, Fe^{3+} 116
  luminescence 215
  NMR of Li, Na, Al 135
  proton magnetic resonance 132, 133
  types of the centers 283
  X-ray spectra 65
Zincite EPR of Mn^{2+}, Fe^{3+} 117
Zircon
  EPR of Ti^{3+}, Fe^{3+}, Nb^{4+}, Eu^{2+}, Dy^{3+},
    Tm^{2+}, Y^{2+}, Gd^{3+} 117
  luminescence 210, 211
  thermoluminescence 238
  types of the centers 283
Zoisite
  EPR of VO^{2+}, Cr^{3+}, Fe^{3+} 117
  NMR of Al^{27} 135

A.S.Marfunin

Physics of Minerals and Inorganic Materials

An Introduction
Translated from the Russian by G. Egorova,
A.G.Mishchenko

1979. 138 figures, 50 tables. XII, 340 pages
ISBN 3-540-08982-9

Contents: Quantum Theory and the Structure of Atoms. – Crystal Field Theory. – Molecular Orbital Theory. – Energy Band Theory and Reflectance Spectra of Minerals. – Spectroscopy and the Chemical Bond. – Optical Absorption Spectra and Nature of Colors of Minerals. – Structure and the Chemical Bond. – Chemical Bond in Some Classes and Groups of Minerals. – References. – Subject Index.

Solid-state theories and spectroscopy account for the third crucial change within this century in our concept of the basis of mineralogy and the inorganic materials sciences. This book is a revised, updated, and supplemented translation from the Russian edition, providing a complete system of recent theories of solids as they apply to minerals and inorganic materials. Both basic principles and sophisticated new theories are presented with the mineralogist and materials researchers in mind. The book contains extensive references for further study and will be a valuable reference work since each chapter is self-contained.

Springer-Verlag
Berlin
Heidelberg
New York

A Springer Journal

Physics and Chemistry of Minerals

In Cooperation with the International
Mineralogical Association (I.M.A.)

Editors

S. S. Hafner, University of Marburg, Institute of Mineralogy,
Lahnberge, 3550 Marburg, Federal Republic of Germany
A. S. Marfunin, IGEM, Academy of Sciences of the USSR,
Staromonetnyi 35, Moscow 109017, USSR
C. T. Prewitt, State University of New York at Stony Brook, Department of Earth and Space Sciences, Stony Brook, New York 11794,
USA

Advisory Board

T. J. Ahrens, Pasadena; A. Authier, Paris; P. M. Bell, Washington;
G. B. Bokij, Moscow; V. Gabis, Orléans la Source; T. Hahn,
Aachen; H. Jagodzinski, Munich; J. C. Jamieson, Chicago;
N. Kato, Nagoya; R. C. Liebermann, Stony Brook; J. D. C.
McConnell, Cambridge, U. K.; A. C. McLaren, Clayton; N. Morimoto, Osaka; A. Navrotsky, Tempe; R. E. Newnham, University
Park; A. F. Reid, Port Melbourne; R. D. Shannon, Wilmington;
I. Sunagawa, Sendai; D. W. Strangway, Toronto; R. G. J. Strens,
Newcastle upon Tyne; V. M. Vinokurov, Kazan; E. J. W. Whittaker,
Oxford; B. J. Wuensch, Cambridge, MA.

Physics and Chemistry of Minerals is an international journal
that publishes articles and short communications about physical
and chemical studies on minerals, as well as solids related to
minerals.
The journal supports competent interdisciplinary work in mineralogy as it relates to physics or chemistry. Emphasis is placed on the
application of modern techniques and theories, and the use of
models to interpret atomic structures and physical or chemical
properties of minerals.
Subjects of interest include:
Relationships between atomic structure and crystalline state
(structures of various states, crystal energies, crystal growth,
thermodynamic studies, phase transformations, solid solution,
exsolution phenomena, etc.)
General solid state spectroscopy (ultraviolet, visible, infrared, Raman, ESCA, luminescence, X-ray, electron paramagnetic resonance, nuclear magnetic resonance, gamma ray resonance, etc.)
Experimental and theoretical analysis of chemical bonding in
minerals (application of crystal field, molecular orbital, band theories, etc.)
Physical properties (magnetic, mechanical, electric, optical,
thermodynamic, etc.)
Relations between thermal expansion, compressibility, elastic
constants, and fundamental properties of atomic structure, particularly as applied to geophysical problems.
Electron microscopy in support of physical and chemical studies.

Sample copies and subscription information upon request.

Springer-Verlag
Berlin
Heidelberg
New York